Genuine new directions in research are unanticipated. It is the unanticipated event which is so important and has to be followed up; a new viewpoint then develops.

Neil Bartlett,

"The Oxidation of Oxygen and Related Chemistry", World Scientific Ed., 2001

Fluorine

Progress in Fluorine Science Series

Volume 1 – *Photonic and Electronic Properties of Fluoride Materials*
A. Tressaud, K. Poeppelmeir, Eds.

Volume 2 – *New Fluorinated Carbons: Fundamentals and Applications*
O. Boltalina, T. Nakajima, Eds.

Volume 3 – *Modern Synthesis Processes and Reactivity of Fluorinated Compounds*
H. Groult, F. Leroux, A. Tressaud, Eds.

Volume 4 – *Fluorine in Life Sciences: Pharmaceuticals, Medicinal Diagnostics, and Agrochemicals*
G. Haufe, F. Leroux, Eds., ISBN: 9780128127339.

Progress in Fluorine Science Series - Volume 5

Fluorine

A Paradoxical Element

Alain Tressaud
ICMCB-CNRS, University Bordeaux,
Pessac, France

ELSEVIER

Elsevier
Radarweg 29, PO Box 211, 1000 AE Amsterdam, Netherlands
The Boulevard, Langford Lane, Kidlington, Oxford OX5 1GB, United Kingdom
50 Hampshire Street, 5th Floor, Cambridge, MA 02139, United States

Cover image credit: From F. J. Zúñiga, A. Tressaud, J. Darriet, The low-temperature form of Rb_2KCrF_6 and Rb_2KGaF_6: the first example of an elpasolite-type structure with pentagonal bipyramid in the B-sublattice, 176, 3607–3614, Journal of Solid State Chemistry, Copyright Elsevier (2006).

Notices

Knowledge and best practice in this field are constantly changing. As new research and experience broaden our understanding, changes in research methods, professional practices, or medical treatment may become necessary.

Practitioners and researchers must always rely on their own experience and knowledge in evaluating and using any information, methods, compounds, or experiments described herein. In using such information or methods they should be mindful of their own safety and the safety of others, including parties for whom they have a professional responsibility.

To the fullest extent of the law, neither the Publisher nor the authors, contributors, or editors, assume any liability for any injury and/or damage to persons or property as a matter of products liability, negligence or otherwise, or from any use or operation of any methods, products, instructions, or ideas contained in the material herein.

Library of Congress Cataloging-in-Publication Data
A catalog record for this book is available from the Library of Congress

British Library Cataloguing-in-Publication Data
A catalogue record for this book is available from the British Library

ISBN: 978-0-12-812990-6

Publisher: Candice Janco
Acquisition Editor: Emily McCloskey
Editorial Project Manager: Katerina Zaliva
Production Project Manager: Omer Mukthar
Cover Designer: Katerina Zaliva

Typeset by SPi Global, India

Contents

Foreword

The history of fluorine becomes lively when Alain Tressaud reports on the Section 1 of the book with the title, "Fluorine: A Paradoxical Element." Indeed fluorine has a Janus head with two faces, and the author elegantly shows its pros and cons. This is a real masterpiece of a book. Sections 2 and 3 are entitled, "Fluorine: A Key Element of the 21st Century," and "Fluorinated Compounds in Our Environment." On the one hand, both chapters demonstrate how important fluorine compounds contribute to the technical development of our daily life. On the other hand, we face the problem of contamination from fluorine-containing compounds. In principle, this has already been known since the Swiss medical doctor Paracelsus (1493–1541) wrote, "All things are poison and nothing exists without poison, alone the amount is responsible for being a poison." Presently, everyone faces daily contamination of the air with increasing amounts of exhaust gases from burning coal and gasoline, or learns from the newspaper about the contamination of water with nitrates and large amounts of plastic in the oceans. Huge consumer demand for products creates an enormous problem today and for the future. Alain Tressaud shows this convincingly in his discussion of the development of fluorine-containing products. Therefore, this book is a very valuable and key contribution and an eye-opener for the researcher, producer, and consumer interested in sustainable development. I hope that this book will not only be read by fluorine experts but also distributed to others who are involved in research and environmental issues.

Herbert W. Roesky
Göttingen, Germany

Introduction

Last few decades, fluorine and fluorinated products have attracted significant attention, because they are greatly involved in our daily life and environment. Although many beneficial breakthroughs have been brought to us thanks to fluorine and fluoride compounds, they are decried today because of some environmental drawbacks. The goal of this book is to provide a better understanding of the many advantages, but also some disadvantages, of these products, focusing on the real place and importance of fluorine and fluorinated products in our current environment and everyday life. Of course, everyone knows that this element is present in toothpastes we use regularly, but who is really aware that it surrounds us in most instances of our life? Fluorine is not a rare element. In fact, it is the 13th most abundant element constituting the Earth's crust, where it is present at about 0.06% to a depth of 16 km. It is the second anion after oxygen. It is most found in minerals such as fluorite, apatite, topaz, and tourmalines, but also in the oceans, groundwater, lakes, and rivers. Much closer to us, it is also an essential constituent of our teeth and our skeleton, as well as those of all mammals. Finally, fluorinated molecules also exist in several plants. Unlike oxygen (O_2), which is the essential constituent of our atmosphere, fluorine gas (F_2) could not be found in nature until very recently, due to its high chemical reactivity. Indeed, due to its electronic structure ($1s^2\,2s^2\,2p^5$), one electron is missing in the 2p electronic layer to complete the valence shell to eight electrons ($2s^2$, $2p^6$), which corresponds to the electron configuration of the nearest noble gas, neon (octet rule). Thus, because of the reaction $\frac{1}{2}F_2 + e^- \rightarrow F^-$, the element oxidizes almost all other elements, leading to generally very stable compounds in which it is present under the form of a fluoride ion, F^-. However, for simplicity, the word fluorine is often used to identify the species present in a product, whatever the charge.

Physical chemical properties of fluorine will illustrate how it is certainly the most paradoxical of the elements, because the majority of its properties derive from a duality: so reactive on one hand that it could not be isolated for centuries, and on the other hand, giving most stable bonds with almost all other elements (particularly with carbon), so stable that the formed products often stay unmodified in the stratosphere. Obviously, we will discuss the contribution of the refrigerant gases containing fluorine, CFC and HCFC, to the depletion of the ozone layer, and perfluorocarbons and other perfluorinated molecules' contribution to the greenhouse effect.

Discoveries and anecdotes about the real history of fluorinated materials throughout centuries will be illustrated using some major examples in the first section. We will present some decisive discoveries and stages of development in fluorinated products that took place during the 20th century following the isolation of fluorine by H. Moissan in 1886, with particular emphasis on the technological advances that have followed. The contributions of fluorinated products these last 50 years will be described in the second section. Issues relative to these products are of primary concern for humans in important areas such as modern technology, medicine, pharmacy, and biotechnology. Examples will be given in various fields: photonics, energy production, new drugs, surfactants, fire emulsions, TV components, fluorinated polymers, alternative materials in cardiovascular surgery, anti-decay products, and medical imaging applications. A third section will deal with the presence of fluorine and fluorinated products around us and their effect on our environment (atmosphere, lithosphere, water, gaseous emissions), focusing on efforts to limit harmful consequences. We will conclude this section on fluorine's impact on man by giving examples of different types of fluoride contaminations (water, food, industrial environmental), together with the physiological effects of fluorides on living organisms; the solutions that have been proposed to solve these problems will be also considered.

Because of the importance of this element, international meetings are regularly devoted to all fields involving Fluorine. The last International Symposium on Fluorine Chemistry was held in Oxford, UK on July 2018 and the next European one will take place in Warsaw on August 2019. It can be added that several "Fluorine Networks" have been created in various countries, including France , USA, UK, Germany, Russia and Japan, to promote activities related to fluorine in various sectors of chemistry, new technologies, life sciences, medicine and the environment. It should be noted that several specialized books have appeared in recent years dealing with many fields of fluorine chemistry and fluorinated materials: organic and inorganic compounds, fluoropolymers, pharmaceutical molecules, industrial and biomedical applications, etc. These volumes, which propose highly specialized information, are collected in a table at the end of this book. But so far, no attempt had been made to produce an English-language work with a comprehensive and global approach to this element's impact on humanity, that is, advanced technologies we use every day, the air we breathe, the water we drink, the land we live on, etc. Rather than listing tedious documents on the various classes of fluorinated products (the features and applications of which can be better found in more specialized books), it was decided that the following pages should propose an original global approach to the topic, illustrating (I hope) the importance of this element on our environment and life. The varied fields involving fluorine, such as history, cosmochemistry, global warming, radiative forcing, natural and anthropogenic emissions, the water cycle, etc., will be thus evoked, but without ignoring some negative aspects of these products. The place of fluorine and fluorinated products in key areas for our future will be also evoked, dealing with waste reduction, global warming, energy independence, microelectronics, nanotechnologies, etc.

In this book, the author wanted to show a personal version of the image of fluorine and fluorinated products, of their importance in our daily lives, as well as all the promising contributions in many fields of science and technology. However, the disadvantages caused to our health and environment by some of these products have not been underestimated. Unfortunately, the author could not consider all properties and applications of these materials, and begs concerned colleagues to excuse him in case of important omissions.

Note: In this book, the word "fluorine" may be taken either for the element or for the F_2 gas (difluorine); on the other hand, the word "fluoride" may be taken either for a compound/material or the anion F^-.

History and milestones of fluorine and fluorinated products through the centuries

1

Chapter Outline

Until the dawn of technological (al)chemistry, the value of many minerals was widely recognized and exploited without suspecting the nature of the element at the origin of these properties, for example, aluminum in alum and fluorine in fluorite (fluorspar or fluβspat). It can be noted that the name of these two elements finally resulted from the properties of these minerals: i.e. the element that flows for F and the metal from clay for Al. The word fluorine (*fluores* in Latin) appeared for the first time in 1530 in a book written by G. Agricola, but long before that, the melting properties (or "flux") of the fluorinated minerals had been identified, and the beauty and reputed magical properties of these crystals had fascinated many ancient civilizations. Some examples will help us to follow the history of fluorine and fluorinated minerals throughout the centuries: fluoride crystals as gemstones, the real history of the murrhine vases of

Fluorine. https://doi.org/10.1016/B978-0-12-812990-6.00001-5

Imperial Rome, fossil fluorapatites used as synthetic turquoises by the monks of the Middle Ages, description of the use of cryolite in smelting by G. Agricola during the Renaissance period. An important place will be devoted to the fierce competition between the greatest scientists of the 18th and 19th centuries to characterize fluorocompounds and isolate the F_2 element, a competition that allowed exceptional discoveries, culminating in the isolation of fluorine by Henri Moissan in 1886.

1 History and uses of fluorinated minerals

In very ancient times, beautiful Egyptian ladies were using fluoride crystals as ornaments, especially fluorite. Carved figurines and beetles have been found during archeological excavations, hundreds of fluorite seals have been found in Aegean sites of the Late Bronze Age,[1] and a multitude of various objects were made over many centuries in ancient China. At Moncy-St-Pierre (France), necklaces with fluorite beads accompanied wealthy Merovingians in their last resting places (6th–7th centuries). In Amerindian civilizations, fluorite crystals were likely cut for religious reasons, as evidenced by the figurine of a sitting man discovered in the Angel pit site in Indiana (United States), which could have been an idol for the main temple.[2] The importance of murrhine vases in Imperial Rome will be later mentioned. As such, it is said that Hadrian would have launched the conquest of northern England mostly to appropriate deposits of "Blue John" fluorspar (Fig. 1), a banded mineral with

Fig. 1 Banded fluorite crystal from Castleton, United Kingdom.
Photo Michael Vickers, Used with permission from Michael Vickers.

[1] A. Dickers, Die spätmykenischen Siegel aus weichem Stein. VLM Verlag Marie Leidoerf GmbH, Rahden, Germany (2001).

[2] C. R. Stafford, Prehistoric People of Indiana, INARNG Integrated Cultural Resources Management Plan A187, Indiana State University (2001).

extremely bright colors that was—and still is—extracted in Cumbria[3] (see – "Etymology: the 'fluid' rather than 'destructive' ").

Etymology—The "fluid" rather than "destructive"

Many minerals have been recognized and used for centuries without suspecting what element was at the origin of their properties. However, the names attributed to these products corresponded most times to the real nature of the element. This is the case for fluorine in fluorspar ores (fluorite, CaF_2), as for aluminum in alum. Thus, the origin of the word "fluorine" comes from the Greek root "fluo-" (φλυω: "sink") and the Latin "fluorine" ("fluere," or "flow"). Saxon miners named fluorite ores "flußspat" (also "flosse," "flüsse," "fluss-spaht," "spatum vitreum," or "calx fluorata") due to the properties of these crystals (from "fluss," or "flow," and "spat," or "spar," that derived from the Old German "spalten," meaning "crack," or "cleave"). They also gave more colorful names like berg-bluhme or metall-bluhme ("bluhmen," or "flower" in Old German), because of the beauty of their blooms. When Agricola translated in Latin these German names for *De Natura Fossilium* and *Bermannus, re metallica of dialogus sive*, he chose "fluores" probably because of the melting properties of these crystals, but one can imagine that he might have also been influenced by the proximity of the Latin words "fluores" and "flores" ("flowers").

The symbol "F" for fluorine appeared in an article in 1814, but until the second half of the 19th century, it was the symbol "Fl" which was most commonly used. It may be noted that in some Anglo-Saxon texts, the misspellings "flourine" or "flourspar" are still used. In his letter to Davy, dated August 26, 1812, Ampere, having realized that hydrofluoric acid contained no oxygen, replaced the word "oxy-fluorique" with a new proposal, the French word, "fluorine." It's Davy who released this name in the United Kingdom the following year by using exactly the word proposed by Ampere. But on a second occasion, Ampere (to account for the aggressive nature of hydrofluoric acid) proposed replacing the word fluorine with "phtore" or "phthore." This term (coming from the Greek *Φθόριος* and meaning "destructive") seemed more appropriate to account for the properties of its derivatives: "I chose the name phtore, which takes, in both noun and adjective form, the meaning 'deleterious,' which has the power to destroy, corrupt."[4]

Although this name was in good agreement with the nomenclature of the other halogens, it was not accepted by the scientific community at the time, partly because of the influence of the journal "Annals of Philosophy," which favored the word fluorine. Today, this term is still mostly used, including, for example, in Japan where フッ素 ("fusso") is associated with the flow concept. However, in a book published in Russia around the same period, *Osnovaniya Obshchei Khimii,* the author, G.H. Hess, proposed the name "ftor," and since that time Russian chemists call the element *Фтор* which is also the case in the Greek-speaking community: *Φθόριου*. Now, all names used worldwide when speaking of this element (i.e., hydrofluoric, fluorine, or phtor) were initially proposed by Ampere. Note that the word "fluorescence" is also derived from the same root.[5]

[3] "Blue John" fluorspar crystals (name derived from the French word "bleu-jaune"), with beautifully colored strips, are extracted since centuries in the Hope Valley, near Castelton, Derbyshire, in the Northern part of England.

[4] A. M. Ampere, "D'une classification naturelle pour les corps simples," Ann. Chim. Phys. 2 (1816) 5–32.

[5] The fact that some fluorite crystals (CaF_2) became luminescent after being heated had already been noticed by J.L. Elsholtz in Vienna in 1676, and this characteristic was studied in 1824 by the German mineralogist F. Mohs (1773–1839). It was in 1852 that the Irish scientist G.G. Stokes (1819–1904) named this property fluorescence: "I am also inclined to coin a word, and call the appearance fluorescence, from fluorspar, as the analogous term opalescence is derived from the name of a mineral."

1.1 Fluoride crystals as gemstones substitutes

Due to their richness and shimmering colors, crystals of fluorite (fluorspar) and apatite, the most well-known of the fluorinated gemstones, were often confused—accidentally or intentionally—with more valuable precious stones. Because the purple, green, or blue varieties of fluorites ("fluorinated violaceo color") can be confused with amethyst, emerald, or sapphire, their names were often given as "false amethyst," "false emerald," or "false sapphire" (in Latin: *pseudo-amethystus, pseudo-smaragdus, pseudo-hiacynthus*)[6] (Fig. 2A). Dark purple fluorspar, called "bull beef" by workmen, may be changed, by heat, to a rich amethyst tint. Nevertheless, their use in jewelry is quite limited, because these stones can be easily scratched, which is not the case with another fluorinated crystal, topaz. In addition, because of their fragility, fluorite crystals are very difficult to process and may require the use of resins during cutting operations (see later).

In addition to fluorites and apatites, some other fluorinated minerals have been used as semi-precious stones (Fig. 2B), such as:

- *Tourmaline*: $NaX_3Al_6(BO_3)_3(Si_6O_{18})(OH, F)_4$ compounds belong to the group of cyclosilicate minerals[7] (Dana class 61.3.1; Dana 1997),[8] in which tetrahedra form $(Si_6O_{18})^{-}_{12}$ rings. These crystals usually exhibit pyroelectric properties. The fluorine content depends on the nature of the cations present in the X site, and also on the Fe/Fe+Mg) ratio. The color of these crystals is strongly influenced by the nature of the X cation: green or transparent for X=Li or Al (indicolite or liddicoatite), brown to transparent for X=Mg (Dravite),[9] or even black for X=Fe or Mn in shorlites. In buergerite, the OH groups are replaced by three O and one F. This leaves a space open for one more positive charge, which is compensated for by 3 Fe^{3+} ions, which leads to the formula, $NaFe^{3+}{}_3Al_6(BO_3)_3Si_6O_{21}F$. This mineral has a color that is bronze to brown and black, with a hardness of 7 Mohs. The best tourmaline gems come from granitic pegmatites (Brazil, Madagascar, Mozambique).
- *Topaz*: Al_2SiO_4 $(F, OH)_2$ which belongs to the nesosilicates group (Dana 52.3.1.1 class) in which SiO_4 tetrahedra are independent and contain some O not bound to Si. The crystals may be transparent or blue, green, yellow, or orange. They are found in eruptive acidic deposits, and their size can reach several tens of kg (Brazil, Russia) (Fig. 2C).
- *Jadeite-nephritis*: $Ca_2(Mg,Fe)_5Si_8O_{22}(F,OH)_2$ which are tremolite varieties (Dana 65.1.3c.1 class). The color of these minerals varies from yellow to green and very dark green. In many civilizations, sculptures were carved in these minerals, such as this Olmec mask (Mexico, 1250–900 BC) (Fig. 2D).

[6]"The 'fluatée' lime (i.e., fluorite crystals) is, among all the stony substances, the one that most lent to the effects of the coloring actions of metal oxides. Also, the illusion that tends to produce the resemblance between the hues that adorn these crystals with those of stones called gems, suggested their names as false ruby, false sapphire, false emerald etc., that the ancient mineralogists had given, according to the diversity of its colors," V. Haüy, Traité de Minéralogie (1822).

[7]F.C. Hawthorne, D.J. Henry, Classification of the minerals of the tourmaline group, *Europ. J. Mineral.*, **11** (1999) pp. 201–215.

[8]R. V. Gaines et al., Dana's New Mineralogy, 8th Edition, John Wiley & Sons, New York (1997).

[9]I. Baksheev, O. Kudryavtseva, Compositions of hydrothermal dravitic tourmalines from Ural, *Geophys. Res. Abst.* **5** (2003) p. 262.

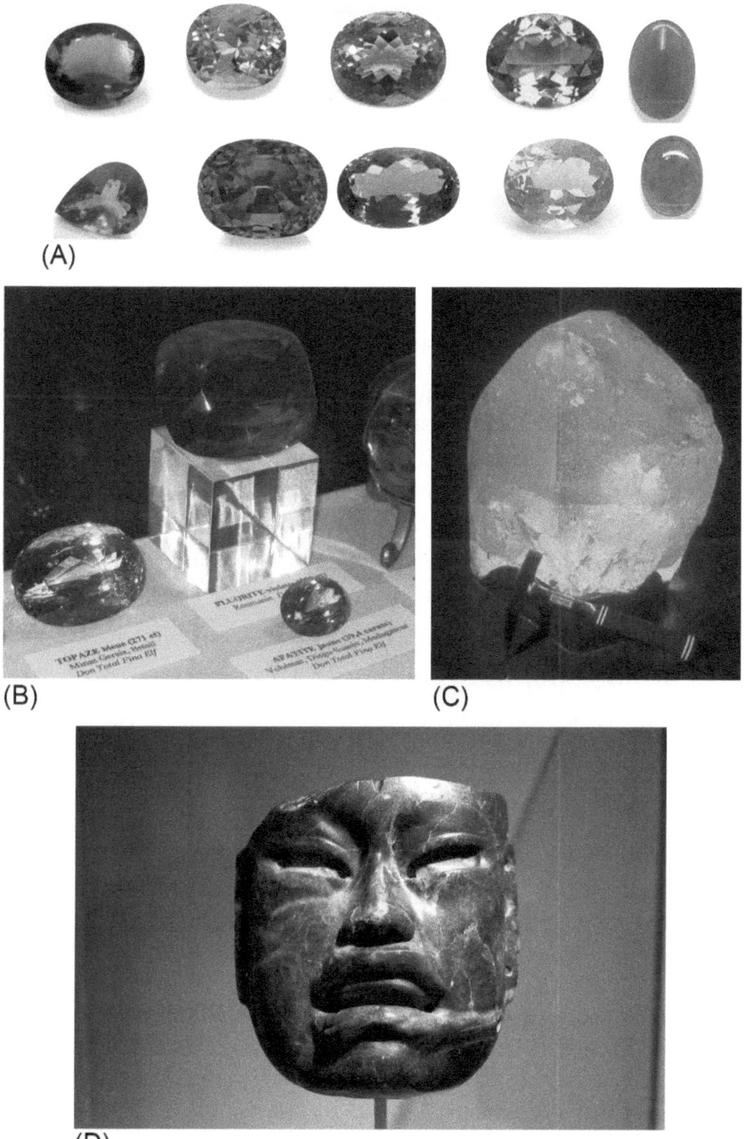

Fig. 2 Fluoride gemstones: (A) fluorite, tourmaline, topaz, apatite, jadeite; (B) topaz, fluorite, apatite from MNHN, Paris; (C) 150kg topaz, Museum Univ., Florence; (D) Jadeite Olmec mask.

1.2 Fluorinated powders as "therapeutics"

It was long assumed that fluorite crystals possessed healing, even magical, powers. The ancient Romans believed that drinking alcoholic beverages in containers cut from such crystals protected the drinker from drunkenness, and until the 18th century, powdered fluorite crystals were ingested in water to reduce the pain associated with kidney stones. In ancient times, it was thought that fluorspar had a marked ability to relax the muscles, could produce a sour taste though simply held in the hand, and sometimes induced a somnambulistic state.[10]

1.3 Uses of fluorspar minerals as pigments

Purple fluorite powder was used in northern Europe as a pigment on polychromic wooden sculptures at the end of the Gothic and early Renaissance periods. Given the places where those pigments were found (Saxony, Tyrol, Switzerland and Silesia), the spectrum of their colors was very wide, including violet and darker colors of fluorspar minerals from Wölsendorf mines. Because of the weak hue of these powders, they were used to reinforce the color effect.[11] Similarly, CaF_2 could be detected by SEM and EPMA in Chinese cloisonné enamels from the 16th and 17th centuries.[12]

1.4 Uses of fluorspar as ornaments and in decoration items

Around Castleton, in the rolling landscape of Derbyshire Peak District in northern England, a unique type of fluorite ore, the "Blue John," has been found for a couple thousand years by miners looking for lead veins. As written by T. Thomson (1773–1852), "The mineral called flourspar, and in this country frequently distinguished by the name of Derbyshire spar, is so common in lead mines, and is so beautiful in consequence of its transparency, its fine colours, and the large size of its cubic crystals, that it must have early attracted the attention of mankind. There can be little doubt that it is mentioned both by Theophrastus and Plinius under the name of false-emerald." This unique type of fluorite is formed of alternating strips of violet, yellow, or gray (see Fig. 1). Today, the mine of Treak Cliff Cavern still produces 500 kg of Blue John fluorite per year. There are up to 14 different types of banded fluorites, and because of their exceptional beauty, they have been used as decoration since the 18th century. Among these, we can cite chimneys by the great architect R. Adam (1728–92), Boulton's work of ormolu, and Vallance's obelisks and vases. Such fluorite objects can also be found in Windsor Castle, the White House, or the Vatican.[13] Note that some

[10] A. R. Bliss, The ancient medicinal uses of gems and precious stones, J. Amer. Pharmac. Assoc. Vol. XXV, (1936) pp. 544–545.

[11] Richter M., Hahn O. Fuchs R., Purple fluorite: a little known artists' pigment and its use in Late Gothic and Early Renaissance painting in northern Europe, Studies in Conservation, **46** (2001) pp. 1–13.

[12] Henderson J., Tregear M., Wood N., The technology of sixteenth- and seventeenth-century Chinese cloisonne enamels, *Archaeometry*, **31** (1989) 133–146.

[13] Ford T. D., Proc. Yorksh. Geol. Soc. **30** (1955) pp. 35–60.

Fig. 3 Fluorite balusters from Voltennes (Burgundy, France) at Opera Garnier, Paris (SAGA, Société amicales des géologues Amateurs, France, https://www.saga-geol.asso.fr/).

balusters of the Opera Garnier in Paris are made from blocks of fluorite coming from the mine in Voltenne (Saône et Loire)[14] (Fig. 3).

1.5 Middle ages forgeries: Fossilized mastodon tusks sold as synthetic "turquoise"

During the Middle Ages, a material of a beautiful turquoise-blue color was commonly used by the monks of Cistercian monasteries in France to decorate religious objects such as cross reliquaries. For example, in the collections of the National Museum of the Middle Ages of Paris, such stones are embedded in the reliquary of the True Cross made in Limoges in the 12th century (Fig. 4). Since the 19th century, it was recognized

[14] Joseph D. Etude géologique et minéralogique des gisements de fluorine de Voltenne et de Maine, 3rd cycle Thesis, Orléans, France (1974).

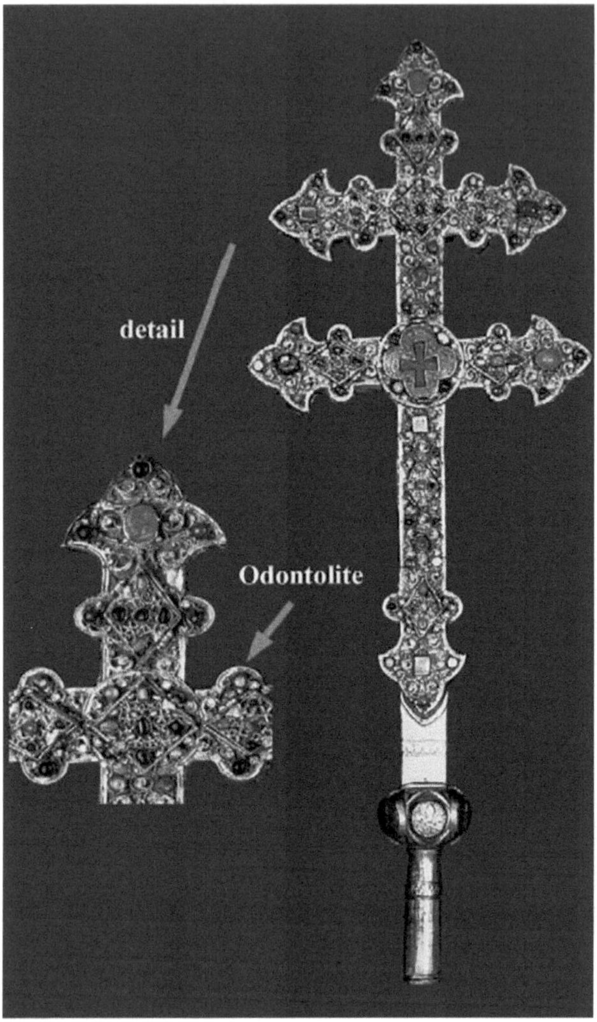

Fig. 4 The True Cross reliquary.
Reproduced with permission of *Musée National du Moyen-Age*, Paris (no. C1.998).

that such materials are totally different from turquoise and are actually made of fossilized ivory, called *odontolite* or *bone turquoise*. Thus, these objects are made of the fossilized bones, teeth, or tusks of mastodons. In addition, odontolite is slightly less hard (Mohs 5) than turquoise (5–6 Mohs).

Back in 1628, the French naturalist G. de la Brosse (c.1586–1641) mentioned that this color was obtained by heating fossil ivory and he proposed that synthetic turquoises were obtained in this way.[15] Later, R.A. Réaumur (1683–1747) experimented

[15]Brosse G. de la, *De la nature, vertu et utilité des plantes*, Paris (1628).

with a furnace for producing "prepared turquoises" which originated from the Simorre area in the Gers region, in the southwest part of France.[16] During the 19th and 20th centuries, various hypotheses were forwarded to explain the origin of the color in these materials, for example, the presence of copper salts, small amounts of vivianite $Fe_3(PO_4)_2{\cdot}8H_2O$, mixed phosphates of Ca and Fe.[17,18] Webster announced that the odontolite was not a true mineral, but "the most fascinating of simulated turquoise impregnated with $Fe_3(PO_4)_2{\cdot}8H_2O$, a mineral which is sometimes called blue iron earth."[19] C.A. Baud, using ESR and IR spectroscopies, found that the blue color of the fossilized bone and ivory was the result of the heating process which introduced carbonaceous components into the material. Above 600°C, a new phase, whitlockite $Ca_3(PO_4)_2$, was supposed to appear.[20] It was not until the early 21st century that the mystery of colorful odontolites was elucidated by I. Reiche et al.[21] These authors studied various mastodon ivory samples (mastodons being elephant precursors who lived in the Miocene, i.e., 26 to 2 million years ago, in some areas of Europe) (Fig. 5). The studied samples were specifically dated from the Middle Miocene (i.e., 12–15 million years BP). Using various techniques such as TEM, SEM-EDX, PIXE, and PIGE, it was shown that turquoise-blue samples (odontolites) used in the Middle Ages had a microcrystalline structure different from that of more recent bone samples, for instance, those that were only 10,000 years old. The first samples consisted primarily of crystallites of fluorapatite (FAP), $Ca_{10}(PO_4)_6(F, OH)_2$, containing a high percentage of fluorine (4%), whereas more recent samples contained <0.3%. Thus, turquoise-blue odontolites are formed of FAP nanocrystallites, 50–600 nm long. Other elements are present such as Mn, Fe, Sr, Ba, S, Cl, Pb, and Y, but copper only scarcely and Co is never found. By comparing various heat treatments carried out in the laboratory, these authors showed that Miocene-fossilized mastodon ivory acted differently from more recent bones and ivory. While the latter samples became beige and black when heated to 400–550°C, with formation of carbon black, then turned gray at 700°C, and finally, white above 800°C, the fossilized specimens turned blue after being air-heated at 550° C for 20 h. During the heating cycle, a peculiar trend was observed in the case of fossil ivory samples: at 550°C, the nanometric (20–30 nm) fluorapatite crystals started to grow as flat-shaped crystals to reach a 100–300 nm average at 700°C, as shown from TEM images in Fig. 6. Comparison of Fig. 6B and C clearly shows that the apatite

[16]Réaumur R. A. de, *Mémoires Acad. Royale* (1715).

[17]Bouillon-Lagrange E. J. B., *Ann. Chim.* **59** (1807).

[18]Fischer M. G., Ann. Chim. **8** (1823) 326.

[19]Webster R., Lapidary J., Feb. issue (1975) p. 1750.

[20]Baud C. A., Tochon-Danguy H. J., Bull. Assoc. Pro aventico **29** (1985) p. 49.

[21](a) Reiche I., Vignaud C., Menu M., Heat-induced transformation of fossil mastodon ivory into turquoise "odontolite", Solid State Sci. **2** (2000) pp. 625–636; (b) Reiche I., Vignaud C., Champagnon G., Panczer G., Brouder C., Morin G., Armando V., Charlet L., Menu M., From mastodon ivory to gemstone: the origin of turquoise color in odontolite, Amer. Miner. **86** (11-12) (2001) pp. 1519–1524; (c) Reiche I., La couleur de la turquoise osseuse, Seminar on Color of Materials, Roussillon, France, mai 2002; (d) Reiche I., Fluorine and Its Relevance for Archaeological Studies, pp. 253–283, in Fluorine and the Environment, Vol. 2, A. Tressaud Ed., Elsevier, Amsterdam (2006).

Fig. 5 Representation of mastodons (Gomphotherium), Miocene period, tertiary era. From https://en.wikipedia.org/wiki/Palaeomastodon#/media/File:Paleomastodon.jpg.

crystals of fossil ivory heated in the laboratory and those of the turquoise-blue odontolite have the same morphology. The modification of the color of the crystals under air-heating conditions confirms the correctness of these assumptions (Fig. 7). Furthermore, given the absence of copper or vivianite, previously proposed assumptions should be abandoned. The riddle of the turquoise-blue fossil ivory could be finally solved by I. Reiche et al. as being consecutive to a heat treatment of fossilized mastodon tusks in oxidizing conditions above 600°C. Two main conditions must be met to obtain this result:

(i) The process of fossilization (or diagenesis) is combined with an anionic exchange (F^-/OH^-) between these fossilized objects and the surrounding environment. A complete fluorination of the apatitic crystalline network of mastodon teeth or tusks is obtained with a simultaneous growth of FAP crystals. In this very long process of dissolution and recrystallization that occurs at ambient temperature during millenaries, nanoscale needle-like crystals are formed, a process which cannot be observed in either a shorter time scale or in recent ivory.

(ii) To explain the change in color, we must admit that, simultaneous to the F^-/OH^- substitution, a segregation occurred between the transition elements (Mn and Fe) that migrated into ivory during diagenesis. The transition elements concentrated in the grain boundaries and underwent oxidation during heating of the materials in air.

In natural, unheated fossil ivory (white), Mn^{II} was identified by TRLIF. The combination of several other techniques, UV/visible/near-IR reflectance spectroscopy, time-resolved laser-induced luminescence spectroscopy (TRLIF), and XANES allowed

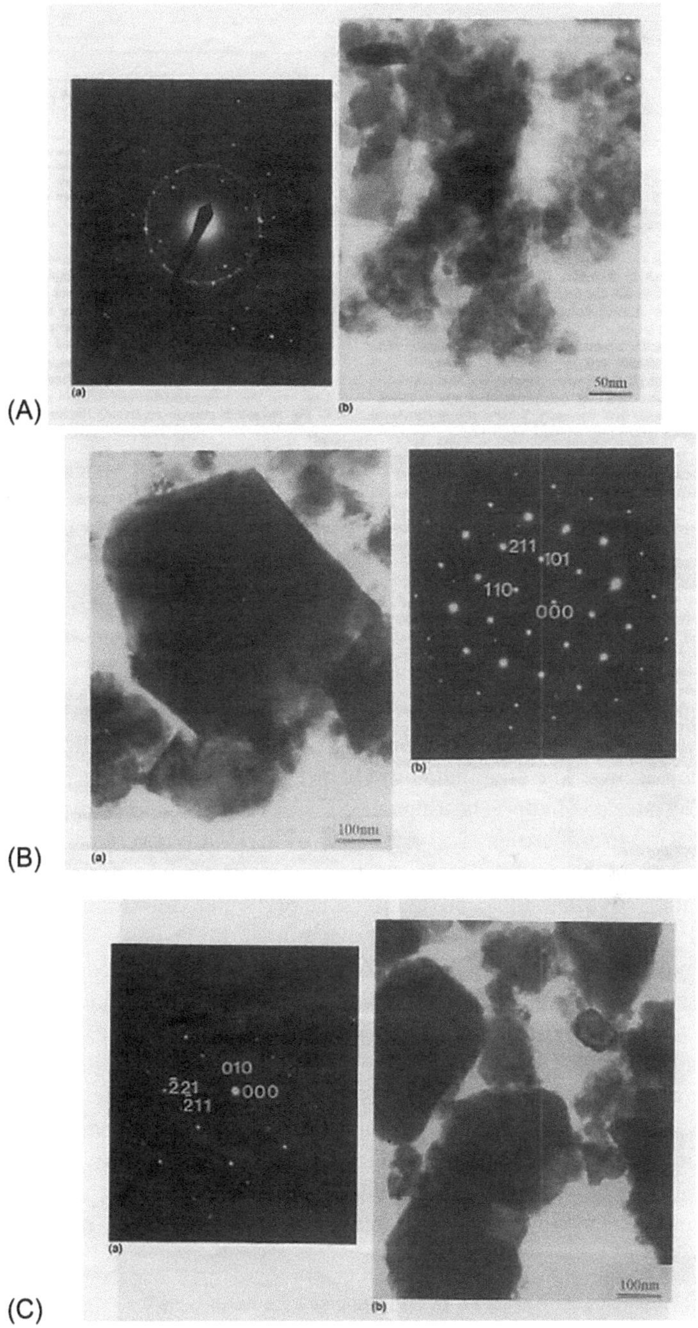

Fig. 6 TEM images of fossilized mastodon ivory: (A) not heated samples, (B) heated 20h at 700°C, (C) turquoise-blue odontolite sample showing FAP crystal along ⟨102⟩ axis (Museum National d'Histoire Naturelle, MNHN, Paris).
Reproduced from I. Reiche et al. Solid-State Sci. 2 (2000) pp. 625–636, with permission from Elsevier, Amsterdam.

Fig. 7 Color modification induced by thermal treatments on a fossilized tusk of mastodon (from Rajegats, France, Middle Miocene: 13–16 Myr).
From Reiche I., Vignaud C., Champagnon G., Panczer G., Brouder C., Morin G., Armando V., Charlet L., Menu M., From mastodon ivory to gemstone: the origin of turquoise color in odontolite, Amer. Miner. **86** (11-12) (2001) pp. 1519–1524.

determination that the blue color in heated fossil ivory materials was due to the presence of Mn^{V} in distorted tetrahedral environments.[22] In particular, X-ray absorption spectroscopy allowed the elucidation of changes occurring in the local environment and valence states of Mn upon heating. XANES and EXAFS spectra were recorded at the K-edge of Mn impurities (200–650 ppm) in various types of samples (Fig. 8). The position and intensity of the pre-edge structure of Mn clearly indicate changes in the structural environment and oxidation state of the element. Unheated (white) fossilized ivory shows Mn^{2+} ions in octahedral coordination (Fig. 8a). By contrast, in the fossilized ivory heated at 600°C (turquoise-blue) and in the two odontolite samples (from Museum collections), the major part of Mn is in 5+ oxidation state, as confirmed by the pre-edge structure observed at 6541.3 eV (Fig. 8b,c). In the XANES spectra, heated fossilized ivory and odontolites are compared with synthetic Mn-chlorapatite (Fig. 8e), a strongly absorbing matrix at this energy (6.5 keV), in which Mn^{5+} ions are known to be substituted to P^{5+} in the tetrahedral sites. In such an environment, Mn^{5+} ions give rise to an intense turquoise-blue color,[23] causing the appearance of this peculiar color in fossilized ivory. These XANES and EXAFS investigations of manganese environments are also in good agreement with previous

[22]Dardenne K., Vivien D., Huguenin D., Color of Mn(V)-substituted apatites $A_{10}(B,Mn)O_4)_6F_2$, J. Solid State Chem. **146** (1999) 464–472.

[23]Reinen, D., Lachwa, H., and Allmann, R., EPR- und ligandenfeld- spektroskopische Untersuchungen an MnV-haltigen Apatiten sowie die Struktur von $Ba_5(MnO_4)_3Cl$. Z. Anorg. Allg.Chem., **542** (1986) pp. 71–88.

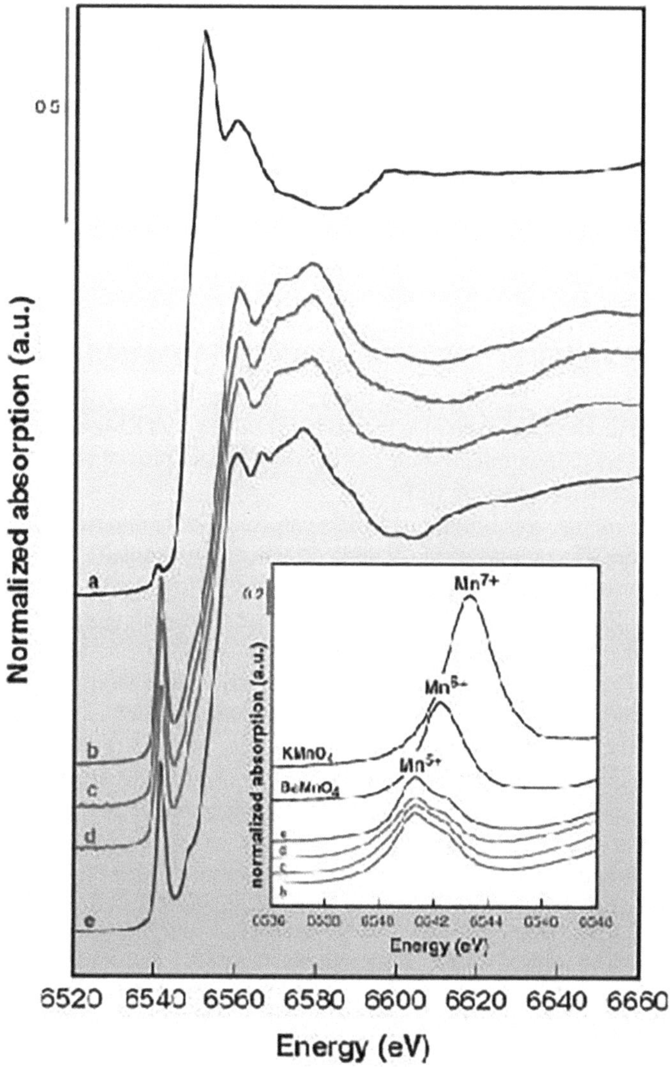

Fig. 8 Mn K-edge XANES spectra of: (a) fossilized ivory; (b) 600°C heated fossilized ivory; (c, d) turquoise-blue museum odontolites; and (e) synthetic apatite $Ba_5(PO_4)_{2.5}(MnO_4)_{0.5}Cl$, taken as reference mineral with Mn^{5+} in tetrahedral coordination. Inset: Pre-edge structure of tetrahedral Mn^{5+} in (b) heated fossilized ivory; in (c, d) odontolites, and in (e) synthetic $Ba_5(PO_4)_{2.5}(MnO_4)_{0.5}Cl$, are observed at 0.9 and 1.9 eV below those of tetrahedral Mn^{6+} in $BaMnO_4$ and Mn^{7+} in $KMnO_4$, respectively.
From Reiche I., Vignaud C., Champagnon G., Panczer G., Brouder C., Morin G., Armando V., Charlet L., Menu M., From mastodon ivory to gemstone: the origin of turquoise color in odontolite, Amer. Miner. **86** (11-12) (2001) pp. 1519–1524.

results on chloroapatites and synthetic Mn^{5+}-doped fluoroapatites, $A_{10}[(B, Mn)O_4]_6F_2$, (with A = Ca, Sr, Ba; B = P, V).[22]

During the heating process, Mn^{II} ions in distorted octahedral sites are oxidized into Mn^{V} in distorted tetrahedral environment. In apatite, short Mn-O distances are indeed found by EXAFS for the nearest neighbors: 1.69 Å. The introduction of a second type of Mn-O distances at 2.12 Å is required to take into account an incomplete transformation of Mn^{2+} into Mn^{5+}. Another contribution at 3.24 Å corresponds to Mn-Ca distances, which means that in the apatitic network, Mn^{5+} is located in oxygenated tetrahedral sites similar to those of P^{5+}. In the pre-edge structure of the Mn K-edge XANES spectra, the maximum of the signature of tetrahedral Mn^{5+} in heated fossilized ivory (b), odontolites (c,d), and synthetic $Ba_5(PO_4)_{2.5}(MnO_4)_{0.5}Cl$ (e) is located at 0.9 and 1.9 eV lower than those of tetrahedral Mn^{6+} (in $BaMnO_4$) and Mn^{7+} (in $KMnO_4$), respectively. It can be noted that spectra (a), (b), (c), and (d) are recorded in fluorescence mode, spectrum (e), and those of $BaMnO_4$ and $KMnO_4$ in transmission mode. The presence of an intense blue coloration was also proved to arise from Mn^{5+} in a tetrahedral environment by ESR.

The history of turquoise-blue odontolite can thus be summarized in four steps. A very long time ago, a herd of mastodons drowned in a flooded river on a site that would become the southwest of France. During the following 14 million years, an alteration took place through exchanges between the surrounding soil and the mastodons' teeth/tusks. The composition and morphology of the apatite crystals were modified, and new transition elements were introduced simultaneously with an F/OH exchange. In the Middle Ages, monks from neighboring monasteries incidentally observed that these peculiar stones became blue after heating. They had the idea to use them in place of much rarer and expensive turquoises, and even started a small business to provide "turquoise" to reliquaries manufacturers. Finally, it was not until the beginning of the 21st century that the enigma's solution was found,[21] involving both a morphological change and a change of composition, with the critical presence within the fossil ivory of fluoride ions and transition elements exhibiting particular electronic structures.

2 Fluorine dating in archeology

During the last several decades, the determination of fluoride contents has been used for various purposes such as assessing diagenetic processes in fossils, in lithic items, and in developing relative chronologies for human, hominid, or animal remains. The penetration of chemical elements into archeological and geological vestiges may provide valuable information on the length of stay of these objects in their discovery sites. The enrichment of halides, especially light fluoride ion, allows relative dating of certain types of materials: bones and lithic objects. Although this method does not allow absolute dating, it may, nevertheless, provide often-decisive information for the relative dating of objects in the same geological stratum. The principle is simple: the objects buried in the ground accumulate fluoride ions present in the surrounding soil. Runoff generally contains an average of $\sim$0.1 μg fluoride ions per g of water and may even contain up to several mg/g. In normal conditions, a concentration profile starts at the surface and continues to the heart of the object. If the diffusion coefficient is known, it is theoretically possible to estimate the duration of exposure (see "A history of fluorine dating").

A history of fluorine dating

Middleton, J. 1844 On fluorine in bones, its source, and its application to the determination of the geological age of fossil bones. *Proceedings of the Geological Society (London) 4*:431–433.

Carnot A. 1893 Recherches sur la composition générale et la teneur en fluor des os modernes et des os fossiles, Annales Mines, Ser. IX, **3** (1893) 155.

Oakley, K. P. and C. R. Hoskins. 1950 New Evidence on the antiquity of Piltdown man. *Nature 165*:379–382, 1950.

Wiener, J. S., K. P. Oakley and W. E. Le Gros Clark. 1950 The solution of the Piltdown problem. *Bulletin of the British Museum (Natural History). Geology 2*(3):139–146.

Parker, R. B., J. W. Murphy, and H. Toots. 1974 Fluorine in fossilized bone and tooth: distribution among skeletal tissues. *Archaeometry 16*: 98–102.

Haddy, A. and A. Hanson. 1981 Relative dating of Moundville burials. *Southeastern Archaeological Conference Bulletin 24*: 97–99.

Haddy, A. and A. Hanson. 1982 Nitrogen & Fluorine Dating of Moundville Skeletal Samples. *Archaeometry 24*:37–44.

Callaghan, R. T. 1986 Analysis of the fluoride content of human remains from the Gray site, Saskatchewan. *Plains Anthropologist 31*(114):317–328.

Schurr, Mark R. 1989 Fluoride dating of prehistoric bones by ion selective electrode. *Journal of Archaeological Science 16*: 265–270.

Schurr, Mark R. 1989 *The Relationship between Mortuary Treatment and Diet at the Angel Site*. Ph.D. dissertation, Indiana University.

Schurr, Mark R. and Sherri L. Hilgeman. 1990 Fluoride Dating and Pottery Chronology at Angel. Midwest Archaeological Conference, Evanston, Illinois.

Berger R., Protsch R. 1991 *Fluorine Dating,* in Scientific Dating Methods, H.Y. Göksu, M. Oberhofer, D. Regulla, Eds, Kluwer (1991).

Ezzo, James A. 1992 A refinement of the adult burial chronology at Grasshopper Pueblo. *Journal of Archaeological Science 19*: 445–458.

Johnsson, K. 1997 Chemical dating of bones based on diagenetic changes in bone apatite. *Journal of Archaoeological Science 24*:431–437.

Schurr, Mark R. and David J. Hally. 1999 Fluoride Dating Burials from Short-Occupation Sites: A Mississippian Period Test of the Technique. Poster presented at the Society for American Archaeology 64th Annual Meeting, Chicago, Illinois.

Gregory, David and Mark R. Schurr. 2000 Fluoride Dating within an Early Agricultural Period Settlement in the Tucson Basin. Poster presented at the Society for American Archaeology 65th Annual Meeting, Philadelphia.

A. A.-M. Gaschen, U. Krähenbühl, M. Döbeli, A. Markwitz, B. Barry. 2005 Fluorine and calcium profiling by PIGE/PIXE for exposure age dating in archaeology, *Int. J. PIXE* 15, 327 (2005).

Döbeli M., A.A.-M. Gaschen, U. Krähenbühl. 2006 Fluorine Analysis by Ion Beam Techniques for Dating Applications, pp. 215–252, Fluorine and the Environment, Vol. 2, A. Tressaud Ed., Elsevier (2006).

A.A.-M. Gaschen, M. Döbeli, A. Markwitz, B. Barry, S. Ulrich-Bochsler, U. Krähenbühl. 2008 Restrictions on fluorine depth profiling for exposure age dating in archaeological bones. *Journal of Archaeological Science* **35**:3, 535–552 (2008).

M. Schurr, I. Kuijt, J. Anfruns, M. Molist. 2011 Timing the Neolithic transition: the application of fluoride dating at Tell Halula, Syria. *Journal of Archaeological Science* 38(7): 1496–1501.
R. L. Lyman, C. N. Rosania, M. T. Boulanger. 2012 Comparison of fluoride and direct AMS radiocarbon dating of black bear bone from Lawson Cave, Missouri. *Journal of Field Archaeology* 37(3).
M.R. Goodrum. 2016 Fluorine Dating, Encyclopedia of Geoarchaeology, Part of the series Encyclopedia of Earth Sciences Series, pp. 275–276 chapter 37, Springer (2016).

Given the complexity of the varied experimental parameters for both the object (nature, compactness, stability in time) and the environment (geological, meteorological, etc.), it is easy to understand that this type of method may often bring some valuable information on a polemical issue, even if does not provide a definitive answer. Once made conventionally, fluorine analysis is now carried out using powerful techniques such as XPS, AES, EPMA, IBA, or RNA.

2.1 Relative dating of bones and skeletal remains

Since the first half of the 19th century, it has been known that bones buried in humid ground are strongly enriched by fluorine. The first paper quoting the use of fluorine for dating bones appeared in 1844. At that time, fluorine was known to be present in bones. In the end of the 19th century, A. Carnot, a French mineralogist, analyzed a large number of bones from different geological horizons, and showed that their fluorine-content increased with geological age. He proposed a mean F-amount in bones, which increases as age increased, but it is now well known that these figures are only averages, and that one cannot date a bone by estimating its F-content only, because depending on the locality, fluorine may be either abundant or rare in the groundwater.

Hydroxyapatite (HAP) $Ca_{10}(PO_4)_6(F,OH)_2$, the main component of bones and teeth, readily takes up fluorine and is transformed into the more stable and water-insoluble fluorapatite (FAP). This process will not end before the total exchange of HAP into FAP is achieved, with a corresponding fluorine concentration reaching 4.8 at.% This value is much higher than the average F-concentrations in both groundwater and soil. During the fossilization process, a progressive exchange of fluorine for hydroxyl groups occurs in the apatitic components of bone or tooth. Using this principle, Eisenbarth and Hille used neutron activation to propose a calibration curve up to 1 million years for the fluorine and nitrogen amounts for bones found in the northeast part of Austria.[24] Their results allow rather precise estimation of the age of remains coming from the same geographic area. Valuable information was obtained by NRA on numerous samples coming from New Zealand and the Pacific islands.[25] Some special applications of the method were pointed out, for instance, when numerous samples have to be studied rapidly, and when remains are only several hundreds of

[24] P. Eisenbarth, P. Hille, A nondestructive method for age determination of fossile bone, J. Radioanal. Chem. **40**, Issue 1–2, (1977) pp. 203–211.

[25] Coote, G. E. & Sparks, R. J., Fluorine Concentration Profiles in Archaeological Bone: An Application of a Nuclear Microprobe, New Zealand J. Archaeol., **3** (1981) 21–32.

years old, instances for which ^{14}C dating is not accurate enough. It would be, however, hazardous to draw such assumptions for any type of archeological remains. The amount of fluorine incorporated in bones and teeth depends indeed on many external parameters such as the speed of circulation of water, the chemical and geological characteristics of the surrounding soil, the nature and structure of the calcified tissue, etc.

2.1.1 The Piltdown hoax and the Tepexpan man

The first example of the use of fluorine analysis was in bringing the Piltdown Man hoax to light. In 1912, Charles Dawson, an amateur paleontologist, found what was thought to be an early human skull and jaw (Fig. 9) in the Piltdown gravel deposit in East Sussex, England. Because it had an ape-like jaw and was found in association with the bones of extinct animals, this "Piltdown Man" was immediately believed to be a very ancient human. At that time, it was popularly acclaimed as "the missing link" in human evolution by recognized scientists, including a young priest, Teilhard de Chardin, and writers such as Arthur Conan Doyle. Many authorities suggested that these remains were about 500,000 years old, and perhaps even one million! At that time, there was a trend among the scientists to think that it was the spirit which was the driving force in evolution laws, and finding a modern human skull associated with an ape-like jaw was grist to their mill. After a long period, the controversy was more or less forgotten, and for over three decades, the scientific community accepted the Piltdown man as an authentic artifact. In 1950–53, the Piltdown bones were finally tested for their fluorine content by K. Oakley,[26] and the fraud became apparent. The fluorine content of the Eoanthropus skull averaged 0.1%–0.2%, whereas in the mandible and teeth, it was as low as <0.03%. In startling contrast, the bones of the older elephant and of the mastodon found in the same deposit, that is, the Lower Pleistocene remains, showed 1.9%–3.1% of fluorine. Thus, the age of Piltdown man was thus evaluated to be closer to 50,000 years than 500,000 years. In addition, the skull and jaw clearly were not from the same time period. The jaw was likely to have come from a modern young adult orangutan. The forgery was further confirmed through X-ray fluorescence examination. The teeth and mandible had been cleverly carved to fit the skull and stained to look ancient. Unfortunately, by 1950, Charles Dawson and all of the other people involved with the Piltdown Man discovery and analysis had died, so we do not know for sure who was responsible for the hoax. At about the same time, F-test was applied to the bones of the Swanscombe skull found deep in a gravel deposit in Kent, England, and accepted by geologists as certainly contemporary with the containing deposit, which is rich in Acheulian hand-ax. It proved to contain 2% fluorine, thus confirming an early Paleolithic age. These conclusions were further confirmed using more accurate physicochemical investigations, leading to an estimated age of 200,000–300,000 years for the Swanscombe man, an archaic *Homo sapiens*.

[26]Oakley, K. P., C. R. Hoskins New Evidence on the antiquity of Piltdown man, Nature **165** (1950) 379–382; Wiener, J. S., Oakley K. P., Le Gros Clark W. E., The solution of the Piltdown problem, Bull. British Museum (Natural History). Geology **2**(3) (1950) 139–146; Oakley K. P., The fluorine-dating method, Yearbook of Physical Anthropology, **5** (1951) 44–52; Oakley K.P., Weiner J. S., Chemical Examination of the Piltdown Implements, Nature **172** (1953): 981–982.

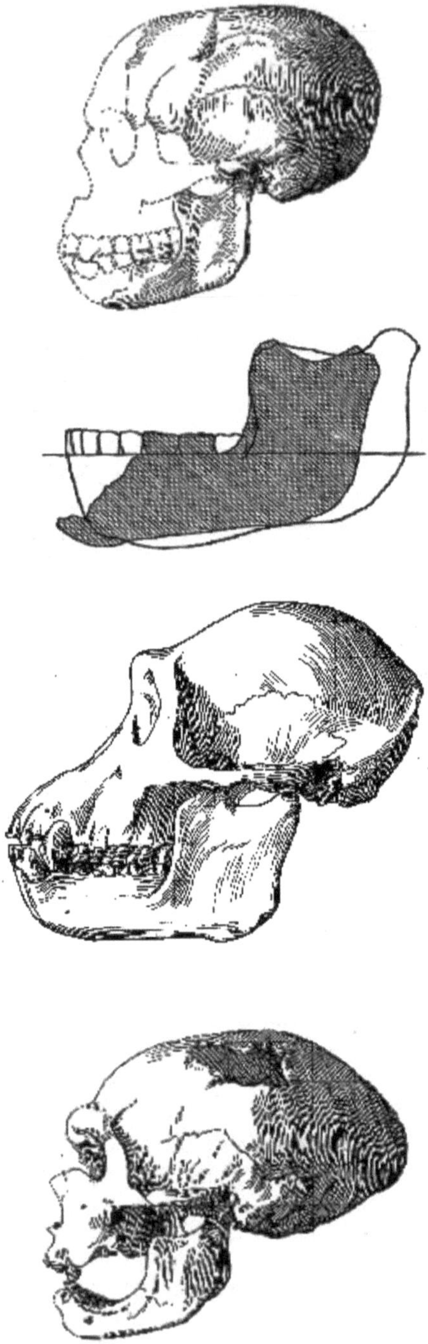

Fig. 9 Original illustrations from Dawson's paper (1913). From top: Skull and mandible of Piltdown's *Eoanthropus dawsoni*; mandible from Piltdown superimposed on that of *Homo heidelbergensis*; orang-utan skull; Neanderthal man skull and mandible from La Chapelle-aux-Saints, Corrèze, France, after M. Boule.

Another important information was obtained through F-dating regarding the occupation of Mesoamerica during the late Pleistocene. In 1949, the skeletons of two mammoths were excavated by H. de Terra near Tepexpan, on the northern shore of Lake Texcoco, 36 km northeast of Mexico City. In the same geologic layer, a human skeleton was found: the Tepexpan "man." It was shown that these mastodons had been killed with spears fitted with lance-like stone points, and butchered on site. Initially, a possible date of about 80,000 BP was suggested for the Tepexpan man. While the association of the human bones with the mammoths was questioned at first, fluorine tests proved them to indeed be contemporary. Now, most geneticists agree that Native Americans are descended from Siberians who crossed over the Bering Strait into America 26,000–18,000 years ago. Further analysis showed that these remains were indeed "only" 10,000–13,000 years old, and that Tepexpan "man" was, in fact, an indigenous American Indian female (*Homo sapiens*). These remains are some of the oldest human archeological evidence found in Mexico.

2.1.2 Early hominid remains

For early hominid remains, fluorine dating is able to produce valuable information for a relative dating. The method is again based upon the principle that fossil bones take up fluorine from surrounding soil. When skeletal remains are buried in the earth, water inundates the bones with a solution of minerals drawn from local soils. The main mineral component of bones is hydroxyapatite, which acts as a trap for fluorine ions present in most ground waters. During the F/OH exchange, the hydroxyapatite crystals are converted into fluoroapatite, which is a more stable mineral with regard to weathering. Because of the porosity of bones, the alternation generally proceeds throughout the center of the material. The amount of absorbed fluorine is dependent upon the amount in the surrounding deposits and the time the bone has been buried. Varying concentrations of fluorine in different deposits preclude the method from being considered absolute, and thus, fluorine dating is chiefly used to determine whether bones or human skeletal remains were buried at the same time as other bones found close to them. In other words, fluorine analysis is primarily used for verifying whether or not two bones or fossils in the same strata were in fact contemporaneous. If they are not, they most likely come from different eras, and then, at least one of them must have been displaced from its original site.

2.1.3 Earliest modern human skeletons in Japan and East Asia

The above situation is often the case for *Homo erectus* fossils found on Java Island. Many parts of the more than 50 different specimens which have been found there have been incidentally discovered on the surface of the ground by inhabitants. A possible combination of circumstances, that is, fossil washed out from an outcrop, rolled down in a stream, etc., may have pushed the remains far from their original site. In the absence of conclusive information from their real environment, it is difficult to establish the true chronology. Kondo et al. have shown recently that fluorine dating, associated with a multi-elements analysis, allows to effectively distinguish the bone stratigraphic sources and to successfully discriminate the provenance of several fossil remains from Java.[27]

The origin of the first inhabitants of the Japanese archipelago remains a subject of controversy, particularly for the occupation before 35,000 years.[28,29] A forgery in one of the key sites has disrupted some previous assumptions.[30] Regarding the occupation of the archipelago by modern men during the Pleistocene, it seems established now that several fairly distinct populations have mixed. Recently, fluorine dating has been successfully used to date the human cranial and postcranial remains (Fig. 10) that have been found from 1967 to 1974, together with many animal bones unearthed in limestone fissure deposits at the Minatogawa quarry near the southern coast of Okinawa Island. Analyses of fluoride concentrations in the different bones (human and animal) and different levels are given in Fig. 11. The analyses have been carried out using conventional analytical procedure, as proposed by Matsu'ura.[31]

As shown in Fig. 11, the fluorine contents of the boar bones from the upper deposit (Level 1–4) are grouped around 0.7%–1% F, whereas those from the lower deposit (Level 5–8) appear to be in two separate groups. It is worth noting that these two types of samples have different colors; whereas the six boar specimens with high fluorine content have a light color on their surface, the other three boar specimens are dark gray or brown. This finding, together with the fluorine analysis results, may indicate that three of the nine fossils which were found at the lower deposit are probably coming from an upper horizon. Unlike other sites of similar period, such as Liujiang and Upper Cave in China,[32] there is no concern over the reliability of the dating of Minatogawa man; the fluorine contents of both the human and non-human (boar, deer) bones within the site suggest that they were contemporaneous.

[27] Kondo M., Matsu'ura S., Aziz F., Sudijono, Fujimori E., Sawatari H., Haraguchi H., A multi-element approach to the relative dating of bone fossils, Abstr. A231, Goldschmidt Conf., Kurashiki, Japan (2003).

[28] Keally C. T. (1991b). A Model for the Origins of the Japanese Paleolithic. Paper presented at the seminar "Ancient Relations between Japan and the Continent," 36th International Conference of Orientalists, May 20, 1991, Tokyo, Japan.

[29] Oda S. Nihon no Kyusekki Bunka (The Japanese Palaeolithic Culture). Tokyo: Doseisha (2003) (in Japanese).

[30] A renowned Japanese archaeologist was caught in October 2000 to hide newer stone tools in the site Kamitakamori (Ancient Paleolithic,dated 600,000 years BP) where he directed the excavations (see Shukan Shincho, "Can the 500,000-year-old site really be believed" March 9, 2000, pp. 53–54 (in Japanese), and the website of CT Keally on Japanese archeology: http://www.t-net.ne.jp/~keally/.

[31] Matsu'ura S., Determination of fluorine in fossil bone with an ion-selective electrode, Bull. National Science Museum, Tokyo, Ser. D, **7** (1981) 7–16; Matsu'ura, S., Fluorine dating of Upper Minatogawa man, J. Anthropol. Soc. Nippon **92** (1984) 111–112; Matsu'ura, S., in "The Minatogawa Man," H. Suzuki, K. Hanihara Eds., Bull. Univ. Museum, Univ. Tokyo **19**, chapter 6 (1982).

[32] Brown P., The first Mongoloids: another look at Upper Cave 101, Liujiang and Minatogawa1. Acta Anthropologica Sinica **17** (4) (1998) 255–275; Brown P., The first modern East Asians? A view from the Late Pleistocene and Neolithic of China and Japan. In K. Omoto, Ed. "Interdisciplinary Perspectives on the Origins of the Japanese", International Research Center for Japanese Studies, Kyoto, (1999) 105–131; Brown P., Chinese Middle Pleistocene hominids and modern human origins in east Asia, in Barham L.S., Robson-Brown K., Eds., "Human roots: Africa and Asia in the Middle Pleistocene," Western Academic & Specialist Press, Bristol (2001) 135–147.

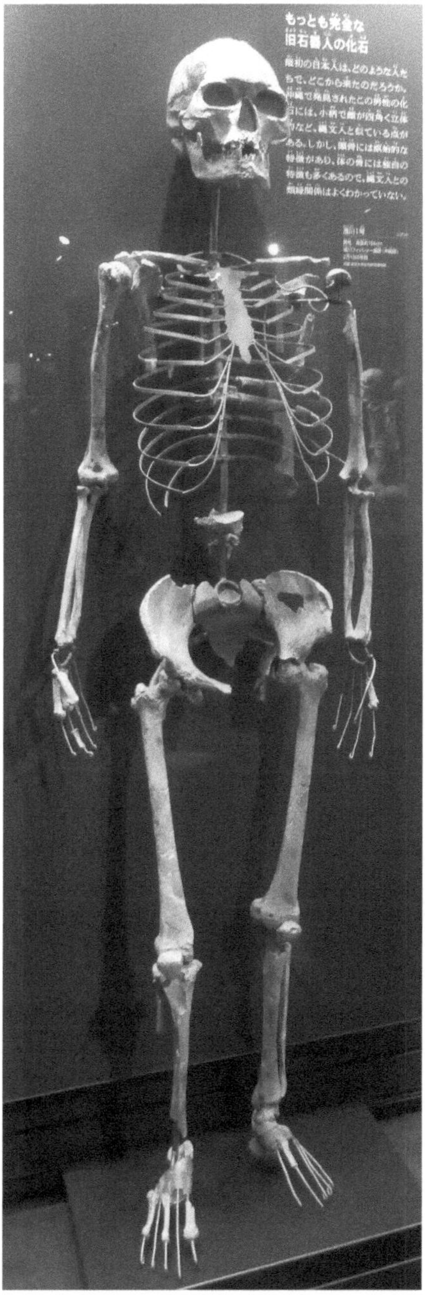

Fig. 10 The Minatogawa man 1 skeleton is that of a relatively short person, approximately 1.53 m tall and the skull is correspondingly small but robust for its size (skeleton housed in the National Museum of Nature and Science, Tokyo, Japan).
Credit: Photaro CC BY-SA 3.0 (https://creativecommons.org/licenses/by-sa/3.0), via Wikimedia Commons.

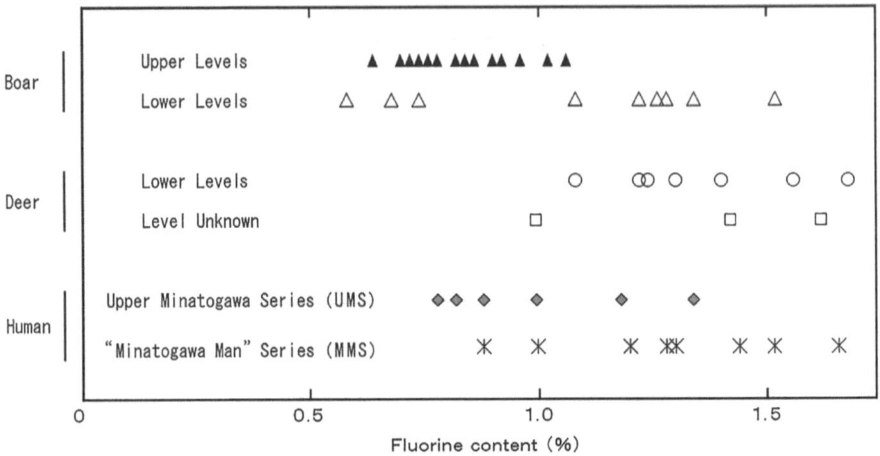

Fig. 11 Fluorine content of fossil bones from Minatogawa site.
Used with permission from Shuji Matsu'ura.

The conclusion that the Minatogawa man lived in the Upper Pleistocene seems to be supported by the bones' high fluorine content (0.9%–1.6%). However, it must be pointed out that the decisive effect of hydrological conditions is not taken into account. Charcoal fragments from the lower deposit inside the fissure of the Minatogawa site have yielded radiocarbon dates of 18,250 ± 650 years BP and 16,600 ± 300 years BP.[33] Assuming that the site was well stratified, that the carbon dates do bracket the skeletons, and that the skeletons were not intrusive, then the Minatogawa remains can be considered to be the earliest modern human skeletons (*Homo sapiens*) in East Asia, with those found in Luijian and Shandingdong sites in China. However, following Brown, it should be added that these Late Pleistocene men can be clearly assigned as Asian aborigines, bigger and skeletally more robust than the actual Japanese men. So, in conclusion, they cannot be recognized as ancestors of any modern East Asian population.[33]

2.1.4 Limits of the method

Kottler et al. have pointed out the limitations of the fluorine diffusion method to yield absolute dating.[34] About 10 bones of 1000 and 6000 years from several archeological sites in Switzerland were investigated using RNA in order to shed more light on the feasibility of the method, more particularly on the consistency between

[33]Kobayashi H., Hirose Y., Sugino M., Watanabe N., *Radiocarbon* **16** (1974) 384.

[34]Kottler C., Döbeli M., Krähenbühl U., Nussbaumer M., Exposure age dating by fluorine diffusion, Nucl. Inst. Meth. Phys. Res. B **188** (2002) 61–66.

measurements obtained from different parts of the same object and from different objects found at the same site. One of the reasons for misinterpretation is the fact that fluorine is not distributed homogeneously throughout a bone, but has its own diffusion profile. Thus, not only the average F-concentration, but also the shape and range of the concentration profile should be directly related to the exposure age. Since their samples (either human or animal) came from different archeological sites and had different structures, they could not surmise a general relationship between the depth of the fluorine diffusion profiles and the exposure age of archeological bones. In bones with a loose structure, for instance, diffusion is combined with convective penetration of the object through cracks and voids with a possible percolative character.

On the other hand, bones that remained for a long time in archeological sediments have been shown to present an elemental composition very different from that of modern bones because of diagenetic alteration.[35] The extent of the modification in element composition depends on the local burial conditions in the archeological layer. Thus, analysis of the surrounding sediments should be carried out systematically in order to assess their contribution to the modification of the bone's chemical composition. In the general case of ancient bone and dentine specimens from various origins, ranging from the Paleolithic period of the middle Miocene (several tens of thousands of years) to the Neolithic period (4000 BD), TEM is shown to be a very appropriate technique to reveal the peculiar recrystallization phenomena occurring during either diagenesis in soil or heat treatments, and it should be associated with other analytical methods.[36] During the heating process, the organic component is eliminated, and the remaining apatite forms a more homogeneous medium; thus, the growth of apatite crystals is possible, leading to the formation of large polygonal crystals of about 100 nm. By measuring the size and shape of apatite crystals, it may be possible to evaluate the temperature at which archeological bones or dentine have been heated. The nature of elements which have been in contact with the sample during the diagenesis process is of decisive importance, because it can be noted that, in most studied cases, the crystals were hydroxyapatite, whereas, in the case of mastodon tusks, it was found to be almost pure fluoroapatite. More recently, comparing fluorine dating measured by neutron activation analysis (NAA) and direct AMS radiocarbon, R.L. Lyman considered that the F-method is valid only in cases when the chronological validity of its results is confirmed with independent chronometric data.[37]

In consequence, it can be concluded that dating bones using fluorine diffusion gives valuable information, indeed, on the relative origin and dating of the

[35] Reiche I., Favre-Quattropani L., Calligaro T., Salomon J., Bocherens H., Charlet L., Menu M., Trace element composition of archaeological bones and post-mortem alteration in the burial environment, Nucl. Inst. Meth. Phys. Res. B **150** (1999) 656–662.

[36] Reiche I., Vignaud C., Menu M., The crystallinity of ancient bone and dentine: new insights by TEM, Archaeometry **44** (2002) 447–459.

[37] R. L. Lyman, C.N. Rosania, M.T. Boulanger, J. Field Archaeol. 2012, **7**, 3, 226–237.

archeological objects, but should be applied only in special cases where the nature of surrounding sediments is perfectly known. The on-site presence of very similar objects of known age for comparison is of great help, and so, there should be a multi-technique approach involving up-to-date tools such as HRTEM, RNA, FTIR, NMR, etc. Estimation of the influence of environmental and material factors on the evolution of the fluorine distribution could be deduced from NRA, PIGE, and PIXE methods.[38]

2.2 Dating archeological lithic artifacts by fluorine diffusion

When prehistorical items contain either carbon or heated ceramics, it is clear that ^{14}C radiocarbon, AMS radiocarbon, or thermoluminescence on fired materials are among the preferred methods. However, in the case of lithic samples, mostly flints, the hydration profiles in aluminosilicates, including the rate of migration of fluorine within the material, are able to produce a convincing relative dating. In these cases, a hydrated layer is formed on the surface of the mineral by reaction with surrounding wet soils, and the determination of the thickness of this layer allows one, in the best case, to deduce the age of the object.[39] Given the very small variations in the level of the nutrient, technical ion beam analysis (IBA Ion Beam Analysis) gives, by its high accuracy, a new impetus to this type of study.[40,41] Fluorine concentration gradients in silica-based objects (obsidian or flint) also provide valuable information about the origin of these materials. This research must be strictly associated with a multi-element analysis to provide converging experimental evidence.[42,43] This method of analyzing the fluorine content has been used to various types of lithic materials, for which different types of scattering profiles have been proposed (see "Analysis of traces of fluorine by ion beam analysis"). Fig. 12 shows the particle accelerator AGLAE ("Accélérateur Grand Louvre d'Analyses Elémentaires") used for such purpose. This accelerator allows producing protons with energy up to 4 MeV and α particles up to 6 MeV.

[38] A-M Gaschen, Relevance of Fluorine Diffusion for Exposure Age Dating in Archaeological Bones and Teeth, PhD thesis, Univ. Bern (2005).

[39] Taylor R.E., Fluorine diffusion: a new dating method for chipped lithic materials, World Archeology **7** (1975) 125–135.

[40] Taylor R.E., Advances in Obsidian glass studies, Noyes Press, Park Ridge, USA (1976).

[41] Jin J., Weathers D. L., Biscar J. P., Hughes B. F., Duggan J. L., McDaniel F. D., Matteson S., Convolution Fitting and Deconvolution Methods for Fluorine Depth Profiling by Resonant NRA, AIP Conf. Proc. **392** (1) (1997) 681; Jin J., Weathers D. L., Picton F., Hughes B. F., Duggan J. L., McDaniel F. D., Matteson S., Nuclear Reaction Analysis study of fluorine uptake in flint, AIP Conf. Proc. **475** (1) (1999) 554.

[42] Bellot-Gurlet, L., Poupeau, G., Dorighel, O., Calligaro, T., Dran, J. C. Salomon, J., A fission track dating/PIXE approach to sourcing studies of obsidian artefacts in Columbia and Ecuador. J. Archaeol. Sci., **26** (1999) 855–860.

[43] Döbeli M., A.A.-M. Gaschen, U. Krähenbühl, Fluorine Analysis by Ion Beam Techniques for Dating Applications, pp. 215–252, Fluorine and the Environment, Vol. 2, A. Tressaud Ed., Elsevier (2006).

Fig. 12 Particle accelerator AGLAE (for Accélérateur Grand Louvre d'Analyses Elémentaires) used for IBA analyses at C2RMF.
© C2RMF / Jean-Louis Bellec.

Analysis of traces of light elements by ion beam analysis

Development of analysis of light elements' traces, such as fluorine, have been greatly helped by the use of powerful Ion Beam Analysis techniques (IBA), in addition to other physical–chemical techniques such as FTIR, ESR, or NMR spectroscopies, XPS, SEM–EDX, TEM, etc. IBA includes Rutherford backscattering (RBS) and Resonant Nuclear Reaction Analysis (NRA) PIGE, PIGME, or PIXE. Among these, NRA methods are particularly suitable for analysis of light elements such as fluorine.[43]

Rutherford backscattering spectrometry (RBS): During the bombardment of a surface with a charged particle beam, some of the particles will be elastically scattered in the backward direction with respect to the incident particle. The particle loses energy along the beam path in the sample, and therefore, its energy is dependent on the depth, t. For example, the thickness of a metallic film on a silicon wafer can be studied down to a few nm.

Particle-induced X-ray emission analysis (PIXE): Generally, a 3 MeV proton beam is used to generate characteristic X-rays in a solid sample. A proton creates a vacancy in the inner electronic shell of the atom. The vacancy is filled with an electron from the outer shells. The energy can be released during de-excitation by the emission of an X-ray or an Auger electron. The X-rays originating from the electronic transition from the L shell to the K shell are usually referred to as K_a, and the transition from the M shell is called K_b, according to the Siegbahn notation. The energy of the emitted X-rays is characteristic for the elements in the sample, and its intensity yields the concentration of the element. Only elements heavier than fluorine can be detected. The method is non-destructive and sensitive because concentrations of elements down to a few 100 ppb can be measured, with a spatial resolution of the beam at 1 mm and the penetration depth in a solid sample at about 100 μm.

Particle-induced gamma-ray emission analysis (PIGE): Similar to PIXE, PIGE (or PIGME) analysis utilizes a nuclear reaction, that is, (p, p′ gamma), (p, gamma), or (p, alpha gamma), for which an emitted gamma ray is detected. The setup used for PIGE analysis is the same as for PIXE, except that the detection of the emitted gamma rays is usually done by large volume Ge detectors instead of the IGP detector. The energies of the emitted gamma rays depend on the nuclide. PIGE is normally used in conjunction with PIXE, and RBS is used to measure concentrations of low Z elements such as Li, F, Na, Mg, Al, N, C, and Si. The detection limit of F element is 2.10 atoms/cm^2. As an example, a 2.7 MeV proton microbeam is used for detecting traces of fluorine via the gamma rays emitted by the reaction $^{19}F(p,\alpha\gamma)^{16}O$, in which a proton reacts with a ^{19}F atom forming a ^{16}O atom, and emitting α and γ particles (see the scheme of the method). This reaction has several resonances in the incident proton range from 0.3 MeV to 1 MeV. The resonance at $E_p = 0.872$ MeV was chosen for this measurement because it has a cross-section of 540 mb and a resonance width of 4.5 keV.

These NRA methods schematized in the figure are currently applied to geological samples for determining the F-content in mineral particles, particularly in meteorites, and may be used in archeology for measuring fluorine diffusion in ancient bones and archeological stone matter.

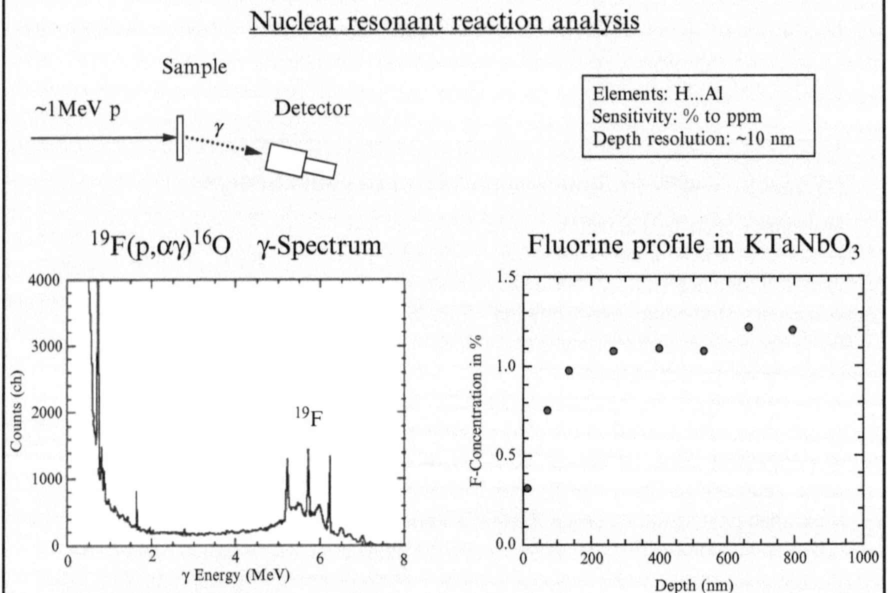

Basic principles of NRA.

The analysis of tools and weapons found in prehistoric sites has allowed us to better imagine the lifestyle of our distant ancestors. In 1975, Taylor extended the method to archeological silicas.[40] More recently, dating by fluorine was used to determine the residence time of meteorites on Earth, especially on the ice of Antarctica. Fluoride analysis is now obtained by powerful spectroscopic methods such as XPS, auger electron spectroscopy (AES), X-ray microprobe, resonant nuclear analysis, or ion beams. Some examples below illustrate the archeological interest in the method.

2.2.1 Dating bronze-age arrowhead flintstones of Fort-Harrouard, France

For instance, using the F-dating method, Walter could solve an interesting problem concerning the prehistoric site of Fort Harrouard near Paris.[44] This site has been occupied continuously from the Neolithic period (4500 BP) until the Roman period (500 AD), and in the same levels, bronze arrowheads were found mixed together with flint arrowheads of similar shapes. The question was whether humans of the Bronze Age used those flint arrows (Fig. 13) they found on site, or did they continue to develop an industry belonging to a past era? P. Walter showed that the penetration of the fluorine in the flint depended on the length of contact between the object and the surrounding environment. The FWHM of a fluorine profile obtained through the nuclear reaction $^{19}F(p,\alpha,\gamma)^{16}O$ is more or less linear with the contact time.[45] Fig. 14 shows fluorine penetration into flint arrows found in various levels of the Bronze Age (200–300 nm) and certify that the people of that time continued the ancient tradition of carving flint arrowheads, and they did not reuse old Neolithic flints, in which a penetration of over 400 nm was found, because of a longer contact time.

Fig. 13 Flint arrowheads found on Fort Harrouard prehistorical site, near Paris.

[44] Walter P., M. Menu M., Vickridge I. C., Fluorine depth profiles as a relative dating method of chipped flints. Nucl. Instr. Meth. Phys. Res., B**45** (1990) 119–122.

[45] Walter P., Menu M., Dran J. C., Dating of archaeological flints by fluorine depth profiling: new insights in the mechanism of fluorine uptake. Nucl. Instr. Meth. Phys. Res., B**64** (1992) 494–498; Walter P., Altération géochimique et datation en archéologie. Prodgs Congrès Atome et Géologie, Mém. Soc. Géol. France, **162** (1994) 249–260.

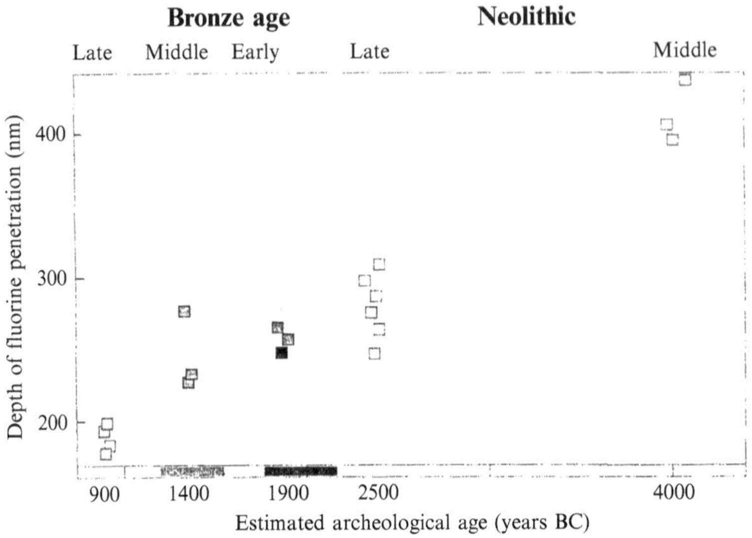

Fig. 14 Dependence of fluorine penetration into the flint samples on the estimated archeological age (900–4000 BC). Results obtained from $^{19}F(p,\alpha,\gamma)^{16}O$ nuclear reaction.

2.2.2 Following trading routes in South Pacific islands using obsidian artifacts

For over 20,000 years, obsidian objects, made of natural vitreous silica, have been the subject of frequent trading between populations living in South Pacific islands, sometimes as much as several hundred kilometers remote one from one another. Melanesians found these obsidians on the slopes of the numerous volcanoes which are located in this area of the globe, constituted by the intersection of several tectonic plates.

Four main groups of islands were defined where stocks of these items were found: Admiralty Islands, New England, Entrecasteaux, and New Hebrides (Banks Islands, Vanuatu). Archeological data confirmed that these exchanges lasted from the Pleistocene era, and are still continuing today. The use of hyphenated techniques PIXE/PIGME allowed a much finer approach to the issue[46,47] (Fig. 15). The multi-element ion beam analysis, from Si to Nb, and more specifically, for Fe by PIXE, and Na, Al, and F by PIGME analysis over a wide range of 1200 objects[48] provided the following information. Obsidian of the Admiralty Islands contains the highest rate

[46]Summerhayes G. R., Bird J. R., Fullagar R., Gosden C., Pecht J., Torrence R., Application of PIXE-PIGME to archaeological analysis of changing patterns of obsidian uses in West New Britain, Papua, New Guinea, Chapter 6, pp. 129–158 in "Archaeological Obsidian Studies," M. S. Shackley Ed., Plenum Press (1998).

[47]Torrence R., Bird J. R., Bailey G., Johnson P., Description of the procedure for PIXE-PIGME analysis at ANSTO, Internat. Assoc. for Obsidian Studies Bull. **23** (1999) 22.

[48]Duerden P., Bird J. R., Clayton E., Cohen D. D., Leach B. F., Provenance studies of New Zealand obsidian artefacts, Nucl. Instr. Meth. **B3** (1984) 419–423.

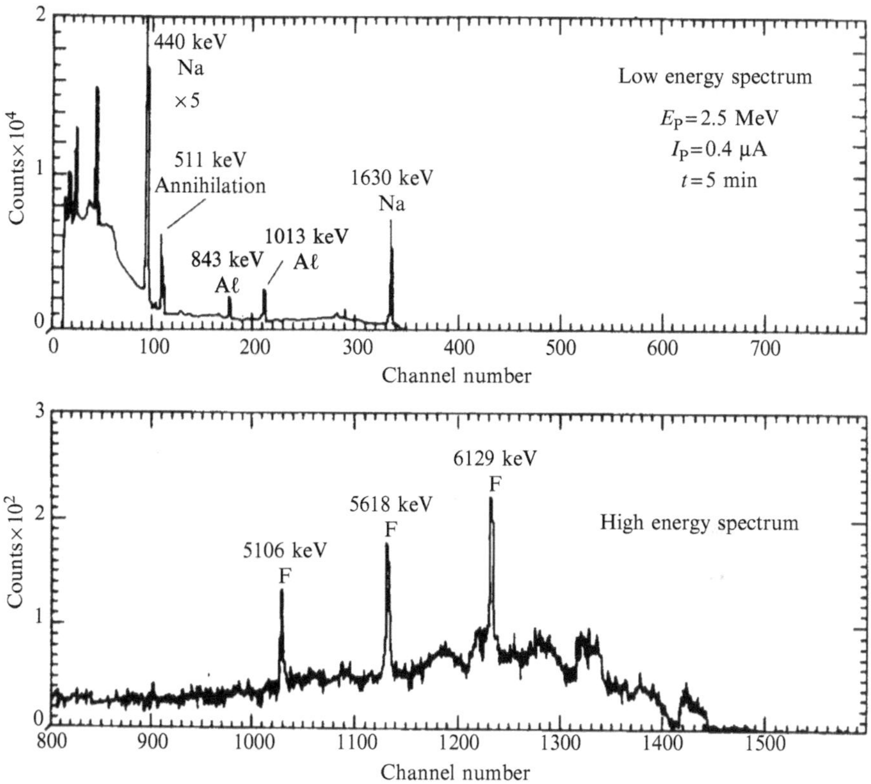

Fig. 15 PIGME spectrum of an obsidian sample obtained after 10 min γ-rays irradiation. Informations relative to fluorine are located at high energies (between 5 and 7 MeV). Reproduced from Duerden P., et al. Nucl. Instr. Meth. B3 (1984) 419–423, with permission from Elsevier, Amsterdam.

of fluorine. That of New England has low rates of F and Na; those of the New Hebrides and Vanuatu have low rates of F and Al. Despite the overlapping of certain data, the chart of Fig. 16 shows that it is possible to find areas where most samples were collected by considering the relationships between the elements F, Al, and Na present in the items. It is now possible to know the original volcano which produced the obsidian dagger, sometimes the date of release, and thus, to trace the Melanesian trade routes for many centuries (Fig. 17).[49]

[49]The hydration speed is around 10–20 μm^2 par millennium and depends on both the variations of paleoclimatic conditions and nature of soils. The knowledge of diffusion coefficients can only be attained with standard samples and allows dating objects between 200 and 100.000 years old. For instance the average error for 2000 years old obsidians can be evaluated about 10%.

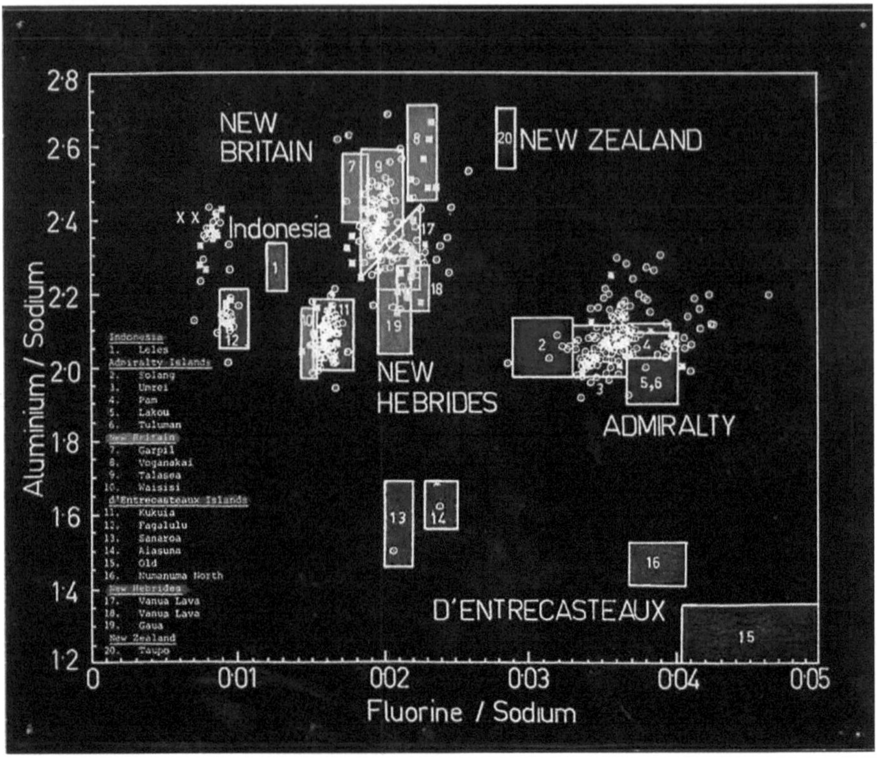

Fig. 16 Correlations between the amounts of Al, Na, and F present in obsidian lithic trades in Papua-New Guinea (Al/Na and F/Na ratios).
Reproduced with permission of D. Cohen, ANSTO, Australia.

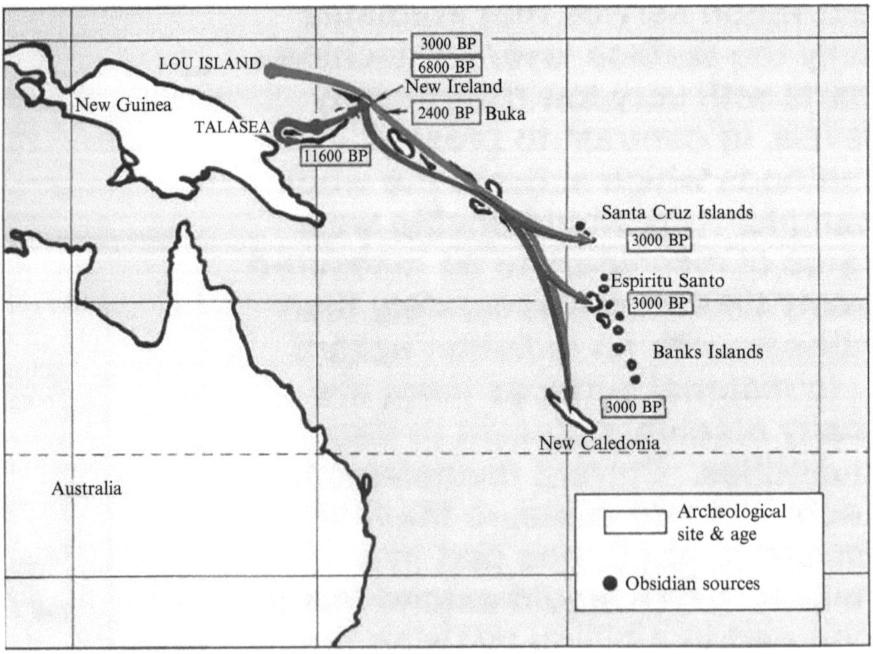

Fig. 17 Archeological trading routes 12,000–2000 BP in New Guinea-Papua area.

3 The true nature of the Roman murrhine through a critical lecture of Pliny the Elder's "Naturalis Historia"

Who knows that the most appreciated gems in Greece and ancient Rome were fluoride crystals? In many Latin texts, the term murrhina ("vasae murrhinae") often appears, which designates a category of large vases and luxury items. Various spellings of the word (murina, myrrhina, murrha, myrrha, murra) occur in many manuscripts. From the descriptions of such items, the aspect of the material that was used to make these vases could be associated with fluorite ores exhibiting very different colorations, but could also present some analogies with iridescent glasses or porcelain wares. Two puzzling questions remained: what could this precious mineral be? What relationship could exist between these wares and myrrh (myrrha), the aromatic gum so important in Biblical times?

It seems now established that murrhine artifacts were indeed made from a mineral—not from a glass or ceramics—and that the ancient Greeks and Romans carved these crystals as a gemstone.[50] In their workshops, they made admirable vases with bright colors that were celebrated in Imperial Rome as murrhine vases. Due to the scarcity of the raw material, many fakes were made of glass, and, in turn, gave rise later to techniques yielding imitation gemstones, such as Venetian millefiori glasses.

The exceptional value of these crystals, which seems to have been recognized in Rome from the end of the republic, is attested by Suetonius in his *Life of Augustus*: "He (Augustus) also showed that he had little passion for luxury, when after the capture of Alexandria, he reserved for himself, from all the kings treasures, only one murrhine vase for his daily use, and he melted all the other golden vessels."[51] Other examples of the high value of murrhine objects are found for example in the *Satires* of Juvenal.[52] But it is especially Pliny the Elder (23–79) (Fig. 18), the early encyclopedist who, in Book XXXVII that highlighted his Natural History, published around 77 AD (*Naturalis Historia*[53]), provides several very pictorial examples of the value that the noble Romans attached to these gems. For Pliny, the most precious of all stones was indeed the exceptional murrhine vases which were introduced in Rome by Pompey the Great in 62, after his victory over Mithridates, and that he presented as an offering to the temple of Jupiter.

But the most important testimony coming from *Naturalis Historia* regards the geographical origin of various gems used during the Roman Empire, including murrhine. Pliny's book makes it possible to draw a map of possible places of production,

[50] A. Tressaud, M. Vickers, Ancient Murrhine Ware and Its Glass Evocations, *J. Glass Studies*, **49** (2007), pp. 143–152.

[51] Suetonius (70–140), Life of Augustus [2,71] LXXI. Latin text, "The Latin Library." French translation, M. Cabaret-Dupaty, Paris, 1893, revised by J. Poucet, Louvain, 2001.

[52] Juvenal (65–128), Satire VI, French Translation, H. Clouard.

[53] Pliny the Elder (23–79), Naturalis Historia, Book XXXVII, Book XXXIII (5). Latin text in "Itinera electronica," Univ. Catholique Louvain, Belgium, http://agoraclass.fltr.ucl.ac.be/concordances/intro.htm#pa.

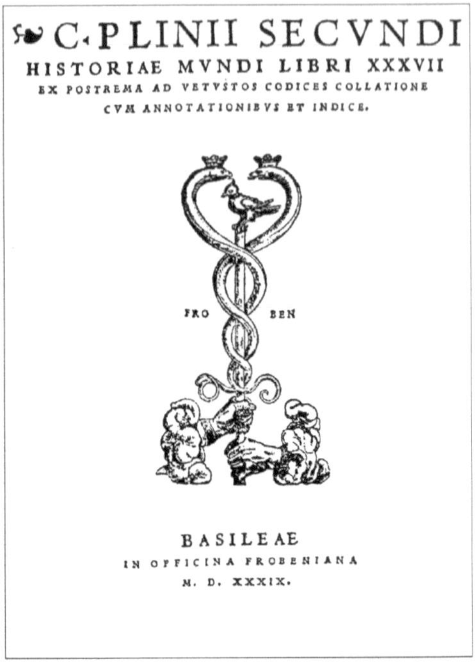

C. PLINII SECVNDI
HISTORIAE MVNDI LIBRI XXXVII
EX POSTREMA AD VETVSTOS CODICES COLLATIONE
CVM ANNOTATIONIBVS ET INDICE.

FRO BEN

BASILEAE
IN OFFICINA FROBENIANA
M. D. XXXIX.

Fig. 18 Statue of Pliny the Elder, Plinius Gaius Secundus (23–79), on the facade of the Como cathedral, Italy (photo: A. Tressaud), and cover of the *Naturalis Historia*, edited by Froben, Basel (1539).

"The murrhine are imported from the East and are found there, in several locations which have nothing remarkable, particularly in the Parthian empire, but the most beautiful are from Carmania."[54] Carmania was between Ariane in the North, the Persian Gulf and Eritrea Sea in the South, Persia in the West, and Gadrosie in the East, and corresponds to the Kerman area in Iran today. Indeed, in imperial times, trade between Rome and the Eastern provinces had greatly increased. The method of extraction and carving operations were also described.

The exact nature of this material has generated many controversies.[55] In a 1919 article on the subject, E. Babelon proposed that chalcedonies, that is sardonyxes or agates, were the component of the murrhine vases.[56] He thus followed the identification that

[54]Ibid. Book XXXVII (8) Latin text: "Oriens myrrhina mittit. Inueniuntur ibi pluribus locis nec insignibus, maxime Parthici regni, praecipua tamen in Carmania. umorem sub terra putant calore densari."

[55]L. de Launay, Mineralogy Veterans, p. 85, Paris 1803; Mr. Rozière, Memory murrhine on vases that once brought into Egypt, and those who made it, or Egypt Description Series observations and research that have been made in Egypt during the expedition of the French army, Volume VI by Edme-Francois Jomard and al. 1822.

[56]E. Babelon, Murrhina vasa Antique Dictionary, p. 2046C. Daremberg, E. Saglio, Hachette, 1919.

Agricola proposed four centuries earlier.[57] Loewenthal and Harden,[58] and a few years later, Bromehead,[59] refuted this assumption point by point, advancing a series of critical arguments. The most useful clues were given through the richness of details given by Pliny in the description of colors, transparency, and light effects characterizing these crystals. These similarities have led many researchers to associate fluorspar or fluorite (CaF_2) to murrhine. If we list the various points described by Pliny, one can deduce: (i) Murrhine is a mineral, often extracted together with rock crystal, which eliminates ceramics and millefiori glasses, as it has sometimes been proposed[60]; (ii) the scarcity of the material limits the size of the items to small containers or drinking cups[61]; (iii) the detailed description of hues and colors lead to the conclusions that antique murrhine had a glossy luster that showed a range of colors ranging from purple to blue and white, often with a reddish tint, and that its diffraction of light could indeed produce a rainbow (*see* Fig. 1); (iv) another feature of the material was its fragility. The material that accounts for Pliny's descriptions is clearly fluorspar (CaF_2):

- First, it is clear that fluorite crystals are extracted from mines; the fact that quartz crystals are often found together in the same veins reinforces this hypothesis.
- Moreover, the dimensions of the fluorspar crystals are generally limited, and therefore, the cut pieces are reduced.
- Purple and blue colors are actually the most characteristic of fluorite, and Pliny's general description gives colorations perfectly matching the characteristics of this mineral.
- Fluorite's ease of cleavage and low hardness (4 in the Mohs scale, i.e., just below that of tooth enamel) is well known compared to the characteristics of quartz, carnelian, onyx, or agate ($\sim$7 of hardness index).
- The high density of murrhine cups, compared to crystal ("crystal cups and heavy murrhine") corresponds to the difference in specific masses existing between quartz ($\sim$2.65) and fluorite (between 3 and 3.25).
- Finally, it is well known that large pieces of all other possible minerals are not so rare, as evidenced by the large number of exceptional objects discovered during excavations. So, it is the extreme rarity of large pieces of fluorite[62] that made such items so valued for Roman emperors and some rich patricians.

Today, all these convergences, range of color, fragility, close aspect with precious stones, etc., provide a whole body of evidence that reinforces the correlation between

[57] G. Agricola De Natura Fossilium, Basle, 1546.

[58] Loewenthal A.I., D.B. Harden (Vasa murrina), J. Roman Studies 39, 1949 31–37, with a mineralogical note of C. E. N. Bromehead.

[59] C. N. E. Bromehead, Antiquity, XXVI (1952) 65–70.

[60] Note that Pliny speaks namely of glass imitations of vases, i.e., Diospolis murrhine (presumably from Thebes in Egypt) that were transported through Red Sea harbors. Furthermore, in Venice, so-called murrhina glasses (vetro murrino) are known since the sixteenth century as a composition based on pasta or colored glass rolls, twisted and melted together to shape various objects (see A. Dorigato, The Murano glass, Citadels & Mazenod, Paris, 2003).

[61] "Naturalis Historia", Book XXXVII (8) "with size no exceeding small vessels (abacus)" [Latin text: "amplitudine numquam paruos excedunt abacos"].

[62] One of the largest fluorite single crystal (30 × 20 × 20 cm), from Dalnergosk, Ural, Russia is exhibilted at MNHN, Paris (specimen No. 193-26).

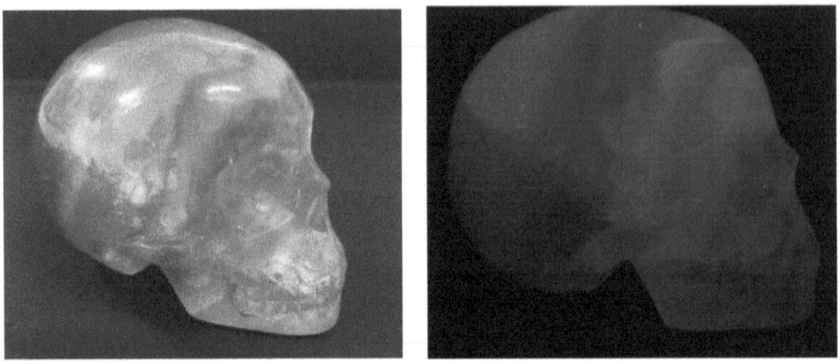

Fig. 19 Skull carved in a fluorite (CaF_2) crystal, before and after UV irradiation at $\lambda = 365$ nm, showing intense blue luminescence (photo A. Tressaud; experiment by Dr. A. Garcia, ICMCB-CNRS, France).

fluorite crystals and murrhine. This mineral has an almost infinite range of colors; when pure, it is colorless and transparent, but it can be yellow, pink, green, blue, orange-red, almost black, banded, a rainbow of multiple colors, or may even present some inclusions of colorful octahedra in a transparent matrix, or vice versa. Why not also consider that the exceptional optical properties of such crystals could have associated them with some magical character? Some varieties, such as the chlorophane are indeed fluorescent, phosphorescent, thermoluminescent, and/or triboluminescent, that is to say, when the crystals are first illuminated, heated, or even ground, they emit a green or blue glow that can last some time (Fig. 19).

One of the most amazing features of these murrhine/fluorspar objects was the smell that seemed to be associated with them: "The smell of this substance also gives it more value."[63] The explanation of this phenomenon may be related to one of the decisive characteristics of fluorite, which is, by the way, at the origin of the name fluorspar, that is, these crystals may be easily cleaved. Together with a low hardness, this makes the substance very difficult to properly cut into pieces. This hypothesis has the further advantage of providing some answers to the unforeseen relationships between the two Latin words *murrhina* (or *murrha)* of these crystals and myrrh (*myrrha* in Latin[64]), the natural resin of great importance in ancient Mediterranean civilizations since Biblical times. The cutting of the original crystal block was likely made using a chisel and a mallet and would have been extremely delicate, given the fragility of the material.

[63]"Naturalis Historia", Book XXXVII (8) "Some fragrance is noticeable" [Latin text: "Aliqua and in odore commendatio est"].

[64]The etymology of the word myrrh, according to Webster's Dictionary (1913): mrr Semitic, murru Akkadian, murr Arabic, mor Hebrew: bitter. Myrrh is a natural gum yellow-brown or amber, with aromatic odor and bitter taste. This substance is extracted from shrubby plants of Abyssinia and Arabia: *Balsamodendron myrrha*. Myrrh of the Bible is supposed to come from these plants and also a species of Cistus. This resin has been used since immemorial time for its aroma and its medicinal properties. Its use during the funeral ceremonies is attested in the Gospel according to John. (On the subject, see the articles by JN Lupia in Johannine Literature and Ethical Dualism of Food in the Gospel of John, by HJ Hodges, Annual Meeting, Society of Biblical Literature, Boston, USA, Nov. 20–23, 1999.)

This drawback could have been circumvented by gradually heating the crystal coated with a substance that could melt, penetrating the cracks of the crystal, thus reinforcing its strength and helping achieve the appropriate cut. Plunging the crystalline block into a gum, which could actually have derived from myrrh, would have helped avoid disintegrating the fluorite block during cutting. Obviously, drinking afterwards in such glasses would have given a particular taste to the wine, as prized by the Romans. However, an Iranian etymology of the word murrhina cannot be excluded, because of the geographical origin of these crystals. Note that for centuries, the cutting processes of "Blue John" fluorspar crystals from Derbyshire, United Kingdom, has included a step of dipping the crystal piece in hot rosin, a solid form of resin obtained from pines.

It can thus be assumed that exceptional specimens of fluorspar have been found in the Parthian kingdom in regions close to Carmania and raised in the treasury of Mithridates at Talaura before being transported to Roma. Fluorspar deposits are still present in this region, especially in the islands of the southern coast of the Persian Gulf, either in the form of translucent purple or light-violet crystals, or as pink, green, purple, or black crystal, that is to say, matching Pliny's description word for word. It should be noted that Iran currently produces about 20,000 tons/year of fluorspar, and the most important mining centers are located either in the central part of the country (Ardakan, Yazd), that is, north of the Kerman area (Carmania, as mentioned by Pliny), or in the southern islands of the Strait of Hormuz.

It is unfortunate that there are almost no remaining items of these ancient wonders of Imperial Rome. Corsi mentioned two antique fluorite vases that he saw in 1845 at the Museo Kircheriano in Rome.[65] Two "Blue-John" vases would even have been identified during the excavations of Herculaneum, but cannot be found in museums, so far. Hopefully, three exceptional pieces have come to us, the "Crawford cup" which has been in the British Museum since 1971, and two other pieces that were described and photographed in the article by Loewenthal and Harden.[58] One is a cup with one handle, whose photograph was provided in 1949 to the authors by its owner, Baron Stoclet, and the other, a cantharus with two handles. The cup likely originated, like the Crawford cup, from a Roman tomb discovered during World War I at the border between Turkey and Syria. It made its entry recently into the British Museum.[66] Carved into a single block of fluorite, the coloration includes green, purple, and cream (Fig. 20). The authors describe the cutting edge of perfect regularity, the polished surface that gives a shiny look, and a rainbow effect when observed under illumination. Such a description perfectly corresponds to that given by Pliny for murrhine vases. The similarities between these pieces and the onyx "Ptolemy" cup kept in the National Library (Paris) allowed these authors to date with certainty these two items as Roman, and more precisely, from the Early Empire. There are many reasons to believe that we are faced with one of those vases of which Pliny the Elder spoke emphatically in his

[65] C. F. Corsi Delle Pietre Antiche Trattato, Roma, 1845.

[66] In 2004, the British Museum acquired this rare Roman fluorite cup for the sum of £ 150,000. One can compare this amount to the one proposed by Emperor Nero, who would have paid 1,000,000 sesterces to acquire the same type of cup, as related by Pliny.

Fig. 20 The "Crawford" cup, Imperial Roman fluorite/murrhina cup (GR 1971.4-19.1). Reproduced with the British Museum permission, London. ©The Trustees of the British Museum.

Natural History. Note that another translucent fluorspar cup is also present in the Ashmolean Museum in Oxford.[67]

Figurines and various objects have continued to be carved in fluorite crystals over the centuries, as evidenced by this pair of urns from the late eighteenth century (Fig. 21). Note that during investigations that led Henri Moissan to isolate fluorine, the scientist might have used an electrolysis cell cut from a large block of fluorite crystal (fluorspar) (Fig. 22). The interest of such a material is that it does not react with fluorine and fluorinated gases. However, it is not proven that Moissan actually used this type of cell, for airtight reasons.

4 Georgius Agricola, the father of mineralogy and creator of the word "fluoride"

The attractive properties of fluorspar ores (CaF_2) as fluxes in metallurgy had begun to be recognized from the Middle Ages. These additives were called "Flußspat" by German miners. It is said that that this use was mentioned for the first time by Basilius Valentinus in the late 15th century.[68] But it was G. Agricola who brilliantly described the importance of these fluorinated ores. Georg Bauer was born in Saxony in 1494. His study of the humanities led him to various European universities, first in Leipzig, then in Bologna and Padua. During this period, he Latinized his name to Georgius Agricola. In a Europe that began to be shaken by the ideas of the Reformation, his wisdom and openness allowed him to approach the great minds of his time, especially D. Erasmus. Very influenced by the ideas of his mentor, Agricola remained loyal to the Catholic Church, but in a spirit of openness and tolerance. From 1526, Agricola

[67]M. Vickers, Hamilton, geology, stone vases and taste, J. Hist. Collect. **9** (1997) 263–274.

[68]Gmelins Handbuch der anorganischen Chemie, 8. Auflage, System-Nr. 5, "Fluor," Verlag Chemie, Leipzig/Berlin 1926, p. 1.

Fig. 21 A pair of fluorite urns (XVIIIth century).
Reproduced with permission of Galerie du Rhône, Sion, Switzerland.

Fig. 22 Electrolysis cell cut from a block of fluorite.
Document Henri Moissan Museum, Faculty of Pharmacy, University Paris V, Photo A. Tressaud.

returned to Saxony, where he was appointed doctor first in Joachimsthal[69] then in Chemnitz, where he ended his days in 1555. The location of the city on the eastern part of the Erzgebirge Mountains, one of richest mining centers of the time, allowed him to visit mines, deposits, and foundries as soon as he could escape his official duties. The discussions he had with experienced miners, particularly Lorenz Bermann or Wermann, provided the material for most of its future books. Besides *Bermannus sive de re metallica dialogus*, a treaty on the mining district,[70] successively appeared *De natura Fossilium*, always at J. Froben, an editor in Basel, *De natura eorum quae effluunt ex terra* and *De ortu and causis subterraneorum*, where he rightly attributed the origin of geological deposits by sedimentation of aqueous origin, and he described in detail the effects of erosion.[71] But the major work of Agricola, *De re metallica* (Fig. 23) was published after his death in 1556. He was the very first to account

Fig. 23 Portrait of Georgius Agricola.
Cover of the original Latin version of *De re metallica*, Basel 1530, with illustration of iron foundry operations using fluoride fluxes.

[69]Today Jáchymov in Czech Republic.

[70]Books of G. Agricola: "Bermannus, sive de re metallica dialogus" (1530); "From ortu and causis subterraneorum" (1546); "De natura effluent eorum quae ex terra" (1546); "From veteribus and novis metallis" (1546); "De natura Fossilium" (1546); "De re metallica" (1556).

[71]The fossils in question are minerals and what we call fossils today were regarded at the time as daring achievements of nature. Although the similarity between fossils and living organisms was widely observed, it was not until the early 18th century that it was recognized that these "stones" had been indeed living organisms. At Agricola's time, it was rather some of their medicinal properties, even magical, that were put forward, such as prescribing *lapis judaicus* powders (Echinoderms of Pentremites type) for stone disease, or fossil oysters to reduce lung inflammation.

for the use of fluorinated minerals as fluxes, such as fluorspar, and he even proposed to use the word "fluorine" ("fluores" in Latin). In his books, he related that miners of the Erzgebirge Mountains were aware that some stones could melt when heated, "Lapides qui facile igni liquescent," facilitating the extraction of metals by fusion of the ores (see "–'Stones that easily melt by heating' Lapides qui facile igni liquescunt"). Thanks to all these works, Agricola is recognized as the father of mineralogy. He found the best way to bring his work to posterity by structuring *De re metallica* into 12 books (or chapters), providing clear explanations of the various stages of techniques and processes, and asking an artist Basilius Wefring to illustrate the variety of situations in the most educational manner possible. We find in this book the foundation of natural science mostly based on experimentation that would be later called geology or earth science. He also added: "I have omitted all things that I had not seen by myself or read or heard from people of confidence." This view sometimes led him to question the classical authors like Aristotle and Pliny the Elder, whose statements were taken at that time as dogma. And if in his preface he quoted the ancient authors, it omitted some more recent scientists such as Avicenna (980–1037), Roger Bacon (1214–1294), Albertus Magnus (1193–1280), and even his contemporary and neighbor, Paracelsus (1493–1541). These omissions can be attributed to a certain distrust of the medieval alchemist tradition of Arab-Latin origin. Agricola's work, like that of Pliny to some extent, is marked by an obvious concern for writing down the memory of science and technology of the early 16th century, thanks to a comprehensive inventory of the human knowledge available on the subject in his time. This first scientific description of casting techniques initially written in Latin was quickly translated into many languages, including Chinese. His contemporaries, like Erasmus and Melanchthon, and later, Goethe, highlighted the decisive position he had held in the field of earth sciences. Closer to us, the first translation into English of *De re metallica* was proposed in the early 20th century by a mining engineer, the future US President Herbert C. Hoover, with the help of his wife, Lou Henri Hoover, a geologist and Latinist.[72] Herbert Hoover gave us this very wise definition of this great precursor: "He was the first to found any of the natural sciences upon research and observation, as opposed to previous fruitless speculation."

"Stones that easily melt by heating" (Lapides qui facile igni liquescunt)

G. Agricola was the first to focus on the use of fluorinated minerals as fluxes, even creating the word "fluorine." In *De re metallica*, such a use is shortly reported; it is said that miners of Erzgzbirge Mountains name certain stones "fluores" (in Latin in the original text), that melt when heated ("Lapides qui facile igni liquescunt") facilitating the extraction of metals by fusion of the ores. These points were developed in many of his previous works, *Bermannus*, *Rerum Natura Metallicarum* and *De Fossilium Interpretation*. In *De Natura Fossilium*, he wrote: "Finally, there is now these stones I call, 'Stones that easily blend into the fire,' because thrown into a furnace, they start to melt...These are fluxes ('fluores sunt' in the original Latin text)."

In his first treatise on mining, *Bermannus, sive de Re Metallica* (dedicated to D. Erasmus in 1530), two scholars, Ancon and Naevius, are featured, in a method often used at the time,

[72]Detailed comments on *De Re Metallica* can be found in the first English translation of the Latin text of 1556 by H. C. Hoover and L. H. Hoover, The Mining Magazine, London (1912).

talking with Bermannus,[73] an engineer particularly skilled in mining. The conversation turns to the minerals found in the Erzgebirge Mountains. The way they describe the various types of stones shows the perfect knowledge of the works of ancient authors, especially Pliny's "Natural History." The appearance of these stones, the similarity with precious stones, and several other features perfectly coincide with the description of murrhine by Pliny. In addition, Agricola first released the melting properties of these rocks that had been known, apparently, for centuries in the development of glass, enamels, and other products. The description given by Bermannus perfectly matches that of fluorite crystals, but it is clear that in his day, it was not possible to differentiate between fluorinated minerals and those derived from quartz and silicates. Furthermore, fluorite deposits were mostly localized with oxides or sulfides. Nevertheless, it can be concluded that Agricola was the first to propose for these minerals the Latin word "fluores," borrowed from the German word "fluss," commonly used by the miners of that time, as shown in the dialog below between Bermannus and Naevius.

Bermannus: A dialogue on mining

NAEVIUS—What is this, Bermannus?

BERMANNUS—These are flows/fluxes ("fluorinated sunt" in the original Latin text). Let me take this word of another. Thus, our miners call it, in my opinion without impropriety. Because it looks like quite a molten and frozen material, just as ice, except that it has diverse and pleasant colors.

NAEV—These stones, Theophrastus says, the ἐχ συρροής, that is to say, formed by earthen runoff. These fluxes—I will use now such words—are red and appear identical in red raw silver, you have told us before.

BERM—At first glance, it is true, but it is not so, because fluxes are often quite translucent.

NAEV—These are carbuncles?

BERM—No, thereof in any case.

NAEV—How can we distinguish them?

BERM—First, because the colors do not shine in the eye as with carbuncles (any red gemstone, most often red garnets), although some parts are transparent. When they are opaque, they also differ in this respect. Finally, the fluxes of all types begin to flow as soon as they feel the heat,[74] while carbuncles do not liquefy.

NAEV—Your distinctions are clear.

BERM—You see here another type, a more diluted purple hue.

NAEV—This looks like the cheap amethyst so abundant in Bohemia.

BERM—This is certainly not dissimilar, so you can easily sell them—and they are sold—for precious stones for people who are not familiar with amethyst, if they are set in rings. But here's a third kind with a white shade.

NAEV—Looks like rock crystal.

BERM—A fourth species is yellowish, a fifth, ashy, and a sixth, almost black.

ANC—But what is the use of flux?

[73] A former companion of Agricola's studies in Italy, Jean Naevius was a doctor in Annaberg, then Joachimsthal; Nicolas Ancon, was also a physician, expert in Arabic medicine. Bermannus (from the German word "bermann," or miner), illustrates the advanced technology of the time and could have been Lorenz Bermann, master of mines, another friend of the author.

[74] Such observation has sometimes been confusing, resulting from relationships (promoted by Pliny the Elder) between heat and crystal colors. In the previous chapter, we explained that the ancients thought that minerals were created permanently, and the heat from the depths of Earth was required for the coloration of crystals. A *carbuncle* is any red gemstone, most often a red garnet.

BERM—It is customary to add them to the extraction of metals with fire. They make the melt much more fluid, indeed, and a stone taken from the "flint," as we have said, which is made not far from here in Breitenborn, near Schwarzenberg.[75] From these fluxes, further pigments for painters can also be drawn.

5 In search of the missing halogen, from the Renaissance to 1886

In the course of the 18th century, numerous observations showed that when an acid, for example, sulfuric acid, was added to a fluorspar sample and the mixture was gently heated, toxic fumes evolved, attacking glass containers. It is said that a Nuremberg glassmaker, Heinrich Schwanhard, had the idea in 1670 to use these properties for the production of glassware. Note that this technique is widely used to this day.

The first report written on this chemical reaction was published in 1725 by John George Weygand, relating to a recipe he inherited from Matthäus Pauli from Dresden, and the first scientific article on flußspat from Saxony was published in 1768 by Andreas Marggraf (1709–82).[76] In 1771, the Swedish chemist Carl Wilhelm Scheele (1742–86) began systematic work to discern the chemical nature of the crystals found in the fluorspar ores and their reaction mechanisms with acids. By adding sulfuric acid to fluorite crystals, he observed that the glass was also attacked by the formation of fumes. The dissolution of these fumes in water led to an acid solution that Scheele named "flußspatsäure," fluorspat acid, or fluoric acid. Added to carbonated water, a white precipitate (CaF_2) was isolated and dried with sulfuric acid.[77] So, C.W. Scheele was the first to identify the nature of fluorspar ore. Besides hydrofluoric acid, this scientist prepared and studied many other toxic gases such as H_2S, and HCN. Presumably, his untimely death was the result of a slow poisoning by these chemicals, because Scheele used to taste his preparations. We can note that this scientist, like all his contemporaries, thought that the acid was part of fluorspar ore. Even if the names proposed for this new acid (e.g., "acide fluorique" in France, "acido fluorico"in Italy) remind us today of the element fluorine, nothing was known at the time about the chemical nature of the constituents of the acid in question. Some critics even questioned Scheele on his results, suggesting that this acid was a derivative of the sulfuric acid used in the experiment. And even when Scheele showed long after his first experiments that the same results could be obtained by reaction of nitric or phosphoric acid on fluorspar, some others suggested that the resultant product was due to a modification of these acids.[78]

[75] In Saxony's Erzgebirge Mountains, south of Chemnitz and Zwickau.

[76] A. Margraff, Observation concerning outstanding volatilization of some of the stone species to which we give the names of Flosse, Flüsse, Flus-Spaht and also that of Hesperos, Memoirs Royal Academy of Sciences of Berlin, translated from German, Volume XXIV, 3–11, 1768.

[77] C. W. Scheele: Sämmtliche physische und chemische Werke. Nach dem Tode of Verfassers deutscher Sprache und in gesammelt herausgegeben Sigismund von Friedrich Hermbstädt, 2 Vols., Reprint of the edition of 1793, Niederwalluf, 1971, Vol. 2, pp. 3–31 (first published in 1771).

[78] B. Pelletier, Donadei L.: "Memoirs of lime phosphate", Ann. Chim. 7 (1790) 79.

The true nature of hydrofluoric acid began to emerge with the 19th century. When muriatic acid (HCl) was identified as a phase containing chlorine, a recently discovered element, it was recognized that oxygen (from the Greek, ὀξύς "generating acid") was not the essential component of acids, as was supposed before. By comparison, fluoric acid was thought to consist of hydrogen and another element with properties similar to those of chlorine. In 1809, when the discovery of sodium and potassium by the English chemist Humphrey Davy (1778–1829) was announced in France, André Marie Ampère (1775–1836) anticipated the existence of chlorine and fluorine as chemical elements, but did not publish his hypotheses. Struck by the analogies between muriatic acid (hydrochloric acid) and fluoric acid (hydrofluoric acid), he concluded that this was due to an element he first called oxyfluorique, and then fluorine, once he was convinced that hydrofluoric acid contained no oxygen. Among the letters he exchanged with Davy during the war raging between the two countries, on November 1, 1810, he even suggested the possibility of isolating the element fluorine by subjecting anhydrous hydrofluoric acid to electrolysis: "The question is whether electricity would decompose hydro-fluoric acid in its liquid form, when as much as possible water would have been removed, carrying hydrogen on one side and oxy-fluoric on the other side. The only drawback to fear is the combination of oxy-fluoric with the conducting piece. But, assuming that the oxy-fluoric was, as the oxy-muriatic, unable to combine with carbon, this latter element might be good enough conductor to be used successfully in such in experience."[79] In response in February 1811, Davy did not accept the existence of a single body proposed by Ampere, but considered it contained oxygen. "In the views that I have ventured to develop, neither oxygen, chlorine or fluorine are asserted to be elements." However, after 3 years of correspondence, Ampere left the fame to Davy as, in 1813, the latter announced the discovery of this new element.[80] An autobiographical note, however, established the priority of his discovery to Ampere, in fact well recognized by Davy: "During the period I was engaged in these investigations, I received two letters from M. Ampere of Paris, containing many ingenious and original arguments in favor of the analogy between the muriatic and fluoric compounds. M. Ampere communicated his views to me in the most liberal manner; they were formed in consequence of my ideas on chlorine and supported by reasonings drawn from the experiments of MM. Gay-Lussac and Thénard."

The quest for this new element occupied numerous researchers throughout the 19th century. A first step was the preparation of pure water-free hydrofluoric acid by Louis-Jacques Thénard (1777–1857) and Louis-Joseph Gay-Lussac (1778–1850). The product they produced spewed thick smoke into the air, rapidly dissolved glass, and caused very stubborn burns, a phenomenon that had been observed for a long time. Later, J.J. Berzelius (1779–1848) characterized ammonium fluoride.[81] During this period, several researchers paid a heavy price in search of fluorine.[82] G. and T. Knox suffered from severe fluorine intoxication, and the Belgian chemist P. Louyet died.[82] Around

[79]Correspondence of Grand Ampere, Gauthiers-Villars, Paris (1936) (containing the correspondence between the two scholars from 1810 to 1825).

[80]H. Davy, Philos. Trans. Royal Soc. London, **102** (1812) 352; ibid. **103** (1813) 263; ibid. **104** (1814) 62.

[81]Berzelius JJ., From the hydrofluoric acid, in Chemistry Treaty, Vol. 2, Brussels, 1833, p. 179.

[82]Rabuteau A. P. A., Toxicology Elements, Paris 1887; Roholm K, Fluorine Intoxication, Kopenhagen—London 1937.

1860, Jean Charles de Marignac (1817–94) synthesized and described in a surprisingly accurate manner the morphological characteristics of many crystals of anhydrous or hydrated salts such as fluorotitanates or fluorozirconate (see "J. C. de Marignac and the first synthetic fluoride crystals").[83] Henri Étienne Sainte-Claire Deville (1818–81) and Edmond Frémy (1814–94) also characterized many inorganic fluorides. In 1869, the great Russian chemist Dmitri I. Mendeleev (1834–1907) perfectly positioned this element in the periodic table that he had conceived.

A decisive step was taken when Frémy succeeded in preparing high-purity anhydrous hydrogen fluoride and KHF_2, also called Frémy's salt, "KFl·HFl" in the notation of that time.[84] G. Gore (1826–1904) determined its physical-chemical characteristics. It should be noted that in 1836 Frémy came close to the solution by electrolyzing salts based on molten calcium or potassium fluoride. Frémy has also long been credited with the paternity of this isolation because in 1856, he had already observed the evolution of a gas decomposing water with production of hydrofluoric acid: "It emerges from the neck of the platinum retort an odorous gas decomposing water to produce hydrogen fluoride, and which displaces the iodine content toward the halide: the gas seems to be fluorine."[85] However, at the temperature at which he operated, Frémy could not go further in the isolation of fluorine, and he never considered using KHF_2, presumably because of technological difficulties related to the high melting temperature of the compound, $T_m = 293°C$. So, in June 1886, Henri Moissan (1852–1907)[86] an alumnus of Frémy, finally succeeded in isolating fluorine, concluding this long chapter in the history of chemistry, as highlighted J. R. Partington (1886–1965), "The long isolation of fluorine was one of the master problems of inorganic chemistry."[87]

Jean-Charles de Marignac and the first synthetic fluoride crystals

The singular character, outside of any school and away from scientific battles taking place in most European capitals, Jean-Charles de Marignac (1817–1894) taught all his life at the Academy of Geneva, later transformed to a university.[88] Marignac performed all his work alone without co-workers or assistants, mostly in the lab he had established in his personal home. Upon returning from his "post-doc" training in the Giessen laboratory of J. Liebig, Marignac

[83] A. D. Galissard de Marignac, Collected Works, Volumes I & II (1840–1887), Physics Society of Geneva, Masson, Paris (1890).

[84] JR Partington, The early history of hydrofluoric acid, Mem. Proc. Manchester Lit. Phil. Soc. **67** (1923) 73.

[85] Frémy E., Ann. Chim. Phys. **47** (1856) 5.

[86] Among the many publications of H. Moissan on fluorine and fluorine compounds, some may be mentioned: Recherches sur l'isolement du fluor, Annales Chim. Phys. 6ème série, t. XII (1887) 472. Nouvelles recherches sur le fluor, Annales Chim. Phys. 6ème série, t. XXIV (1891) 224. Fluorure double de chrome et potassium, Annales Chim. Phys. 7ème série, t. II (1894) 66. Le bifluorure de platine anhydre, C. R. Acad. Sci. **109** (1889) 807. Le Fluor et ses Composés, G. Steinhel Ed., Paris (1900) /Traité de Chimie Minérale, Masson Ed., Paris (5 volumes, 1904–1906).

[87] J. R. Partington, General and Inorganic Chemistry, MacMillan, London (1947).

[88] Information on the life and work of Marignac can be found in the "Book of the Centenary of the Ecole Polytechnique," Gauthier-Villars, Paris (1897), including a description of his scientific activities by E. Grimaux.

began his research on complex fluoride salts with the purpose of determining the atomic weights, and the similarities between simple bodies. He wanted to bring experimental evidence to the discussion of Prout's hypothesis (that is to say, that atomic weights are multiples of that of hydrogen) and Mitscherlich's law of isomorphism. This work drew praise from Berzélius: "These experiences seem to have been executed with particular accuracy, and were repeated with a patience worthy of praise. It is hoped that chemists who will undertake a revision of atomic weights should add to the great accuracy of Marignac, his patience and scrupulous conscience."

It is by examining the crystalline forms of the complex inorganic fluoride salts that he sought to resolve the issues he had raised. His first job was to note that fluosilicates and fluostannates are isomorphic, and silicon, for which he definitely fixed the atomic weight at 28 (contrary to the opinion of Berzelius), can replace tin without altering the morphology of the crystals. The study of fluotitanates and fluozirconates allowed Marignac to confirm the atomic weight of titanium and zirconium. Transforming the "potash fluoxyniobate: $NbOF_3$, 2KF" into a fluoniobate "NbF_5,2KF" by the action of hydrofluoric acid provided Marignac an argument to prove that the body once considered to be niobium was actually an oxide of this metal. Similarly, in the controversy that pitted him against M. Hermann, he aptly demonstrated that white and yellow chlorides of this element, quoted as simple chlorides "$NbCl_3$" and "$NbCl_6$," were actually $NbOCl_3$ and $NbCl_5$. Note that in 1886, Marignac also proposed a method of separating niobium and tantalum, a process that remained valid for more than one century. This method took advantage of the property that in diluted HF, relative insoluble K_2TaF_7 was formed, whereas, niobium led to the very soluble hydrate $K_3NbOF_5 \cdot 2H_2O$.[89]

Marignac works on complex fluoride salts also concerned fluotantalates, fluoxymolybdates, fluoxytungstates, and represent a considerable amount of work.[90] His studies are marked with rare precision and accuracy. Even today, looking through his laboratory notes it is possible to reproduce exactly the results of his experiments. For example, the crystal growth of $NH_4CuTiF_6 \cdot 4H_2O$ reproduced according to Marignac's procedure, demonstrated in 1967 the one-dimensional nature of this phase, with chains formed of elongated octahedral $Cu(H_2O)_4F_2$ and regular TiF_6 interconnected by their opposite vertices and separated by sequences of $(NH_4)^+$ and F^- ions.[91] The F^- ions, which are not directly bound to the transition elements, are tetrahedrally coordinated by 4 H_2O via short bonds OH…F of 2.52 Å.[92]

[89]Note that now, a solvent extraction technique is used. For example, tantalum is extracted with methyl isobutyl ketone from aqueous solutions of HF and niobium can be removed by acidification of the aqueous phase. After conversion to pentoxides, metals are obtained by reduction or by electrolysis of molten fluorides.

[90]Complete Works of A. D. Galissard de Marignac—Volumes I & II (1840–1887), Société de Physique de Genève, Masson, Paris (1890), containing his major titles: Fluosilicates et fluostannates, Biblioth. Univ. Archives, t. II, pp. 89–101 (1858). Recherches sur les formes cristallines, Annales Mines, t. XV, pp. 221–290 (1859). Fluozirconates, Annales Chim. Phys. 3e Série, t.LX, pp. 257–307 (1860). Tungstates et fluotungstates, Annales Chim. Phys. 3e Série, t.LXIX, pp. 5–86 (1863). Combinaisons du tantale (Fluotantalates), Biblioth. Univ. Archives, t. XXVI, pp. 89–116 (1866). Recherches sur le niobium et l'ilménium, Réponse à M. Hermann, Biblioth. Univ. Archives, t. XXVII, pp. 373–378 (1866). Fluosels de l'antimoine et de l'arsenic, Biblioth. Univ. Archives, t. XXVIII, pp. 5–19 (1867).

[91]A. De Cian, J. Fisher, Acta Cryst. **22** (1967) 338; J. Fisher, G. Keib, R. Weiss, ibid. **22** (1967) 340.

[92]W. Massa, Z. Anorg. Allgem. Chem. **436** (1977) 29.

6 The isolation of fluorine on June 1886 by a pharmacist and chemist: Henri Moissan

The memorable experience of the isolation of fluorine gas (F_2) took place on June 28, 1886, in the preparation laboratory that professors H. Debray and C. Friedel had generously made available to Henri Moissan, near the School of Pharmacy, in temporary quarters established to replace the old Sorbonne, which was then being demolished. Although an associate professor at the School of Pharmacy since 1882, H. Moissan did not yet have his own laboratories. After several unsuccessful attempts starting from phosphorus and arsenic fluorides, H. Moissan succeeded in isolating fluorine by subjecting anhydrous hydrofluoric acid to low-temperature (−50°C) electrolysis, making the substance into a conductor by adding molten acid potassium fluoride, or fluohydrate of potassium fluoride KF·HF (or KHF_2). The procedure and precautions outlined by E. Frémy, his former boss, had been very useful for obtaining pure and anhydrous hydrofluoric acid, as well as for preparing KHF_2. Electrolysis was carried out in a platinum U-tube (copper subsequently replaced platinum) with platinum electrodes (Fig. 24). The resulting gas, fluorine (F_2), barely tinted yellow, was very toxic and proved to be a powerful oxidizing agent, igniting all organic materials in contact therewith, and directly

Fig. 24 Electrolytic cell that allowed the isolation of fluorine and 1906 Nobel diploma presented to Henri Moissan.
Document Henri Moissan Museum, Faculty of Pharmacy, University Paris V, Photo A. Tressaud.

reacting (often violently) with almost all elements. One of the most difficult problems of inorganic chemistry had been resolved.[93]

This great discovery, which earned France the 1906 Nobel Prize in Chemistry (Fig. 25), was the culmination of centuries of hide-and-seek with an element whose reactivity was so important that, although its properties had long been recognized, no one had ever been able to definitively synthesize it. H. Moissan's decisive idea was to use KHF_2 in solution in anhydrous HF (aHF) as electrolyte. Indeed, aHF could not operate alone because its electrical conductivity was too low (1.4×10^{-5} - Scm^{-1} at $-15°C$). Moissan chose platinum for building the cell because he thought that the metal would resist fluorine well, but Moissan lowered the reaction temperature to $-50°C$ to limit corrosion. In the cell, the F^- ions were oxidized at the anode where they produced F_2 (see Fig. 24). Moissan announced a light yellow-green gas. Note that some authors raised some doubts about this observation, connecting this color to the presence of impurities. A 1986 re-examination of the color of this gas in a 5 m long silica tube confirmed the accuracy of Moissan's observation[94] (Fig. 26); the yellow color of liquid F_2, cooled down 77 K, is shown on the same figure.

On the Monday following the discovery, H. Debray presented the Academy of Sciences with a short note from Moissan, "Action of an electric current on the anhydrous hydrofluoric acid," in which he wrote: "Allow me not definitive conclusions of this action of the current on hydrofluoric acid. We can, indeed, various assumptions about the nature of evolved gas; The simplest would be that one is in the presence of fluoride, but it would be possible, for example, that it was a hydrogen perfluoride or even a mixture of hydrofluoric acid and ozone active enough to explain the action if energetic as this gas exerts on the crystallized silicon. I'm now only indicate these first results ..." Moissan's extreme caution was justified by the failures of his predecessors, combined with the importance of the discovery. It remains true that for Moissan, the certainty of having actually obtained fluorine gas was more firmly established than this note seemed to show. Because of the high stability of silicon fluoride, this non-metal should combine rapidly with fluorine, as Moissan actually noticed from the gas obtained in these experiments. Following that note, the Academy appointed a commission of renowned chemists, Berthelot, Debray and Frémy, making them responsible for observing the accuracy of Moissan's work. For the occasion, they prepared a highly pure hydrofluoric acid with great care. The experience turned to failure because the current refused to cross the acid, and the commission returned strongly disappointed. Moissan quickly found the explanation for this failure: during his very first experiment,

[93] Valuable information on the life and work of H. Moissan (1852–1907) were presented on various occasions: Centenary of the Discovery of Fluor, Paris 1986: "Life and work of H. Moissan" by C. Viel, J. Flahaut, J. Fluorine Chem. 33 (1986) 27–44; Centenary of the Ecole Nationale Supérieure de Chimie de Paris, including H. Moissan as one of the promoters: "Henri Moissan, chemist experimenter" Collongues R., F. Galtier, For Science, 280 (1996) 46; Centenary of the 1906 Nobel Prize in Chemistry awarded to Henri Moissan. A. Tressaud, J. Flahaut, "National Celebrations," French Ministry of Culture, 2006; "Henri Moissan, chemist, first French Nobel Prize in Chemistry," F. Viel, Pharmathèmes, Paris, 2006.

[94] J. Burdon, Emson B., AJ Edwards, Is fluorine gas really yellow? J. Fluorine Chem., 34, Issues 3–4, January (1987), 471–474.

Fig. 25 Henri Moissan (1852–1907).
Document Henri Moissan Museum, Faculty of Pharmacy, University Paris V, Photo A. Tressaud.

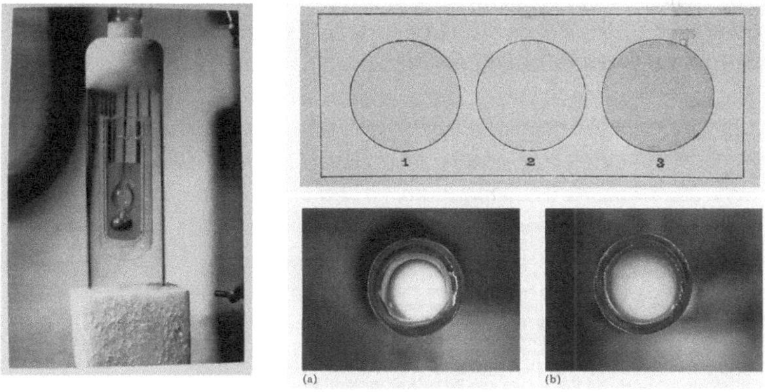

Fig. 26 The color of fluorine. Top right: visual observations noted by Henri Moissan. (1) air-filled tube; (2) F_2-filled tube; (3) Cl_2-filled tube. Bottom right: experiment conducted in a 5 m long quartz tube and filled with air (left) and F_2 (right) by J. Burdon et al., confirming the color described during the 1886 experiment (J. Fluorine Chem., 34, 471–474 (1987)). Photo left-Liquid F_2 *(yellow)* at liquid nitrogen temperature, 77 K.
Private gift by Michel Bargues, ORANO-AREVA.

a small amount of potassium fluoride hydrofluoride was formed, making the acid a conductor. The acid prepared for the experience before the commission of the Academy was pure and anhydrous, and so, an insulator; it was appropriate to add a small amount of potassium fluoride to render the system conducting. Moissan made it so and again convened the commission so that this time, they could see fluorine release and admire the vivid combustion of silicon, iron, manganese, arsenic, and antimony, as well as the inflammation of alcohol, ether, benzene, and turpentine that reflected the extraordinary chemical reactivity of the isolated element. On July 19, 1886, Moissan published a second note confirming the observations of the first one: "The gas obtained in our experiments is fluorine or hydrogen perfluoride"; and it is only in a third note, dated July 26, after a series of experiments, that he finally concluded that the gas emitted during electrolysis was fluorine: "The gas that is released from the electrolysis of anhydrous hydrofluoric acid or molten fluoride hydrofluoride is therefore fluorine."[95] Edmond Frémy, who had always claimed the paternity for the isolation of fluorine (even writing in his 1890 curriculum: "So, I isolated fluorine ..."), generously supported Moissan's nomination at the Academy of Sciences by a June 1, 1889 letter addressed to their peers: "You know that the isolation of fluorine by M. Moissan is one of the greatest discoveries of inorganic chemistry; it may have for science incalculable consequences." Frémy's remarks were prophetic when considering the growing importance of fluorinated compounds, both in inorganic and organic fields, importance that Moissan did not suspect when he described the experience he had just realized.

Until 1891, Moissan's research was carried out exclusively on the properties of fluorine and its derivatives, in collaboration with his students, P. Lebeau, Mr. Meslans, and C. Poulenc,[96] then with the help of distinguished colleagues such as H. Becquerel and M. Berthelot. Numerous reaction products between fluorine and elements such as C, Si, Ag, Pt, Au, transition elements, etc., were discovered and characterized, opening the way for inorganic fluorine chemistry in the early 20th century. Then, having designed an electric furnace to reach 2500°C, he oriented his research toward high temperatures. This research on fluorine isolation and high-temperature chemistry was rewarded by the 1906 Nobel Prize in chemistry. The scientist did not benefit long from this glory, and 2 months after having received the prize, Henri Moissan died of acute appendicitis on February 20, 1907. At the opportunity of the centenary anniversary of Moissan's Nobel Prize, a 20-min video was created on the scientist's life and achievements.[97]

It was not until 1986 that K.O. Christe proposed a new way of synthesizing fluorine, this time via a chemical route. It is well known that a weaker Lewis acid can be displaced from its salts by a stronger Lewis acid such as SbF_5:

$$K_2MF_6 + 2SbF_5 \rightarrow 2KSbF_6 + [MF_4]$$

[95]H. Moissan, C. R. Acad. Sci. **102** (1886) 1543; ibid. **103** (1886) 202; ibid. **103** (1886) 256.

[96]C. Poulenc, Contribution à l'étude des fluorures anhydres et cristallisés, Annales Chim. Phys. 7th series, vol. II (1894) 5.

[97]D. Bour, J. Trouchaud, Henri Moissan, Pharmacist, First French Nobel Prize for Chemistry, A3V-Studio, daniel.bourlevrai@free.fr, and https://www.youtube.com/watch?v=M2P6ckMK_rE.

If the liberated MF_4 is thermodynamically unstable, it will decompose to give a fluoride with lower oxidation and evolution of elemental fluorine. Manganese was chosen, and the following reaction yielded to evolution of fluorine gas[98]:

$$K_2MnF_6 + 2SbF_5 \rightarrow 2KSbF_6 + MnF_3 + \tfrac{1}{2}F_2$$

7 Physical chemical characteristics of fluorine

In standard temperature and pressure conditions, fluorine is a yellow gas (see Fig. 26) of irritating odor, difficult to liquefy. It is the most reactive element of the entire periodic table; it reacts with almost all substances except for some special alloys. Neil Bartlett brilliantly demonstrated that fluorine could even react with noble gases[99] (see "Neil Bartlett invalidating the dogma of inert gases"), except for the two lighter ones, helium and neon. A stable compound of argon, HArF, having been recently characterized by IR.[100] Its electronegativity on the Pauling scale is the highest of elements: $\chi = 4$ (O, $\chi = 3.45$; Cl, $\chi = 2.8$). Its van der Waals radius (0.135 nm) is close to that of oxygen (0.14 nm) and allows substitution and exchanges. On the other hand, its dissociation energy is very low (36 kJ/mol) compared to that of oxygen (118 kJ/mol) or even chlorine (57 kJ/mol). In all fluorinated compounds, it takes the valence −1 (or 0 in F_2). It provides very high binding energies F-C (490 kJ/mol), F-H (570 kcal/mol), or F-S (330 kJ/mol). Its boiling point is −188°C (Table 1). There is still controversy on the isolation of a hypervalent positively charged F^+ ion. Although indirect experimental evidence was presented for the transient generation in solution of a fluorine atom symmetrically bonded to two carbon atoms in fluoronium ions, it was inferred, through a critical examination of these interpretations, that this symmetrically bonded fluorine atom does not carry a positive charge and is not hypervalent or hypercoordinate. The terms "*fluorine with decreased negative charge and increased electrophilicity*" should be preferred. The F^+ affinity in the N_xF_y series has been shown to increase from N_2 to N_2F_4 by 63 kcal/mol, while that in the O_2F_y series it decreases by 18 kcal/mol from O_2 to O_2F_2.[100a] Its electronic structure $1s^2\,2s^2\,2p^5$, with an atomic weight of 18.998. It contains only one natural isotope, ^{19}F, and one artificial isotope, ^{18}F. It is the 13th most abundant element in the earth's crust at about 0.06%.

[98]K. O. Christe, Chemical synthesis of elemental fluorine, Inorg. Chem., 1986, **25** (21), pp. 3721–3722.

[99]Bartlett, N., and Lohmann, D. H., Fluorides of the Noble Metals. Part II. The Dioxygenyl Hexafluoroplatinate (V) $O_2^+PtF_6^-$. *Proc. Chem. Soc.* **277** *(1962)* p.525*)*; N. Bartlett, Xenon hexafluoroplatinate (V), Proceed. Chem. Soc. (1962) p. 218.

[100]L. Khriachtchev, Mr. Pettersson, N. Runeberg, J. Lundell, M. Räsänen, A Stable compound of argon, Nature **406**, (2000) 874–876. M. Räsänen, Argon out of thin air, Nature Chemistry, 6, 82 (2014).

[100a]K. O. Christe, R. Haiges, M. Rahm, D. A. Dixon, M. Vasiliu, Misconceptions on fluoronium ions and hypervalent fluorine cations, J. Fluorine Chem., 204 (2017) 6–10; M.D. Stubble, M.T. Scerba, M. Sieger, M.T. Lectka, Evidence for a symmetrical fluoronium ion in solution, Science 340 (2013) 57–60; D.J. Grant, T.H. Wang, M. Vasiliu, D.A. Dixon, K.O. Christe, F^+ and F^- Affinities of Simple N_xF_y and O_xF_y Compounds, Inorg. Chem. 50, 5, (2011), 1914–1925.

Table 1 Physical–chemical data of fluorine

Melting point (°C)	−219.62
Boiling point (°C)	−188.14
Transition temp. (solid) (°C)	−227.61
Critical temp. (°C)	−129.00
Critical pressure (atm)[a]	55
Volumic mass (g/cm^3) solid	1.9
Volumic mass of liquid at boiling point (g/cm^3)	1.508
Density (gas)	1.31
Refraction index of liquid at boiling point	1.2
Superficial tension of liquid at:	
193°C (dyn/cm)[b]	14.81
207°C (dyn/cm)	18.85
Liquid viscosity at:	
188°C (cPo)[c]	0.257
204°C (cPo)	0.414
Gas viscosity at 0°C, 760 mmHg (cPo)	0.0218
Dielectrical constant at	
190°C	1.517
215°C	1.567
Gas thermal conductibility at 0°C, 760 mmHg (cal cm^{-2} cm^{-1} °C^{-1} s^{-1})	5.92×10^{-5}
Heat of transition, solid (cal/mol)[d]	173.90
Heat of fusion (cal/mol)	121.98
Heat of vaporization at boiling point (cal/mol)	1561.3

[a] 1 atm = 1.013 bar.
[b] 1 dyn cm^{-1} = 10^{-3} N m^{-1}.
[c] 1 cPo = 10^{-5} Pl.
[d] 1 cal mol^{-1} = 4.18 J mol^{-1}.

Neil Bartlett invalidating the dogma of inert gases

For a very long time, it had been assumed that noble gases (He, Ne, Ar, Kr, Xe, Rn) did not react with any other elements, and so, they were named "inert" gases. We can quote the exchanges between H. Moissan and W. Ramsay reporting the absence of reactivity between F_2 and those gases.

The discovery of O_2PtF_6 brought in light the extreme oxidizing power of PtF_6. A young British chemist, Neil Bartlett (1932–2008), discovered that when PtF_4 was heated under diluted F_2 gas in a Pyrex tube, a deep-red vapor was observed, occurring simultaneously with the attack of the glass tube. The resulting compound $O_2^+PtF_6^-$ illustrated the unexpected oxidizing power of the PtF_6 since this molecule was able to remove an electron from O_2, to precipitate the first O_2^+ salt.[99]

As quoted by Bartlett, "I then deduced that PtF_6 should oxidise xenon, since the first ionisation potential of xenon is almost the same as that of O_2. Once $O_2^+PtF_6^-$ was established, the production of $XePtF_6$ was therefore a predictable event."

	Xe	O_2	N_2
First ionization potential (eV)	12.13	12.2	15.58
Oxidized species	Xe^+	O_2^+	N_2^+

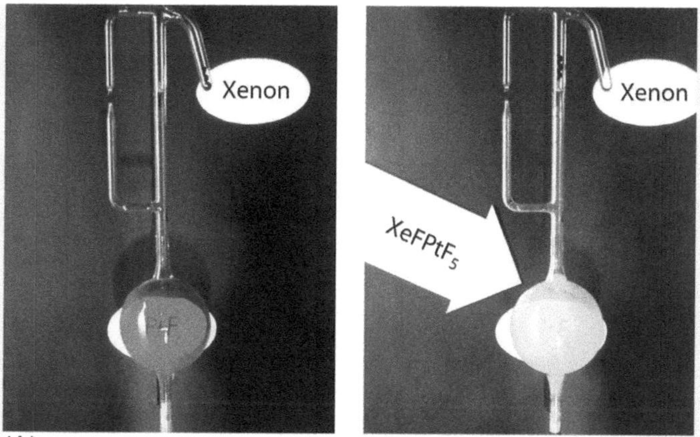

(A) (B)

Fig. 27 Synthesis of the first compound of noble gases by N. Bartlett: (A) Xe and PtF_6 gases prior to reaction; (B) resulting $XePtF_6$.

On March 23, 1962, when platinum hexafluoride, a red gas, was allowed to mix with a large molar excess of xenon (Fig. 27, left), the immediately formed product was a yellow solid with the composition $Xe^+[PtF_6]^-$ (right), the first synthesized compound of a noble gas.[99]

This discovery was acclaimed[101] as one of the "ten most beautiful experiments in chemistry," because[102] it opened the way for the noble gas chemistry and high oxidation species chemistry, as illustrated by the various species isolated in Neil's Berkeley Group: first example of Os +VII: $OsOF_5$; $NOWF_7$, $(NO)_2WF_8$, $NOReF_7$; Isolation of ONF_3: $(ON)_2NiF_6 \rightarrow ONF + ONF_3 + NiF_2$; synthesis and oxidizing properties of $ReOF_5$ and $OsOF_5$; salts of quinquevalent gold: $M^+AuF_6^-$, $NO^+AuF_6^-$, $Xe_2F_{11}^+AuF_6^-$, $XeF_5^+AuF_6^-$, etc. Quoting P. Ball, we can say that "Neil Bartlett's experiment tells us that in chemistry, the wonders never cease."[103]

Today, industrial production of fluorine is still based on the method reported by H. Moissan, electrolysis of a molten mixture of potassium fluoride and hydrofluoric acid, KF·2HF. The considerable growth of the uses of fluorinated organic and polymeric compounds and the use of fluoride in the nuclear industry for the separation of

[101] Philip Ball, "Elegant Solutions: Ten Beautiful Experiments in Chemistry," RSC, 2005. I. Hargittai, "Neil Bartlett and the first noble-gas compound," Struct. Chem. (2009) 20:953–959.

[102] "The oxidation of oxygen and related chemistry," N. Bartlett ed., World Scientific, 2001.

[103] In his famous novel: "The Periodic Table," the Italian writer Primo Levi (a former chemist) tried to convey to non-chemists the idiosyncratic characters of the chemical elements. P. Levi began his book by the chapter "Argon" where he draws analogies between his ancestors and the inert gases, and among them, the Alien (xenon): "There are the so-called inert gases in the air we breathe. They bear curious Greek names of erudite derivation which mean 'the New,' 'the Hidden,' 'the Inactive,' and 'the Alien.' They are indeed so inert, so satisfied with their condition, that they do not interfere in any chemical reaction, do not combine with any other element, and for precisely this reason have gone undetected for centuries. As late as 1962, a diligent chemist, after long and ingenious efforts, succeeded in forcing the alien (xenon) to combine fleetingly with extremely avid and lively fluorine, and the feat seemed so extraordinary that he was given a Nobel prize." From "The Periodic Table," Primo Levi, translated by Raymond Rosenthal for Schocken Books, Inc. © 1984.

Fig. 28 Fluorination facilities at ARC, Inc.
Courtesy Dr. Dayal T. Meshri, ARC Founder and CEO.

uranium isotopes ^{235}U and ^{238}U resulted in a strong increase in fluorine production throughout the century. Several improvements have allowed optimization of the fluorination level (by applying a higher anodic potential to the KF-2HF bath), or enhancing the size of fluorine bubbles during the evolution reaction.[104] A laboratory-scale fluorine reactor was also modeled for proposing fully coupled, fundamental electron, heat, mass, and momentum transfer (two-phase) equations for delivering a transient simulation. Most parameters were taken into account (current density, electric field, temperature, reactive species concentration, gas and liquid velocity profiles, as well as gas fraction distribution within the reactor) in order to simulate theoretical processes and compare them to the experimental systems.[105]

Now, fluorine gas is produced following the electrochemical process, as shown, for instance at Fig. 28 for the F_2 synthesis facilities of Advance Research Chemicals, Inc. (ARC Inc., Catoosa, OK, United States). An important technical point is that hooded areas should be specially equipped for fluoride manufacturing. All work should be done in hoods, 40,000 cfm fresh air being pulled in to avoid depression in the building.

8 Some breakthroughs in the development of fluorine and fluorinated products throughout the 20th century

From its isolation in 1886, fluorine has given rise during the 20th century to remarkable technological progress in a wide range of fields. Among these great advances, we can cite chronologically:

[104] H. Groult, F. Lantelme, M. Salanne, C. Simon, C. Belhomme, B. Morel, F. Nicolas, Role of elemental fluorine in nuclear field, Journal of Fluorine Chemistry128, (4), (2007) pp. 285–295. Groult, H.; Lantelme, F.; Devilliers, D.; Belhomme, C.; Morel, B.; Nicolas, F.; Caire, J.P., Electrochemical synthesis of fluorine from 1886 to 2006: Fluorine, a key-element for nuclear energy, Actualite Chimique; 301–302; (2006) pp. 40–46.

[105] Pretorius R, Crouse PL, Hattingh CJ. A multiphysics simulation of a fluorine electrolysis cell. S Afr J Sci.;111(7/8),(2015) Art. #2014–0082.

- Use of fluorinated fluxes that allowed the synthesis of aluminum with all the dazzling technological advances that we know today in transport, aeronautics, and food packaging, for example.
- The rise of freons as industrial refrigerants that replaced harmful ammonia, followed by the questioning of PFC and CFC in several serious environmental problems. It should be noted that, currently, ozone-depleting substances such as chlorofluorocarbons (CFCs) and hydrochlorofluorocarbons (HCFCs) are now banned, and new alternatives with lower global warming potential (GWP) such as hydrofluorophenols (HFO) are proposed.
- Development of organic fluorine chemistry that led to new molecules with highly selective properties (fluorosulfide compounds used as steroids, triflic acid derivatives, sugars, vitamins), the use of superacid media for the production of fuels with a high octane number.
- The revolution of Teflon followed by the discovery of fluorinated plastics, fluoropolymers and surfactants that gave rise to many improvments in the industrial sector and in everyday life: clothing treatment, biomaterials for cardiovascular surgery, kitchen equipment, fabrics that "breathe" after treatment with Gore-Tex, protection of works of art. The advantages created by the functionalization of the surface properties using fluorinated species (oleo- and hydrophobia for example) served the protection of our cultural heritage, anti-graffiti and anti-reflection layers, UV absorbers, car protection, etc.
- The last part of the 20th century was marked by the appearance of multiple fluorinated molecules with beneficial aspects for agriculture, pharmacy, medicine, and biomedical sciences, with more than 600,000 compounds containing at least one fluorine atom. These drugs have anti-cancer, anti-inflammatory, analgesic, or antidepressant properties, and are used in areas as varied as the treatment of cancer, malaria, and the early detection of Alzheimer's disease, but also in medical imaging (^{18}F PET scan), etc. Three major examples of discoveries and technological improvements that have marked the 20th century will illustrate the rapid development of fluorine and fluorinated products, for example, the rapid development of the aluminum industry, the rise and fall of CFCs, the revolution of PTFE (Teflon), and fluoropolymers development in industrial sectors and in everyday life.

8.1 The rapid growth of the aluminum industry

8.1.1 1886, a great year for both fluorine and aluminum: The Hall-Héroult process

1886 was definitely a great year for fluorine. Simultaneously and completely independently, two young men, both 23 years old, Paul Louis Toussaint Héroult in France and Charles Martin Hall in the United States—both born and deceased the same year (1863–1914)—discovered a new electrochemical aluminum synthesis process that revolutionized the development of the new metal (Fig. 29).[106] The process consisted in electrolyzing alumina dissolved in molten cryolite at about 1000°C. Cryolite (Na_3AlF_6) was extracted from the deposit of Ivigtut, Greenland

[106] Pascaud Claude F., Le développement du procédé Hall-Héroult et son accompagnement par la propriété industrielle, (1886–1994), in: Cahiers d'Histoire de l'Aluminium, n°20, été 1997, published by the l'Institut pour l'Histoire de l'aluminium, Paris 1994. Website on the history of aluminum industry - http://www.histalu.org.

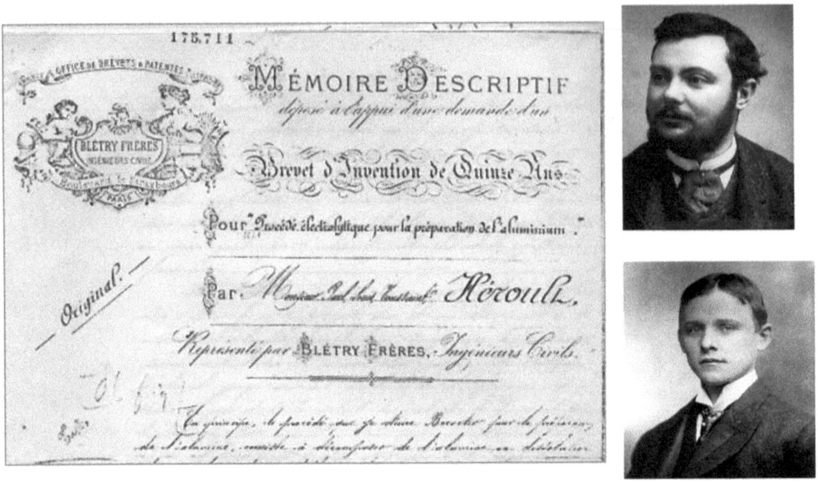

Fig. 29 P. L. Héroult and C. M. Hall. The Héroult Patent, June 26, 1886.

(see “Life and death of the fluoroaluminate deposit at Ivigtut (Greenland)”). This technique extended the expertise of Middle Ages smelters, as described by G. Agricola. Since that time, the process has been called the Hall-Héroult process.[107] This discovery marked the end of the of Deville process's monopoly (H. Sainte-Claire Deville (1818–81)). Producing aluminum by electrolysis of alumina enabled a drastic reduction in cost and opened the way for the development of this new industry. Unlike Moissan's discovery, the importance of which the world immediately recognized, manufacturers who used the Deville process were not immediately convinced of the new method's benefits with relation to a new energy source: electricity. In particular, A.R. Pechiney, who once said, “I do not like electricity,” refused to embark on the adventure. P. Héroult was obliged to ask Swiss industry to find money, and he began to produce aluminum in Neuhausen, the ancestor of Alusuisse, one of the major European producers of aluminum. It took more than 10 years before Pechiney, through the purchase of other companies, implemented the process discovered by the French. Initially, this new metal was seen as an object of curiosity, a semi-precious metal whose price was around that of gold. It had not yet become the key metal of the 20th century. Aluminum ingots were even presented alongside the crown jewels at the Paris Universal Exhibition of 1855. In 1889, K.J. Bayer invented a process allowing large-scale production of alumina from bauxite that is still in use today. In the early 20th century, P. Héroult and C.M. Hall both became the head of gigantic enterprises, having their own yachts and private railway cars; the first one created a huge manufacturing plant at Whitney, North Carolina, and the second founded the Aluminum Company of America, better known under the name of Alcoa.

[107] Héroult P.: French Patent No. 175711 of 23 April 1886; C. M. Hall, US Patent 400.664, July 9, 1886.

Life and death of the fluoroaluminate deposit at Ivigtut (Greenland)

Discovered in 1806 by K.L. Giesecke, the Ivigtut deposit in Greenland has been the world's leading source of fluoroaluminates for two centuries[108]. The main component of this deposit was cryolite (Na_3AlF_6), but most of the known fluoroaluminate minerals were also present. Besides these fluorides, 90 types of other minerals: carbonates, sulfates, siderite, silicates, occurred in the granite veins, associated with the mass of cryolite[109]. This natural stock of cryolite became tremendously beneficial to the growth of the aluminum industry, but backlash caused the total destruction of the site. The operation started very early, in 1850, the ore was then used for the manufacture of alum, fertilizers and even aluminum by reduction of pure cryolite. But the discovery of Héroult and Hall blew the production. Between 1886 and 1960, 3.5 million tons were extracted, until the cessation of operations for reasons of total exhaustion of the vein. The final chapter of the Ivigtut story was written in 1987 with the last load of cryolite (Fig. 30).

Fig. 30 Life and death of the fluoaluminate deposit of Ivigtut (Greenland): (A) Charles Edmond, "Travel in the northern seas aboard the corvette Queen Hortense, 1863," drawing by Karl Girardet, Collection Ecole des Mines, Paris; (B) the mine in full operation in the 1940s; (C) end of activities in 1960; (D) the ghost town of Ivigtut and flooded mine today. Copyright © 1999, 2004 Grønnedalsforeningen haf-kasseren@mail1.stofanet.dk af 1985.

[108] Giesecke K. L., "On cryolite; a fragment of a journal", Edinburgh Philosophical Journal 6 (1822) 141–144. Professor and collector of mineralogy, KL Giesecke (1761–1833) was one of the most famous pioneers who helped to publicize the natural history and geology of Greenland. Showing a very strong personality, he was also an actor and the librettist of Mozart's Magic Flute.

[109] O.V. Petersen, K. Secher, "The Minerals of Greenland," The Mineralogical Record, 24 (2) (1993) 1–67, Tucson, USA. (NB—We can find these fluorinated minerals in most geological collections (In France at Pierre and Marie Curie University, School of Mines, National Museum of Natural History in Paris and at the Geological Museum in Amsterdam.))

8.1.2 *Improvement of the production process of aluminum by addition of fluorinated fluxes*

The electrolysis set up still uses the Hall-Héroult process in which alumina is dissolved at 960–980°C in a molten cryolite bath (Na_3AlF_6) in a graphite-jacketed steel container. The carbon anode is also made of petroleum coke or pitch. The electrical current passes from the carbon anodes via the bath to the carbon cathode cell lining. The electrolytic bath is crossed by a current of 150,000 A under low voltage. The schematized global electrolysis reaction can be written: $2Al_2O_3 + 3C \rightarrow 4Al + 3CO_2$. Possible additions of AlF_3 and CaF_2 allow increasing the efficiency of the electrolytic cell and lowering the temperature. Today, in the absence of sufficient amounts of cryolite, it is the fluorite mineral (as "acid fluorspar") which provides most fluxes.

The growth in global aluminum production increased dramatically from 8000 tons in 1900. It exceeded 500,000 tons before WWII to more than 24 Mt at the end of the century.[110] Today, aluminum can be considered a "young" metal when compared to other non-ferrous metals, because it has been in use just over 150 years, whereas the use of copper, lead, and tin goes back several millennia. However, annual aluminum production was 58 Mt. for 2015 (from the International Aluminum Institute), and far exceeds those of its non-ferrous competitors (Cu, Pb, Sn), that is, 14.3, 4.6, and 0.4 Mt, respectively (Data 2011). It is bauxite, present in Europe but also in the tropical and subtropical zones, which is the most common ore (Australia, China, Brazil, India, and Guinea contain three-fourths of the Earth's total reserves, that is, 25 billion tons).[111] Production units are generally close to cheap energy sources. Four to 6 tons of bauxite produce 2 tons of alumina, which ultimately give 1 ton of aluminum.

8.1.2.1 Production processes

The production units are comprised of four main sectors: a power station transforming the alternating current into direct current for electrolysis, manufacturing carbon anodes, electrolysis rooms equipped with gigantic tanks containing an alumina and cryolite bath, and the foundry that can sink the metal in various forms. The process is large consumer of electricity, but technical improvements have reduced electricity consumption; in 1950, 21 kWh were required to obtain 1 kg of aluminum from alumina, and today, it takes 13 kWh, according to Alcoa. There are two main types of aluminum smelting technology, Søderberg and prebake. The principal difference between the two is the type of anode used. Prebake technology uses multiple anodes in each cell which are baked in a separate facility and attached to rods that suspend the

[110] Lanthony J., "The Aluminium and light alloys," Collection "Que sais-je? No. 543," PUF Presses Universitaires de France, Paris, 1952. Morel P. (under the direction of …) History of Technology Production of Aluminium, Presses Universitaires de Grenoble, PUG, Grenoble 1991. Sheller M., Aluminium Dreams, MIT Press (2014).

[111] World production of fluorspar, is about 6.3mtpa (2012 data) is dominated by China (50%) and Mexico (18%), followed by a smaller production in Mongolia (7%) and South Africa 3%), Other countries, such as Russia, Namibia, Spain, Kenya, Morocco produce each $<2\%$. Over 90% of the production are used for the manufacture of aluminum, the rest for the glass industry, cement and ceramics (6%). Websites—http://www.world-aluminium.org/http://primary.world-aluminium.org/.

Fig. 31 Prebake potroom, Dubai Aluminum Company Limited (DUBAL). The tanks are covered by hoods to recover emissions of fluorine and fluorinated gases (2006).

anodes in the cell. Most primary aluminum production facilities use a variant of prebake technology called point fed prebake, which uses multiple in-cell feeders and other computerized controls for precise addition of alumina to the pot, which improves energy efficiency while reducing emissions, dust, and raw material use (Fig. 31).

8.1.2.2 The anode effect and PFC formation

Small quantities of perfluorocarbons (PFCs) are produced at the anode; this is the anode effect.[112] This effect occurs when the alumina content becomes too low in the electrolysis vessels. It causes a rise in electric current and a decomposition of fluorinated bath with fluorine formation. The gas then reacts with the carbon anode to produce PFC, essentially CF_4 and C_2F_6. The prebake anodes technology (CWPB, for Center Worked Prebake Technology) allows a limitation of these emissions. In modern plants, the gases are captured by an improved ventilation system and are cleaned in processing centers where fluorides bind to alumina. This fluoride-enriched alumina is recovered and returned to the electrolytic cells. Fluoride emissions, in the form of gaseous hydrogen fluoride with sodium and aluminum fluorides and unused cryolite as particules, are the major undesirable fume component produced in the aluminum smelting process. Such emissions can be reduced through the use of fume control systems: an operational good practice and improved technology. In 2014, using the

[112] Mergault P., B. Gaudreau, Brunet-Lefort C., CR Acad. Sci. Paris, **266** (1968) 1332; Watanabe, N., Anode and Graphite Intercalation Compounds Effect—A New Elucidation Addition of Agent Effect in Carbon Anode Reaction of Molten Fluoride Cells. Proc. 1st. Intl. Symp. Molten Salt Chem. Techn., Kyoto, 21 (1983).

Table 2 Fluorine and fluoride emissions from blast furnaces from 1940 to present day (per produced Al ton)

1st generation	1940–1955	12–15 kg
2nd generation	1955–1975	2–6 kg
3rd generation	1975 to present day	0.3–1 kg

prebake technology, for 50,000 kt Al produced, 40 ton of fluorine were emitted, that is 0.8 kg F per ton of Al (2015 data from http://www.world-aluminium.org/). These centers for electrolysis gas treatment are effective over 99%. So, within 60 years, by improving the technology, fluoride emissions have been reduced more than 15 times (Table 2) and annual amounts of fluorinated residues have decreased from 1500 ton after WWII to 60 ton today.

8.1.3 Application fields for aluminum

From a technological standpoint, the lightness of the aluminum parts is an asset wherever mass is a handicap, as in the aeronautics, space, and automotive industries.[113] Aluminum is used for the packaging of foodstuffs: beverages, cans, aerosol cans, hoses, caps, lids, sheet protectors, tetrabriks. It is also sold in roll form, in sheets, trays, other containers, and aluminized fabrics. Aluminum has achieved important breakthroughs such as the first spacecraft (Sputnik), the aluminum foil that the crew of Apollo 11 deployed on the Moon to capture solar wind, the moucharabiehs of the World Arab Institute in Paris, the grandiose geodesic domes of R.B. Fuller, and surface deposit on billions of CDs.

8.2 The rise and fall of CFC

Today, with the natural tendency of our society to catalog and rationalize problems, we tend to differentiate between "good" chemicals and those with problematic issues. Thus, Teflon and most fluoropolymers, perhaps because these materials are used daily in many branches for healthcare or daily life, belong to the first category, whereas CFCs attract opprobrium. But we should remember that these two product families arose from the same reflections which led scientists to find solutions to very critical issues that faced the society in the first half of the 20th century.

8.2.1 A new type of "too perfect" refrigerants

In the early 20th century, the increasing use of refrigeration resulted in the need to find substitutes for refrigerants (at that time, ammonia, propane, or SO_2) that exhibited a considerable risk for users. In the early 1930s, General Motors (via its Frigidaire

[113] I. Grinberg, Aluminium and Aluminium Alloys: Their Properties, Uses, and Methods of Working Them, Scholar's Choice Editions (2015) with the collaboration of Florence Hachez Leroy, J.P. Plateau "The aluminum so light metal" Editons Gallimard Discoveries No. 435, Paris 2003.

subsidiary) was looking for much more stable alternative products, that is, hydrocarbons in which part or all of the hydrogen atoms were replaced by fluorine, chlorine (or bromine); these products were called freon. Two chemists in charge of the program, A.L. Henne and Thomas Midgley, contacted a DuPont group at the Jackson Laboratory in Deepwater, New Jersey, who possessed the physicochemical skills on this type of products. A joint venture was established between the two companies, called Kinetic Chemicals, to develop the synthesis and production of these refrigerants. Among the products sold then, Freon 114 (CF_2ClCF_2Cl) was notable. Since all operating rights for Freon-114 were reserved for Frigidaire, DuPont launched its own program on new molecules. Note that R.J. Plunkett (see below the discovery of Teflon) was present among the chemists involved in this program. In the 1930s, the range of CFCs had spread; $CFCl_3$ (F11), CF_2Cl_2 (F12), $CF_2ClCFCl_2$ (F113), $ClCF_2CF_2Cl$ (F114), and CHF_2Cl (F22) were produced commercially by DuPont.

Very quickly, the new refrigerants synthesized by DuPont experienced a remarkable development. Their fields of applications exceeded the domestic and industrial refrigeration markets to win those of aerosols, insulating foams, microelectronics, and micromechanics. During the following decades, the world market for chlorofluorocarbons (CFCs) scaled up to 1.2 Mt in 1986. A set of qualities made these materials "almost perfect," without competitors: high chemical resistance, flame retardancy, low toxicity, low thermal conductivity in the gas phase, melting temperatures between −80°C and +50°C, depending on the compound, but often near ambient temperature. Between 1945 and 1986, the four most-used refrigerants of all CFCs were CFC-11, CFC-12, HCFC-22, and R502, an azeotrope of CFC-115 (CF_3CF_2Cl) and HCFC-22 (see "Inset: Nomenclature of CFCs, HCFCs and HFCs"). Throughout this period, production, emissions, and properties of the main CFCs and HCFCs were followed very thoroughly.[114]

[114] Among the huge literature on the topic, we can select: P. H. Gamlen, B. C. Lane, P. M. Midgley, J. M. Steed, The Production and Release to the Atmosphere of CCl_3F and CCl_2F_2 (chlorofluorocarbons CFC 11 and CFC 12), Atmos. Environ., **20**, (1986) 1077–1085. L. E. Manzer, The CFC—ozone issue: Progress on the development of alternatives of CFCs, Science **249** (1990) 35. D.A. Fisher and P. M. Midgley. The Production and Release to the Atmosphere of CFCs 113, 114 & 115, Atmos. Environ., **27A**, (1993) 271–276,. D.A. Fisher, P.M. Midgley. The Production and Release to the Atmosphere of Chlorodifluoromethane (HCFC-22), Atmos. Environ., **27A**, 2215–2223, 1993. D. A. Fisher, T. Duafala, P. M. Midgley and C. Niemi. Production and emission of CFCs, halons, and Related Molecules, in Report on Concentrations, Lifetimes, and Trends of CFCs, halons, and Related Species, NASA Reference Publication 1339 JA Kaye, SA Penkett and FM Ormond (eds.), (1994). P. M. Midgley and A. McCulloch. Properties and Applications of Industrial Halocarbons, International Regulations are Halocarbons, and Production, Sales and Emissions of Industrial Halocarbons, in The Handbook of Environmental Chemistry, Vol. 4 Part E: Reactive Halogen Compounds in the Atmosphere, pp. 129–153, 155–190, 203–221, P. Fabian and O.N. Singh (eds.), Springer-Verlag, Heidelberg, (1999). J. W. Elkins, chlorofluorocarbons, The Chapman & Hall Encyclopedia of Environmental Science, DE Alexander and RW Fairbridge Eds., pp. 78–80, Kluwer Academic, Boston, (1999). A. McCulloch, Ashford P., P. M. Midgley. Emissions of Historic fluorotrichloromethane (CFC-11) Based on a Market Survey, Atmos. Environ., **35**, 4387–4397, (2001). A. McCulloch, P. M Midgley, P. Ashford. Releases of Refrigerant Gases (CFC-12, HCFC-22 and HFC-134a) to the Atmosphere, Atmos. Environ., **37** (7), (2003) 889–902.

Most information on the properties of these gases can be obtained via the Alternative Fluorocarbons Environmental Acceptability Study (AFEAS) that provides emission data through 2003 (*see AGAGE website, https://agage.mit.edu/data/afeas-data*). The evolution of the annual production of various CFCs, HCFCs, and HFCs between 1980 and 2002 is given in Fig. 32A, whereas the evolution of ozone reduction potential (ODP) due to CFCs and HCFCs is reported in Fig. 32B. Note that the evolution is calculated from production values given by AFEAS, and ODP from "Production and Consumption of Ozone Depleting Substances," UNEP Report, 2017. (HFCs are not included in the figure because they do not contain chlorine, and their ODP values are zero.)

8.2.2 *Phasing out CFCs and HCFCs*

8.2.2.1 Ozone depletion, the Montreal and Kyoto Protocols

Already in the 1970s, Lovestock et al. had characterized the presence of CFCs 11 and 12 in the atmosphere[115] and during a CFC producers' meeting in 1972, R. McCarthy from DuPont announced that "fluorocarbons are accidentally or intentionally released into the atmosphere at an annual rate of several hundred thousand tons." CFC producers then decided to constitute a committee, the "Fluorocarbon Panel," to study the impact of CFCs on the environment. It is in this context in 1974 that a famous article appeared by Molina and Rowland on the destruction of the stratospheric ozone layer by CFCs.[116] This layer provides a screen for our planet, protecting it from the harmful effects of ultraviolet radiation (UV-B), which can be particularly devastating for our systems and our environment.[117] On this occasion, the global chemical industry, sometimes criticized for its lax attitude concerning environmental damage, showed an unfailing voluntarism.[118] Using the committee which had just been created, almost all of the companies decided to jointly fund a research program monitored by independent experts, while pursuing their course of their own research in these fields. The results of the program were consistent with the predictions of Rowland and Molina, but at that time, the techniques were not able to detect a loss of atmospheric ozone. Computing calculations indicating a decrease of 3% per decade, it was decided that a ban of CFCs should be considered, but over a longer period. However, some countries such as the United States decided in 1980 to unilaterally stop the use of CFCs. It was not until 1985 that the British Antarctic Survey announced the observation of stratospheric ozone loss over Antarctica: the so-called

[115] J. E. Lovelock, Nature **230** (1974) 379; J. E. Lovelock, R.J. Maggs, E. R. Allard, Anal. Chem. **43** (1971) 1962; JE Lovelock, RJ Maggs, RJ Wade, Nature **241** (1973) 194.

[116] Molina MJ, Rowland FS, Stratospheric sink for chlorofluoromethanes: chlorine atom catalysed destruction of ozone, Nature **249** (1974) 810.

[117] United Nations Environment Program (UNEP), Assessing the Environmental Impacts of Consumption and Production: Priority Products and Materials (2010).

[118] A. McCulloch, R. L. Powell, in: R. E. Banks, (Ed.), "Millennium Fluorine Chemistry," Elsevier, Oxford, 2000, pp. 357–360.

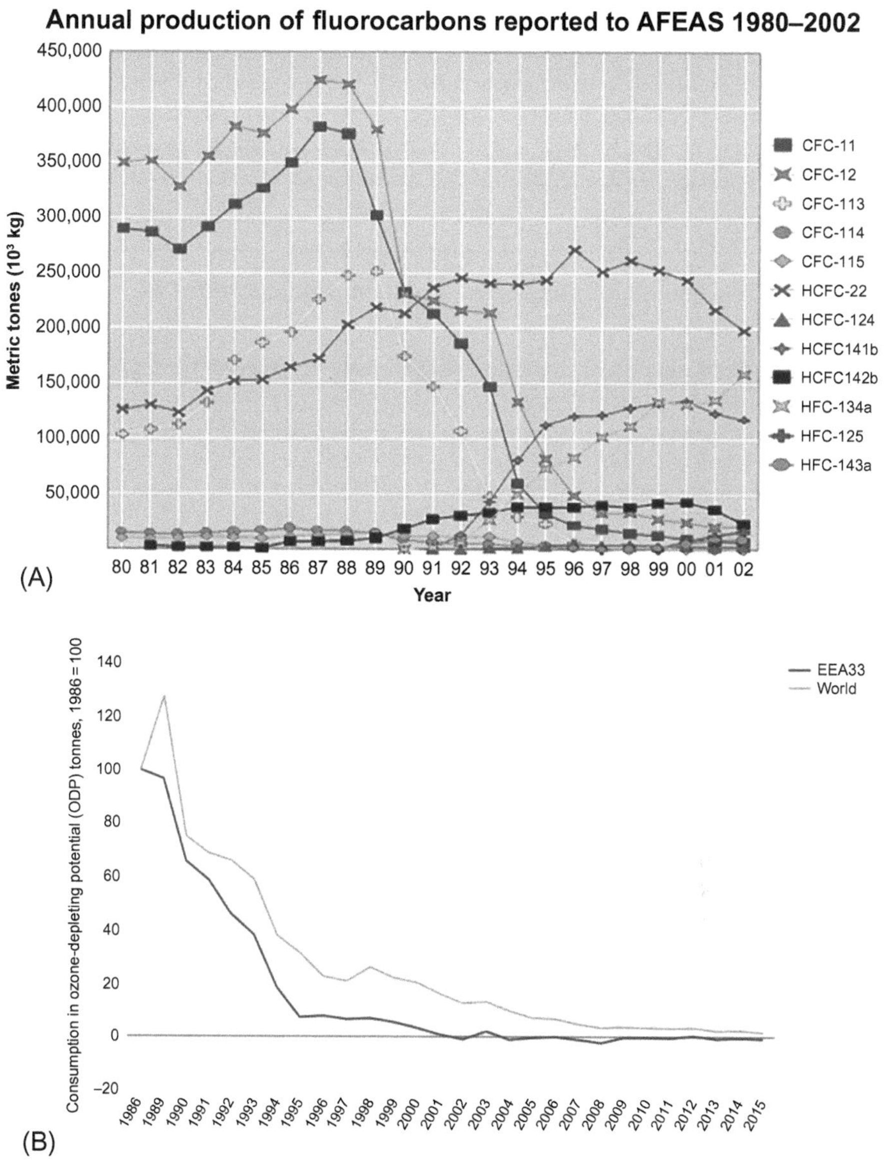

Fig. 32 (A) Evolution of the annual production of various CFCs, HCFCs, and HFCs between 1980 and 2002; (B) evolution of ozone reduction potential (ODP) of CFC and HCFC. A value of 100 is given for ODP-1986.
Credit: https://www.eea.europa.eu/data-and-maps/indicators/production-and-consumption-of-ozone-2/assessment-2.

"ozone hole."[119,120] Chlorine was immediately pointed out in the catalytic mechanisms occurring in the upper stratosphere, at 35–45 km (see "Destruction scheme of stratospheric ozone: Nomenclature of halocarbons").[121] It has been clearly shown that fluoride is not in any way responsible for the degradation of the ozone layer. Because of their high reactivity, the F^- ions and F· radicals are very rapidly converted into HF and $CF_3^{\bullet}$ radicals in CF_2O, and then into CO_2 and HF species that do not affect O_3.

Destruction scheme of stratospheric ozone: Nomenclature of halocarbons

Destruction scheme of stratospheric ozone involving chlorine from CFCs and HCFCs

Decomposition of a CFC:	$CF_2Cl_2\ h\nu \rightarrow Cl^{\bullet} + CF_2Cl$
Stratospheric reactions:	$2 \times (Cl^{\bullet} + O_3 \rightarrow ClO + O_2)$
	$ClO + ClO + M \rightarrow Cl_2O_2 + M$
	$Cl_2O_2\ h\nu \rightarrow Cl^{\bullet} + ClO_2$
	$ClO_2 + M \rightarrow Cl^{\bullet} + O_2 + M$
	(M being another species present in the stratosphere)
Gobal result:	$2O_3 + h\nu \rightarrow 3O_2$

Nomenclature and examples of halocarbons

This system was proposed in 1929 by T. Midgley, Jr., and A.L. Henne, then was subsequently refined by J.D. Park. Hydrochlorofluorocarbons (HCFCs) and hydrofluorocarbons (HFCs) have been further added to initial chlorofluorocarbons (CFCs). The meaning of numbers and letters is as follows, for example, CFC-01234a:

0 = Number of double bonds (not listed if zero)
1 = Number of carbon atoms minus one (if not mentioned zero)
2 = Number of hydrogen atoms plus one
3 = Number of fluorine atoms
4 = Number of chlorine atoms (for bromine the prefix "B" is added)
a = The letter identifies the various isomers.

All the details on this nomenclature can be found from the American Society of Heating, Refrigerating, and Air-Conditioning Engineers (ASHRAE) (2016), American National Standards Institutes (ANSI) and https://www.ashrae.org/resources–publications/free-resources/ashrae-terminology.

Examples of CFCs and HCFCs cited in the chapter are given below.

[119] J. C. Farman, B. G. Gardiner, JD Shanklin, Nature **315** (1985) 207.

[120] S. Solomon, Progress Towards a quantitative understanding of Antarctic ozone depletion, Nature **347** (1990) 347–353.

[121] A. McCulloch, CFC and Halon replacement in the environment, J. Fluor. Chem. **100** (1999) 163–173; Fluorocarbons in the global environment, J. Fluor. Chem. **123** (2003) 21–29.

Formula	Name	Appellation	Dbl bonds	C −1	H +1	F	Br	Isomers (a)	Cl deduced
CCl_3F	Trichlorofluoromethane	CFC-11	–	0	1	1	–	–	3
CCl_2F_2	Dichlorodifluoromethane	CFC-12	–	0	1	2	–	–	2
CCl_2F-$CClF_2$	1,1,2-Trichlorotrifluoroethane	CFC-113	–	1	1	3	–	–	3
CF_2ClCF_2Cl	Tetrafluorodichloroethane	CFC-114	–	1	1	4	–	–	2
$CHCl_2$-CF_3	2,2-Dichloro-1,1,1-trifluoroethane	HCFC-123	–	1	2	3	–	–	2
$CHClF$-$CClF_2$	1,2-Dichloro-1,1,2-trifluoroethane	HCFC-123a	–	1	2	3	–	a	2
CH_3-$CFCl_2$	Dichlorofluoroethane	HCFC-141b	–	1	4	1	–	b	2
CF_3CH_2F	1,2,2,2-Tetrafluoroethane	HFC-134a	–	1	3	4	–	a	–
CH_2F_2	difluoromethane	HFC-32	–	0	3	2	–	–	–
CF_3CH_3	1,1,2-Trifluoroethane	HFC-143a	–	1	4	3	–	a	–
CF_3CF_2H	1,1,1,2,2-Pentafluoroethane	HFC-125	–	1	2	5	–	–	–
CF_2BrCF_2Br	1,2-Dibromotetrafluoroethane	Freon-114B2 or Halon 2402	–	1	1	4	2	–	–

By the late 1970s, a search for CFC substitutes with low toxicity and non-flammability was initiated. Thus, HCFCs were identified, particularly CF_3CH_2F (HFC-134a), CH_2F_2 (HFC-32), CF_3CH_3 (HFC-143a), and CF_3CF_2H (HFC-125), for their low boiling point. A pilot plant was scheduled to implement HFC-134a, but internal political and economic circumstances prevented its implementation. In 1981, the "Vienna Convention for the Protection of the Ozone Layer" simply advised to follow the issues of CFC attentively, but it was not the same for the 1985 convention, because at that time, the depressed Antarctic ozone had been updated. The 1985 Vienna Convention proposed comprehensive measures to enable the control of both the production and emissions of chlorine-based gases.

The first international treaty with purpose of protecting our global environment led to the Montreal Protocol in 1987. The techniques of observation and calculation had progressed considerably and assessed a stratospheric ozone decrease much faster than those provided a decade prior. The agreement provided a term elimination of all CFCs and HCFCs, banning all CFC production as of January 1, 1996, with the exception of some pharmaceutical applications. The European Union forestalled this measure by making it applicable from January 1, 1995. A revision process was originally scheduled on the basis of results which might possibly have been obtained in the meantime. Conferences on maintaining the balance of our planet led in 1997 to the Kyoto Protocol on greenhouse gas (GHG) emissions. A number of categories of the gases responsible for this effect have been identified, including CFCs and derivatives that might be presenting risks due to their very high long-term stability.[122,123]

Examples of ozone-depleting molecules are given in Table 3. The comparison between the lifetime and the global warming potential (GWP) of selected gases appears in Table 4. The maximum historical depletion over the Southern Hemisphere (September 2000) and over the Northern Hemisphere (March 2011) are given in Fig. 33A, and the evolution of the depleted ozone area in the Southern Hemisphere from 1979 is given in Fig. 33B.

The impact of the Montreal Protocol on the abundance of effective stratospheric chlorine (combined effects of chlorine and bromine) is illustrated in Fig. 34. The reduction is a result of restrictions on the production and consumption of synthetic ozone depleting substances. The projections of the future abundance of effective stratospheric chlorine are shown for the following scenarios: no amendments to protocol regulations, regulations under the original 1987 Montreal Protocol, and additional regulations from ongoing and adjustments. The "zero emissions" line shows stratospheric abundance for the case where all emissions are reduced to zero in 2003. (Note that the Kyoto Protocol is based on emissions and not on production.) Following these protocols, a drastic reduction of refrigerant emissions has been reported between 1980 and 2002, as shown with the evolution of the GWP_{100} of CFCs, HCFCs, and HFCs in Fig. 35. This chart is calculated from AFEAS production data and based on the IPCC reports "Intergovernmental Panel on Climate Change," the 45th session of which (IISD—IPCC 45) took place in March of 2017, in Guadalajara, Mexico.

It should be added that the strong increase in the presence of stratospheric chlorine during the second half of the 20th century is anticipated to be reversed in the next decades and might return to pre-1980 values around 2050 in mid-latitudes, and around 2065 in polar regions.

Table 3 Examples of ozone-depleting substances controlled by the Montreal Protocol

Name	Examples	Uses
Chlorofluorocarbons (CFCs)	CFC-11 ($CFCl_3$) CFC-12 (CF_2Cl_2) CFC-113 (CCl_2FCClF_2)	Refrigeration Foam blowing Aerosol propellants
Solvents	CCl_4 CH_3CCl_3	Cleaning
Halons	Halon-1301 (CF_3Br) Halon-1211 ($CBrClF_2$)	Fire extinguisher
Hydrochlorofluorocarbons (HCFCs)	HCFC-22 (CHF_2Cl) HCFC-123 ($CHCl_2CF_3$)	CFCs substitutes

[122] Martyn P. Chipperfield, Global Atmosphere – The Antarctic Ozone Hole, in *Still Only One Earth: Progress in the 40 Years Since the First UN Conference on the Environment*, (2015), pp. 1–33.

[123] Global Ozone research and monitoring Projects, WMO, Reports 1–48 (1977–2015). https://www.wmo.int/pages/prog/arep/gaw/ozone_reports.html.

Table 4 Atmospheric lifetime of ozone-depleting substances and GWP for various selected gases

Gas name	Chemical formula	Lifetime (year)	Global warming potential (GWP) relative to CO_2 for given time horizon (year)		
			20	100	500
Carbon dioxide	CO_2	30–95	1	1	1
Methane	CH_4	12	84	28	7.6
Nitrous oxide	N_2O	121	264	265	153
CFC-11	CCl_3F	52	6730	4750	1620
CFC-113	CCl_2FCClF_2	93	6540	6130	2690
CFC-12	CCl_2F_2	100	10,800	10,200	5200
HCFC-22	$CHClF_2$	12	5280	1760	549
HCFC 123	$CHCl_2CF_3$	1	273	77	24
Halon 1301	$CBrF_3$	72	8480	7140	2760
HFC-23	CHF_3	228	11,900	14,200	10,700
Methyl chloroform	CH_3CCl_3	5	506	146	45
Bromomethane	CH_3Br	1.5	19	5	2
Tetrafluoromethane	CF_4	50,000	4880	6630	11,200
Hexafluoroethane	C_2F_6	10,000	8210	11,100	18,200
Sulfur hexafluoride	SF_6	3200	17,500	23,500	32,600
Nitrogen trifluoride	NF_3	500	12,800	16,100	20,700

Credit: https://en.m.wikipedia.org/wiki/Greenhouse_gas.

Looking for CFCs and HCFCs substitutes: For 15 years, many molecules have been tested as substitutes for CFCs.[124] After the phase-out of CFCs and HCFCs, the first generation of alternative refrigerants included HFCs (hydrofluorocarbons), which had no ozone depletion potential. The first choice of HFC was drowned by the simple observation that the C—H bond in an organic molecule may be more easily attacked by $OH^\bullet$ hydroxyl radicals present in the troposphere than are C—Cl or C—F bonds, thus reducing the stability in time of these species, and thus, their GWP. Regarding the main properties that guide this choice are physicochemical properties in good agreement with a respect for the environment, in particular, a limited stability of long-term molecules; flammability; non-toxicity; thermodynamic characteristics similar to those of CFCs, including the maximum operating pressures.[125]

[124] R. L. Powell, CFC Phase-out: have we met the challenge? J. Fluor. Chem. **114** (2002) 237–250.

[125] A. Sekiya "Evaluation and selection of CFC alternatives" in "Fluorine and the Environment," Vol. 1, Tressaud A. (Ed.), Elsevier, Amsterdam (2006).

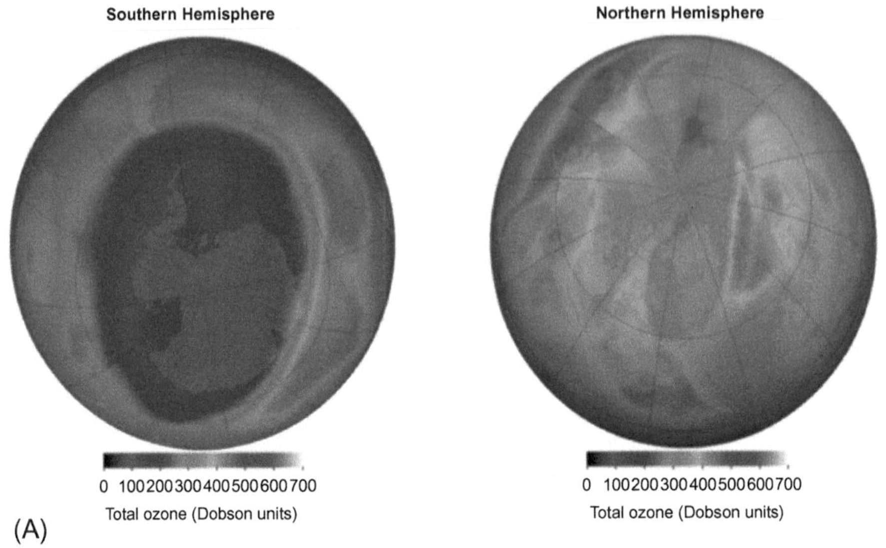

Fig. 33 (A) Maximum historical depletion over the Southern Hemisphere (September 9, 2000) and over the Northern Hemisphere (March 22, 2011); (B) evolution of depleted ozone area (Southern Hemisphere).

Credit: https://www.eea.europa.eu/data-and-maps/indicators/production-and-consumption-of-ozone-2/assessment-2.

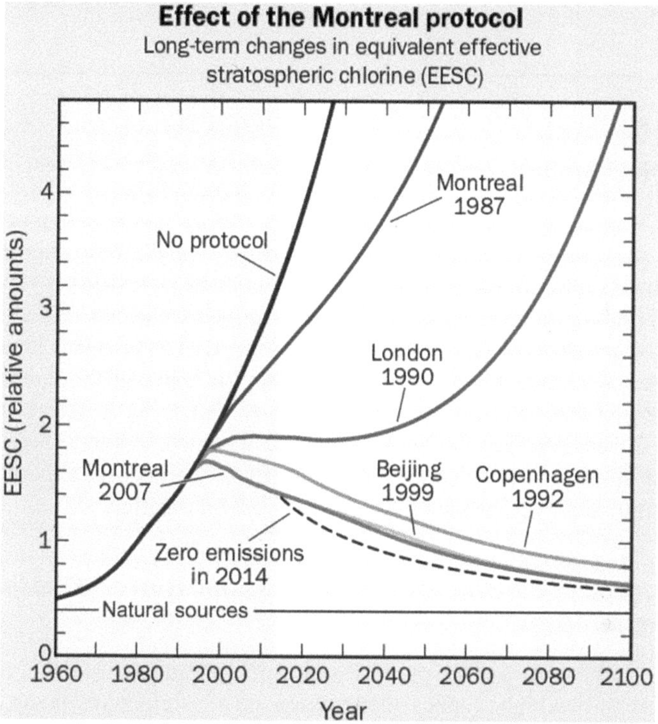

Fig. 34 Impact of the Montreal Protocol on the abundance of effective stratospheric chlorine. From "Twenty Questions and Answers about the Ozone Layer: 2010 update: scientific assessment of ozone depletion, 2010", Global Ozone Research and Monitoring Project Report No. 52, Geneva, Switzerland, 2011 (published by WMO and UNEP), Figure Q15-1. Effect of the Montreal Protocol. Used with permission from WMO.

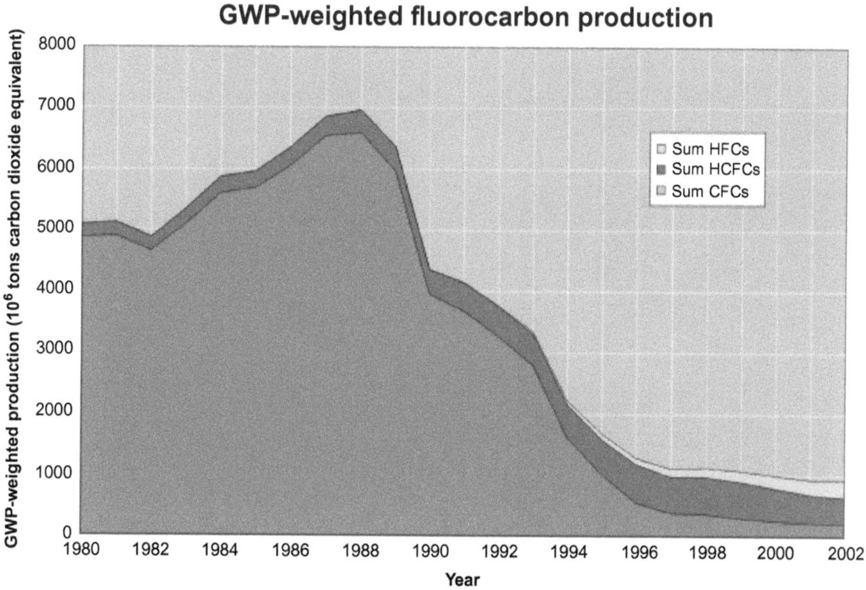

Fig. 35 Evolution of the global warming potential (GWP_{100}) of CFCs, HCFCs, and HFCs between 1980 and 2002.

8.3 The story of Teflon and fluoropolymers

8.3.1 The discovery of Teflon

It was by accident that Teflon was discovered, but this discovery took place in a company that immediately understood the immense impact that this exceptional material could present. On April 6, 1938, Roy J. Plunkett (1911–94), a young chemist who was hired 2 years before by DuPont, assisted by Jack Rebok, failed to properly open a container in which a gas, tetrafluoroethylene (TFE: C_2F_4) was in contact with hydrochloric acid. When they finally had access to the inside, the researchers discovered a white powder, polyfluoroethylene [PTFE $(C_2F_4)_n$], better known today by its brand name Teflon (Fig. 36). R. Plunkett immediately understood the type of reaction that had just taken place, noting the same day in his notebook of experiences, "The white solid obtained was probably a product of polymerization."

8.3.2 Fluoropolymers: From military applications to hospital and kitchen uses

The exceptional properties of the product were quickly characterized: thermoplasticity, lubricating properties, high chemical inertness toward moisture, acids, bases, and various solvents; these led 1 year later to a patent on PTFE-based polymers. A major difficulty in synthesis, nevertheless, remained to be solved, because a spontaneous polymerization of

Fig. 36 The discovery of Teflon by R.J. Plunkett (right) and his team (1938).

the monomer could lead to an explosion due to the high exothermicity of the reaction. DuPont chemists found a solution to the problem, but it was relatively expensive, and a potential market had to be clearly identified. WWII was ravaging the planet, and in the United States, a program called The Manhattan Project was launched in secret.

The project's final goal was to build an atomic bomb based on uranium isotope ^{235}U, that is the only nucleide that could be rendered fissile with thermal neutrons. Among the different uranium compounds at highest oxidation state (VI+), the hexafluoride UF_6 can be obtained by burning tetrafluoride UF_4 in fluorine atmosphere gas (F_2) at high temperature. It is a solid at room temperature and sublimates at 56.5°C. An enrichment in ^{235}U could thus be possible via gaseous separation of the two fluorides $^{235}UF_6$ and $^{238}UF_6$. In such reactive conditions, the various parts of the equipment, valves, seals, etc., should be particularly inert. The DuPont Teflon corresponded exactly to these specifications. It thus became a highly secretive product, and renamed K-416. Teflon's entire production was reserved for various military projects and two-thirds of the Arlington production unit were allocated to the Manhattan project, the rest for other uses such as nose cone proximity bombs (thanks to its electrical insulating property and transparency to radar), fuel tank coatings (thanks to better resistance to low temperatures, etc.). It was only in 1944 that DuPont gave the name of Teflon to the PTFE product.

After the return of peace, the truly exceptional qualities of the material were quickly brought at the service of industry, medicine, and public well-being. Many new fluoropolymer families exhibiting specific features had appeared. As an example, the hydrophobic properties of fluoropolymers had given rise to new lines of clothing such as Gore-Tex-treated gear. The coating of extremely varied surfaces, fabrics, paper, leather, etc., by a film of these fluoropolymers rendered these materials hydrophobic (granting water impermeability) and/or oleophobic (spot-resistant). Many other applications are also concerned. Indeed, in the polymer chains, the substitution of fluorine for the hydrogen atoms improves many characteristics, increasing heat resistance and chemical resistance, delaying or inhibiting the spread of flame, lowering the critical surface tension, and increasing the fabric's insulating character. The effect is more marked as the degree of substitution of H by F is higher, and it reaches its maximum for polymers that contain only carbon and fluorine.

The main types of fluoropolymers are grouped in Table 5 with some of their applications. Their most outstanding properties, that is, high chemical resistance, excellent resistance in aggressive environments, lack of adherence, nonflammability, low dielectric constant, low temperature flexibility, high temperature stability, and low friction coefficient, make them unique materials for all branches of industry, such as chemicals, building, aeronautics, space, biotechnology, clothing, and medicine.[126] These exceptional properties explain why today, the range of fluoropolymers' uses is very wide in the aircraft and automobile industries, surface treatment of clothing, cardiovascular surgery, kitchen equipment, wire insulation for computer networks, and semiconductor manufacturing equipment (Fig. 37). In order to protect metal structures against atmospheric pollution, the treatment of the aluminum parts of the Louvre pyramid or the renovation of the Statue of Liberty in New York have used fluorinated polymers.

[126] Ameduri B., B. Boutevin, Well-Architectured Fluoropolymers, Elsevier, (2004); Ameduri B., Les (co) polymères fluorés,l'Actualité Chimique, N° **241**, 24–30 (2017).

Table 5 Main types of fluoropolymers with applications

Fluoropolymer	Formula	Commercial brands	Major types of applications
PTFE—polytetrafluoroethylene	$[-CF_2-CF_2-]_n$	Teflon, Fluon, Hostaflon, *Neoflon (ex-Kel-F)*	– Bearings, seals, high temperature electrical insulators, non-stick coatings and linings/coating cookware/sealing tape (renovation, Statue of Liberty)/solid lubricant/deposits on razor blades/rainproof clothing, Gore Tex/insulating fittings for electrical/high-voltage wiring/computer mouse
PVF—polyvinyl fluoride	$[-CH_2-CHF-]_n$	Tedlar	– Protecting films – Gas tanks, solar panels, printed circuit
PVDF—polyvinylidene fluoride	$[-CH_2-CF_2-]_n$	Kynar, Solef, *Hylar*	– Thermoplastic, highly resistant – Piping, fittings and tanks, chemical industry/wire and piezoelectric elements insulation in electronics/protection of metal parts in buildings (Louvre Pyramid) – Biotechnology
PFA—perfluoroalkoxy	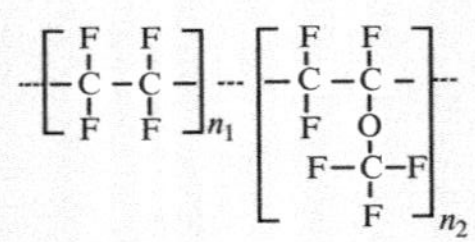	*Hyflon*	– Endurance to high bending/very good mechanical resistance to high temperatures – Special tubings, moldings, coatings in chemical engineering equipment – Transfer molding and extrusion
FEP—fluorinated ethylene propylene	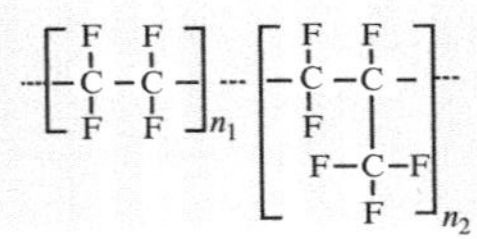	Hostaflon	– Uses in extruded, molded, or injected forms

ETFE—tetrafluoroethylene and ethylene copolymer	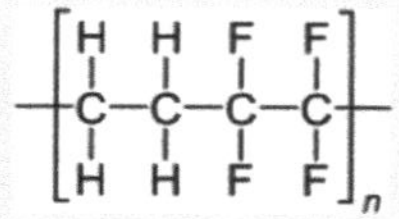	*Tefzel, Fluon, Neoflon ETFE, Texlon*	– Use for very large surface panels: "Water Cube," Beijing, China, 2008 Olympics/20,000 m^2 window for "Tropical Islands" project, Brandenburg, Germany – Protection of electrical equipment in extreme conditions/ resistance to high energy radiations – Aerospace, nuclear energy
ECTFE—ethylene chlorotrifluoroethylene polymer	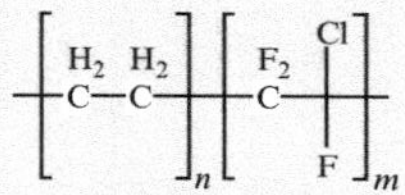	Halar	– Electronics and electrical chemical industries – Cryogenic and aerospace applications
PCTFE—*polychlorotrifluoroethylene polymer*	$(–C\ F_2–CFCl–)_n$	Voltalef	– Crystalline material (with crystalline levels ranging from 30% to 70%/high mechanical strength and low elongation. Cryogenics, polymer stable between −255°C and +150°C/extremely low permeability to gases – Barrier material/electrical insulation in hostile climates: hot and humid climates, marine conditions

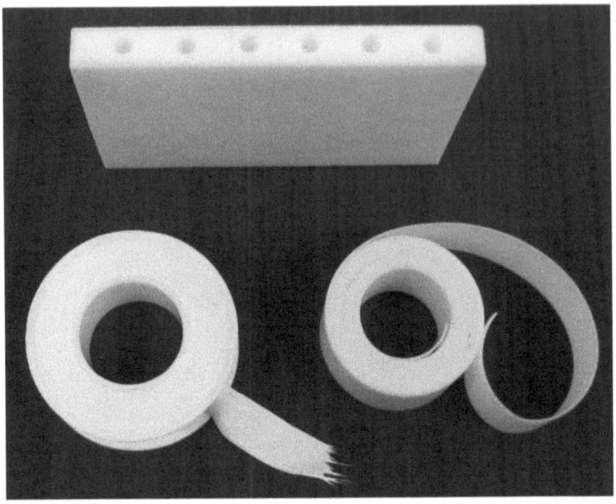

Fig. 37 Various PTFE items.
Taken from https://hi.wikipedia.org/wiki/%E0%A4%9A%E0%A4%BF%E0%A4%A4%E0%A5%8D%E0%A4%B0:Teflon_items.jpg.

8.4 Fluorine, a decisive element in the nuclear cycle

8.4.1 The Manhattan project

During WWII, most discoveries involving fluorine and fluorine products had become top secret. So was Teflon, the Simons perfluorination electrochemical process which would be published in 1949, and of course, research on fluorine derivatives of uranium. The origin of the Manhattan Project back in 1939, when Niels Bohr came from Denmark to spend several months at Princeton University. With his American colleagues, Bohr discussed of the hypothesis of Lise Meitner and Otto Robert Frisch that the absorption of a neutron by a uranium nucleus would cause the separation of the latter into two approximately equal parts and would release a huge amount of energy, a phenomenon they called "nuclear fission." From discussions between Bohr and the Italian-born physicist Enrico Fermi, who studied ionizing radiation produced by the fragments of the U nucleus at Columbia University, came the idea that neutrons might be emitted during the process, leading to a nuclear-chain reaction. On August 2, 1939, Albert Einstein wrote a letter to President Roosevelt to sensitize him to the risks of Nazis developing this type of nuclear reaction in order to make destructive bombs. This correspondence prompted the president to set up a consultative committee for uranium, which, in 1942, yielded the Manhattan Project, led by R. Oppenheimer. This project was strengthened by the first nuclear reaction, self-sustaining chain conducted at the end of the year by the Fermi team at the University of Chicago. Independently, on May 1939, F. Joliot-Curie filed a patent for the use of nuclear energy to obtain explosive chain reactions.[127]

[127] F. Joliot Curie, H. Halban, L. Kowarski, Perfectionnement aux charges explosives, Invention Patent n° 971–324, Caisse Nationale de la Recherche Scientifique, Paris, May 4, 1939.

The realization of the atomic bomb required an enrichment of fissile ^{235}U isotope, the only U nucleide that could be rendered fissile with thermal neutrons. Natural uranium contains 99.28% ^{238}U isotope, with only about 0.71% ^{235}U. With an enrichment of natural uranium into compositions containing between 3.5% and 4.5% ^{235}U could be obtained via gaseous separation of $^{235}UF_6$ and $^{238}UF_6$ uranium hexafluorides. Thus, a critical step was separating 235 and 238 isotopes. UF_6 can be obtained by burning tetrafluoride UF_4 in fluorine atmosphere gas (F_2) at high temperature. It sublimes above 56.5°C at atmospheric pressure. So, the diffusion of the gases through a porous material made it possible to differentiate the two types of $^{235}UF_6$ and $^{238}UF_6$ molecules of different sizes and masses. The lighter molecules diffused faster than the heavier, thus enriching the starting material to fissile isotope $^{235}UF_6$. The bomb ("Little Boy") dropped on Hiroshima August 6, 1945, was made of uranium 235. The uranium isotope separation can be achieved mainly by gaseous diffusion of uranium hexafluoride, but also via other techniques such as thermal diffusion and electromagnetic separation. The main work of separation took place in Oak Ridge. The uranium bomb was using a mechanism called a gun to obtain the critical mass of $^{235}UF_6$, that is, a mass of ^{235}U (the "bullet") was launched in a tube toward another ^{235}U mass, creating the mass critically needed for explosion.

8.4.2 *Importance of fluoride materials in the nuclear cycle*

Today, 60% of the world's production of F_2 is devoted to the production of UF_6 to fuel nuclear power plants. Therefore, elemental fluorine is part of what is commonly called the "nuclear fuel cycle," and thus constitutes an essential link for the preparation of electricity via power plants.

$$U_3O_8 \xrightarrow{HNO_3} UO_2(NO_3)_2 \xrightarrow{T^0} UO_3 \xrightarrow{H_2} UO_2 \xrightarrow{HF} UF_4 \xrightarrow{F_2} UF_6$$

$$UO_{2.0} \xrightarrow{+4HF,\,-2H_2O,\,400°C} UF_4 \xrightarrow{F_2,\,T>400°C} UF_6$$

For the manufacture of nuclear fuel, the fluorinated products occur at two levels: hydrogen fluoride, HF, transforms the ore concentrates into UF_4, and finally, it is an oxidation of UF_4 by fluorine gas which results in the production of uranium hexafluoride, UF_6, whereby the $^{235}UF_6$ enrichment can take place. In France, these processes are implemented by orano-AREVA, which ranks first in Europe and second in the world for the preparation of fluorine and number one in the world for the production of UF_6. Thus, thanks to its fleet of 58 nuclear reactors, France has achieved an energy independence rate close to 50%. The nuclear share of electricity generation is currently around 80%. The French nuclear fleet has 34 units of 900 MW, 20 units of 1300 MW, and 4 units of MW 1450; the last unit was commissioned in 1999 to Civaux.[128] Fig. 38 shows thc place of fluorine at different steps of the nuclear fuel cycle:

[128] H. Groult, F. Lantelme, D. Devilliers, C. Belhomme B. Morel, F., JP Nicolas, J.P. Caire, Fluoride, Elec trochemical synthesis of fluorine from 1886 to 2006: fluorine, a key-element for nuclear energy, Special issue devoted to Fluor and Fluorinated Materials, Actualité Chimique 144–147, (2006), pp. 40–46.

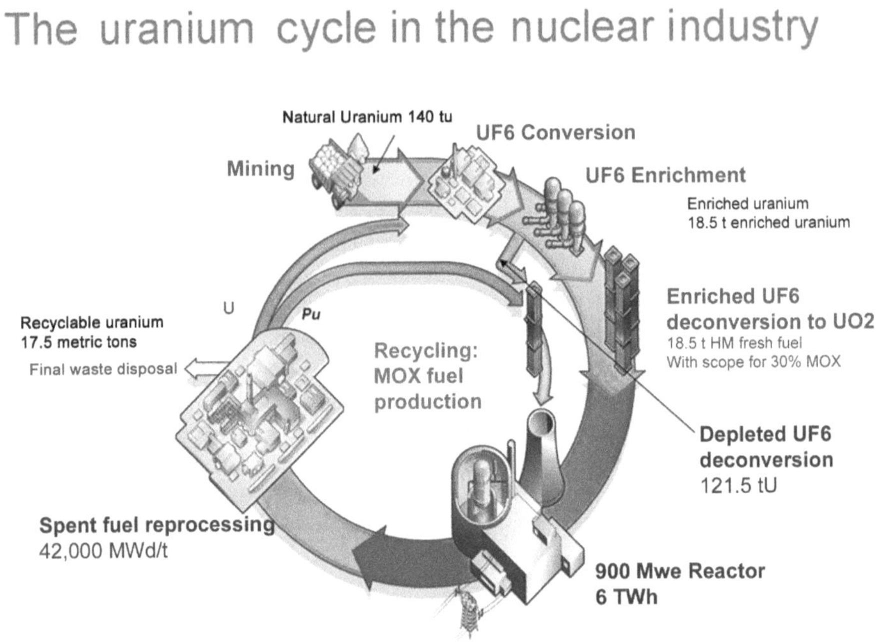

Fig. 38 Contribution of fluorine and fluoride components in the nuclear cycle [*credit orano-AREVA*].

conversion to UF_6 and enrichment, enriched UF_6 converted into UO_2, and depleted UF_6 deconversion. It should be pointed out that civilian nuclear programs—although sometimes criticized—allow us today to limit the depletion of fossil energy resources.

A think tank was initiated several years ago to propose new types of nuclear reactors: the Generation IV nuclear reactors. A "Generation IV International Forum" (GIF) was initiated in 2000.[129] It is an international collective representing governments of 14 countries where nuclear energy is significant now and also seen as vital for the future. Most countries are committed to joint development of the next generation of nuclear technology.[130] The motivations for the project include goals such as improved safety, sustainability, efficiency, and cost.

Among the various types of Generation IV nuclear reactors, are thermal reactors, very-high-temperature reactors (VHTR), molten-salt reactors (MSR), supercritical-water-cooled reactors (SCWR), gas-cooled fast reactors (GFR), sodium-cooled fast reactors (SFR), and lead-cooled fast reactors (LFR).

[129] Locatelli, Giorgio; Mancini, Mauro; Todeschini, Nicola (2013-10-01). "Generation IV nuclear reactors: Current status and future prospects". Energy Policy 61: 1503–1520. "GIF Portal—Home—Public." www.gen-4.org.

[130] The original members of GIF were Argentina, Brazil, Canada, France, Japan, South Korea, South Africa, the UK, and the USA. The agreement was subsequently signed by Switzerland, Euratom, China, Russia, and Australia.

In molten salt reactors, the nuclear fuel may be solid or dissolved in the coolant. The thorium molten salt reactor used mixtures of LiF-BeF_2-ThF_4-UF_4 as fuel. The initial composition was very low in U, with a composition of 77.5% LiF, ThF_4 18.7%, and 3.8% UF_4. One of its advantages was that a mixture of uranium-238 was implemented in order to make sure that recovered uranium would not be weapons grade. The most well-known molten salt breeder reactor was built at the Oak Ridge National Laboratory and used LiF-BeF_2-ThF_4-UF_4 (72-16-12-0.4 (%)) mixtures as fuel. It was moderated by graphite with a 4-year replacement schedule. The secondary coolant was NaF-$NaBF_4$. Its peak operating temperature was around 705°C.[131] Despite its success, the MSR program closed down in the early 1970s in favor of the liquid metal fast-breeder reactor (LMFBR), after which, research stagnated in the US. Regular GIF reports are available, including the 2016 report.[132] On January 2016, the US Energy Department announced new investments in advanced nuclear power reactors.[133]

[131] Energy From Thorium's Document Repository: The Use of Thorium in Nuclear Power Reactors. WASH 1097, ORNL.gov (http://www.energyfromthorium.com/pdf/ http://moltensalt.org/references/static/downloads/pdf/WASH-1097.pdf.

[132] H. Boussier, S. Delpech, V. Ghetta et al.: The Molten Salt Reactor (MSR) in Generation IV: Overview and Perspectives, GIF Symposium Proceedings, 2012 Annual Report, NEA No. 7141, pp. 95 [3]. GIF Portal—Annual Reports https://www.gen-4.org/gif/jcms/c_92757/gif-annual-report-2016.

[133] US Department of Energy. Retrieved January 16, 2016. https://www.energy.gov/articles/energy-department-announces-new-investments-advanced-nuclear-power-reactors-help-meet.

Fluorine, a key element for the 21st century

2

Chapter Outline

Fluorine. https://doi.org/10.1016/B978-0-12-812990-6.00002-7

In this section, various examples of current applications of fluorine and fluoride compounds illustrate their involvement in the many and varied fields of chemistry, engineering, materials science, health, and biotechnologies. The impact of properties and applications of fluorocompounds on our daily life has been the subject of several books over the last two decades, and for more specialized research, the reader should refer to these books (*see a selection of books (2000–2018) at the end of the present book*).

1 Advances in organic fluorine chemistry

1.1 Toward new functional molecules

During the first decades of the 20th century and following the works of H. Moissan, the chemistry of fluorine was mostly concerned with inorganics. However, several events rapidly triggered the development of new sectors of fluorine chemistry[134,135,136,137,138] which are still in use today. Concerning the CFCs saga, following the synthesis of these new refrigerants by T. Midgley, the halogen exchange fluoridation method using anhydrous hydrogen fluoride still remained the major tool for the industrial production of numerous fluoro-organics, including (trifluoromethyl) arenes. The field of fluoropolymers was also very active. Following the launch of the Manhattan Project, intensive technological research on fluorine gas generators was conducted by G.H. Cady, L.A. Bigelow, and W.T. Miller. This research led to significant development of fluorination reactions of organic molecules by F_2 gas and by cobalt trifluoride, CoF_3, which was still used in the 1980s. In 1941, J.H. Simons discovered electrochemical perfluorination[139]: the electrolysis of a mixture of hydrocarbon substrate and hydrogen fluoride, a process that is still very much used today for the production of high-performance surfactants. Finally, since the 1960s, the number of new fluorinated molecules has increased dramatically because of the surprising biological changes introduced by the introduction of one or more fluorine atoms in the molecule. Thus, the organic chemistry of fluorine entered the fields of fine chemistry, pharmacy, biomedical, medical imaging, medicine, and agrochemistry, which today constitute some of the most powerful advances in fluorine chemistry.[140]

[134] Fluorine in Medicinal Chemistry and Chemical Biology, I. Ojima, (Ed.), Wiley-Blackwell, Oxford (2009).

[135] Fluorine Chemistry, a thematic issue, V. Gouverneur, K. Seppelt, Eds. Chemical Reviews, 115 (2015).

[136] Fluorine & Life Sciences, G. Haufe, F. Leroux. Eds. Vol. 4, Progress in Fluorine Science series, Elsevier (online 2018).

[137] B. Langlois, S. Ratton, J.M. Paris, Tendancies and Challenges of organic fluorine chemistry, Special issue of "L'actualité chimique" devoted to Fluorine and fluoride materials—N° **301–302,** pp. 56–66 (2006).

[138] R.E. Banks, J.C. Tatlow, Synthesis of C—F bonds: the pioneering years, 1835–1940, in Fluorine, the first hundred years (1886–1986); R.E. Banks, D.W.A. Sharp, J.C. Tatlow, Eds.; Elsevier Sequoia, Lausanne, 1986; chap. 4, pp. 71–108 and references herein.

[139] J. H. Simons, The Electrochemical Process for the Production of Fluorocarbons, J. Electrochem. Society, **95** (1949) pp. 47–66.

[140] J.P. Bégué, D. Bonnet-Delpon, "Chimie Bioorganique et médicinale du fluor", CNRS Editions/EDP Sciences, Paris, 2005. J.-P. Bégué, D. Bonnet-Delpon, Bioorganic and Medicinal Chemistry of Fluorine, (2008) John Wiley & Sons, USA.

1.1.1 The carbon-fluorine bond

Fluorine atom possesses several remarkable behaviors which differ from those of all other substituents encountered in organic molecules. As a result, the physical and chemical properties of organofluorines can be distinctive in comparison to other organohalogens.[141]

(i) The C–F bond is one of the strongest in organic chemistry (an average bond energy around 480 kJ/mol). This is significantly stronger than the bonds of carbon with other halogens (an average bond energy of, e.g., C–Cl bond, is around 320 kJ/mol) and is one of the reasons why fluoro-organic compounds have high thermal and chemical stability.

(ii) A van der Waals radius, intermediate between those of hydrogen and of oxygen (F: 1, 47 Å, H: 1.20 Å, O: 1.52 Å), which will not cause drastic changes in the molecule's volume. This, together with short bond length, is the reason why there is no steric strain in polyfluorinated compounds. In addition, the fluorine substituents in polyfluorinated compounds efficiently shield the carbon skeleton from possible attacking reagents. This is another reason for the high chemical stability of polyfluorinated compounds.

(iii) A much lower polarizability than that of hydrogen and oxygen (F: 0.56 $Å^3$, H: 0.67 $Å^3$, O: 0.82 $Å^3$) which is associated with the highest electronegativity of all the elements (F: 3.98, O: 3.44, Cl: 3.16) polarizes the CF bonds. Protic functions close to the fluorinated group are acidified; for example, imidazole has a pK_a of 7.1, whereas, its fluorinated homologues at the two or four position have a pK_a of 2.4.

(iv) A very strong hydrophobicity provided by fluorinated groups to the molecules improves their bioavailability by facilitating their transmembrane transfer. This hydrophobicity is quantified by the Hansch parameter Π_R, attached to each substituent R and resulting from the measurement of the partition coefficient between water and *n*-octanol of Ph-R aromatic relative to that of benzene (Table 6). The difference in fluorine behavior depending on the medium are schematized in Fig. 39.

In organic chemistry, the high electronegativity of fluorine causes a large electronic effect on closer carbon atoms. This creates an appreciable effect on the molecule's dipolar moment and the acidity or basicity of other groups.

Table 6 Hansch constants de (Π_R) for various fluorinated substituents (Π_R = log P(PhR) – log P(PhH) (eau/*n*-octanol); $\Pi_R > 0$: hydrophobic substituents; $\Pi_R < 0$: hydrophilic substituents)

H: $\Pi_R = 0.00$	F: $\Pi_R = +0.14$
CH_3: $\Pi_R = +0.56$	CF_3: $\Pi_R = +0.88$
OCH_3: $\Pi_R = -0.02$	OCF_3: $\Pi_R = +1.04$
SCH_3: $\Pi_R = +0.61$	SCF_3: $\Pi_R = +1.44$
SO_2CH_3: $\Pi_R = -1.63$	SO_2CF_3: $\Pi_R = +0.55$
$NHSO_2CH_3$: $\Pi_R = -1.18$	$NHSO_2CF_3$: $\Pi_R = +0.92$

[141] B.E. Smart, Characteristics of the C—F Systems, in Organofluorine Chemistry, Principles and Commercial Applications; R.E. Banks, B.E. Smart, J.C. Tatlow, Eds (1994) Plenum Press; New York, London.

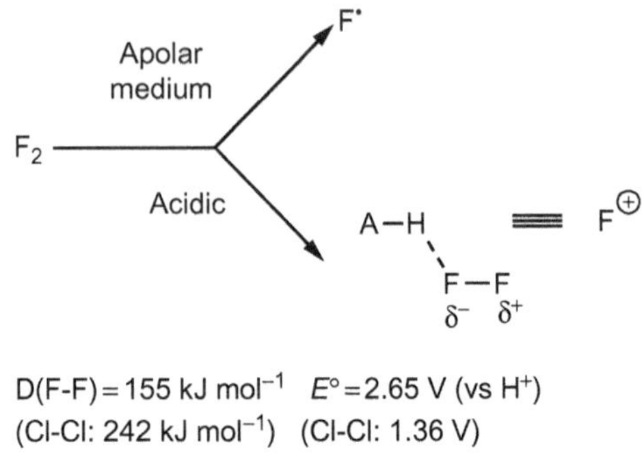

Fig. 39 Difference in fluorine behavior depending on the medium [from 137].

The carbon-fluorine is very strong, polarized, and weakly polarizable. The introduction of a fluorine atom (and especially a fluorinated group) can strongly increase the lipophilicity of the molecule that carries it. This is almost always true when the fluorinated group is carried by an aromatic ring, but must be nuanced when it is an aliphatic chain. Among other notable consequences of the presence of this halogenated pattern, the deep conformational and electronic changes induced not only by the fluorine atom, but also by perfluoroalkyl groups, are noteworthy. In addition, as far as drug molecules are concerned, fluorine is usually more lipophilic than hydrogen, which makes fluorinated compounds more fat-soluble. This allows them to be more easily integrated into membranes, thus exhibiting higher bioavailability. Among significant developments of the last 40 years in terms of synthesis of fluorinated organic compounds, we can cite:

- Direct fluorination, either by elemental fluorine F_2 diluted in a neutral gas, by electrophilic process, by hydrofluoric acid, or by halogen exchange using KF, by nucleophilic process.
- Synthesis by synthons, providing fluorocarbon bricks for building more complex networks.
- Direct fluoroalkylation method, in particular trifluoromethylation with direct introduction of radicals. CF_3, or equivalents of the $^+CF_3$ cation or equivalents of the $^-CF_3$ anion.
- Classical reagents such as gaseous fluoroform or the Ruppert-Prakash's reagent for efficient and practical fluorination processes.
- Discovery of new entities such as the electrophilic trifluoromethylation and trifluoromethylthiolation reagents.
- Interplay with new types of catalysis, such as dual catalysis and/or photocatalysis.
- Industrial advances have gone through the development of new reagents such as DAST (Et_2NSF_3) or its analogue Deoxofluor, tetrasubstituted difluoromethylenediamines which have made it possible to replace alcohol functions by fluorine or to convert carbonyl groups into CF_2 units, making it obsolete the use of very toxic SF_4. It should be noted that until the 1990s, molecular fluoride F_2 was scarcely used because of the difficulties of industrial implementation, and CoF_3 was often preferred. R.D. Chambers opened up new perspectives by demonstrating that it was possible to use fluorine as a pure electrophile by polarizing it ($F^{\delta+} - F^{\delta-}$) in a strong acid medium[142] (Fig. 40). However, the use of electrophilic fluorination reagents, more expensive but easier to be industrially implemented are still preferred to fluorine, such as those provided in Fig. 40.

[142] R. Chambers, Fluorine in Organic Chemistry, Blackwell Publ. (2004).

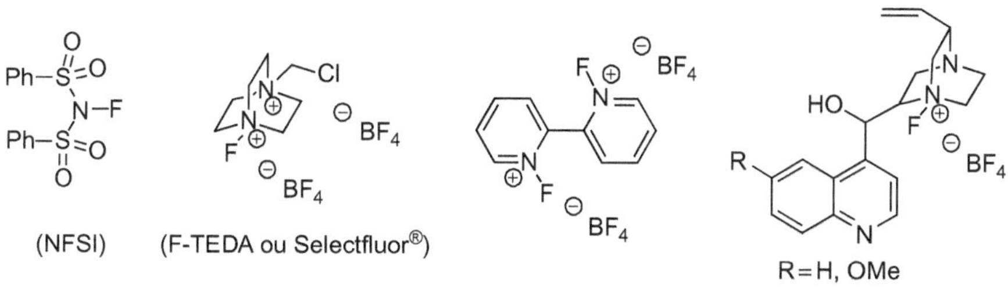

Fig. 40 Some reagents used in electrophilic fluorination.

1.2 Some examples of fluoro-organic chemistry

1.2.1 Electrophilic fluorination

Electrophilic fluorination relies on sources of F^+. Often such reagents feature N–F bonds, for example, 1-chloromethyl-4-fluoro-1,4-diazoniabicyclo[2.2.2]octane bis(tetrafluoroborate), or Selectfluor, which is one of the best reagents used as a fluorine donor. This compound is a derivative of the heterocycle DABCO. This colorless salt was first described in 1992 by E. Banks and has been since commercialized for use in electrophilic fluorination.

Asymmetric fluorination, whereby only one of two possible enantiomeric products are generated from a prochiral substrate, relies on electrophilic fluorination reagents. The preparation of a precursor to anti-inflammatory agents is illustrative of this approach.

1.2.2 Electrofluorination

A specialized but important method of electrophilic fluorination involves electrosynthesis. The method is used to perfluorinate, replacing all C–H bonds by C–F bonds. The hydrocarbon is dissolved or suspended in liquid HF, and the mixture is electrolyzed at 5–6 V using Ni anodes. The method was first demonstrated with the preparation of perfluoropyridine (C_5F_5N) from pyridine (C_5H_5N). Several variations of this technique include the use of molten potassium bifluoride or organic solvents.

1.2.3 Nucleophilic fluorination

The nucleophilic fluorination use reagents that are sources of "F^-" for displacement of typically chloride and bromide. Metathesis reactions employing alkali metal fluorides are the simplest, for example:

$$R_3CCl + MF \rightarrow R_3CF + MCl \ (M = Na, K, Cs)$$

Alkyl monofluorides can be obtained from alcohols and Olah reagent (pyridinium fluoride) or another fluoridating agent. Although hydrogen fluoride may appear to be an unlikely nucleophile, it is the most common source of fluoride in the synthesis of organofluorine compounds. Such reactions are often catalyzed by metal fluorides such as chromium trifluoride. 1,1,1,2-Tetrafluoroethane, a substitute for CFC's, is prepared industrially using this approach.

$$Cl_2C=CClH+4HF \rightarrow F_3CCFH_2+3HCl$$

1.2.4 Deoxofluorination

Deoxofluorination agents replace hydroxyl and carbonyl groups with one and two fluorides, respectively. One such reagent, useful for fluoride for oxide exchange in carbonyl compounds, is sulfur tetrafluoride:

$$RCO_2H+SF_4 \rightarrow RCF_3+SO_2+HF$$

Alternates to SF_4 include the diethylaminosulfur trifluoride (DAST, NEt_2SF_3) and bis(2-methoxyethyl)aminosulfur trifluoride (deoxo-fluor). These organic reagents are easier to handle and more selective.

1.2.5 Fluorinated building blocks

Many organofluorine compounds are generated from reagents that deliver perfluoroalkyl and perfluoroaryl groups. (Trifluoromethyl) trimethylsilane, $CF_3Si(CH_3)_3$, is used as a source of the trifluoromethyl group, for example. CF_3X (X = Br, I), C_6F_5Br, and C_3F_7I are also fluorinated building blocks. These species form Grignard reagents that then can be treated with a variety of electrophiles. The development of fluorous technologies (see below, under solvents) is leading to the development of reagents for the introduction of "fluorous tails."[143]

A special but significant application of the fluorinated building block approach is the synthesis of tetrafluoroethylene, which is produced on a large-scale industrially via the intermediacy of difluorocarbene. The process begins with the thermal dehydrochlorination of chlorodifluoromethane at 600–800°C:

$$CHClF_2 \rightarrow CF_2+HCl$$

$$2CF_2 \rightarrow C_2F_4$$

1.2.6 Fluorous chemistry

Highly fluorinated substituents, for example, perfluorohexyl (C_6F_{13}), confer distinctive solubility properties to molecules, which facilitates purification of products in organic synthesis.

"Fluorous chemistry," an important area of "green chemistry,"[144] is related to the concept of "like-dissolves-like" in the sense that fluorine-rich compounds dissolve preferentially in fluorine-rich solvents. Because of the relative inertness of the C–F bond, such fluorous phases are compatible with even hard reagents. This theme has

[143] P. Ramaiah, R. Krishnamurti, G. K. Surya Prakash. "1-Trifluoromethyl)-1-cyclohexanol". Organic Syntheses, **72** (1998) 232.

[144] J. A. Gladysz, D. P. Curran, I. T. Horváth (Eds.) Handbook of Fluorous Chemistry, Wiley–VCH, Weinheim (2004).

spawned the techniques of "fluorous tagging" *and* "fluorous protection," for instance, the use of fluoroalkyl-substituted tin hydrides for reductions, the products being easily separated from the spent tin reagent by extraction using fluorinated solvents. Hydrophobic fluorinated ionic liquids, such as organic salts of bistriflimide or hexafluorophosphate, can form phases that are insoluble in both water and organic solvents, producing multiphasic liquids.

1.2.7 Click Chemistry of Sulfur Fluoride Exchange (SuFEx)

Click chemistry is a term that initially described reactions giving high yield and selectivity products by carbon-hetero bond formation reactions. The word "click" referred to easily joining molecular building blocks as two pieces of a seat belt buckle. In 2014, K.B. Sharpless, 2001 Chemistry Nobel laureate for the discovery of "Click Chemistry," extended this type of reactions to fluorine chemistry.[145] Owing to the balanced properties of fluoride bonding to hydrogen and sulfur (VI), sulfur fluoride exchange (SuFEx) was identified as the next click chemistry, leading to a variety of sulfonyl fluoride reagents. These molecules can find wide use in applications such as chemical synthesis, materials science, chemical biology, and drug development.

1.2.8 Superacid Media

R. Gillespie gave such a definition in 1971 for any acid with an H_0 value lower than that of 100% sulfuric acid (~ -12). By mixing antimony pentafluoride and fluorosulfonic acid, G.A. Olah prepared the so-called magic acid, named after its ability to attack hydrocarbons.[146]

Fluoroantimonic acid results from combination of hydrogen fluoride (HF) and antimony pentafluoride (SbF_5). In this system, HF releases its proton (H^+) with the simultaneous binding of F^- by the antimony pentafluoride. The resulting anion (SbF_6^-) is both a poor nucleophile and a weak base. The proton from superacids is said to become "naked," accounting for the extreme acidity of this system.

At 140°C, FSO_3H–SbF_5 protonates methane to give the tertiary-butyl carbocation, a reaction that begins with the protonation of methane:

$$CH_4 + H^+ \rightarrow CH_5{}^+$$

$$CH_5{}^+ \rightarrow CH_3{}^+ + H_2$$

$$CH_3{}^+ + 3CH_4 \rightarrow (CH_3)_3C^+ + 3H_2$$

Common uses of superacids include providing an environment to create, maintain, and characterize carbocations. Carbocations are intermediates in numerous useful reactions such as those forming plastics and in the production of high-octane gasoline.

[145] J. Dong, L. Krasnova, M. G. Finn, K. B. Sharpless, Angew. Chem. Int. Ed. **53** (2014) 9430-9448.

[146] Olah, G. A. Crossing Conventional Boundaries in Half a Century of Research. Journal of Organic Chemistry **70** (7) (2005) 2413–2429.

2 Advances in new fluorination routes, surface functionalization, and nanosized fluorocompounds

2.1 Frontiers in inorganic fluorine chemistry

2.1.1 Some routes to high oxidation states of transition elements

Since the discovery of noble gas compounds, a lot of investigations on new synthetic routes have been proposed to have access to high (or unusual) oxidation states of inorganic fluorocompounds. In a number of cases, noble-gas species have been used as synthons to generate high-oxidation state transition metal species and carbocations.

Following the synthesis of XeF_6, the kinetics of the reaction between xenon and fluorides allow researchers to find conditions suitable for synthesis of xenon fluoride in larger quantities; the catalytic effect of first-row transition metal fluorides on the rate of reaction between xenon and fluoride was also investigated. A photochemical method for the production of O_2F_2 by irradiation of a liquefied mixture of oxygen and fluorine was developed. This method was also used for the preparation of large quantities of spectroscopically pure fluorine.

O_2^+ salts dissolved in anhydrous HF at room temperature or below may oxidize numerous d transition elements: $(PtF_6)^{2-}$ to $(PtF_6)^-$, $(AuF_6)^{2-}$ to $(AuF_6)^-$, AgF_2 to $(AgF_4)^-$, $(NiF_6)^{3-}$ to $(NiF_6)^{2-}$.[147]

The highly reactive chemistry of bromine high oxidation state has been extended by G. Schrobilgen. By oxidative fluorination of BrF_5 with the strong oxidant cations, KrF^+ and $Kr_2F_3^+$ in liquid BrF_5, $Br^{VII}F_6^+$ cation was first synthesized. Other compounds of this extremely high oxidation state include $BrO_3F_2^-$, the others being BrO_4^-, BrO_3F and BrF_6^+.[148]

2.1.2 On new polynitrogen $[N_5]^+$ species

For decades, apart from dinitrogen, N_2, other nitrogen species such as $N_3^\bullet$,N_3^+, and N_4^+ have been observed only as free gaseous or matrix-isolated ions or radicals. The absence of data might have resulted from their high endothermicity, yielding instability, and possible explosiveness. The bond energies of N–N single and double bonds are of 160 and 418 kJ mol^{-1}, respectively, that is much less than 1/3 or 2/3 of the N_2 triple bond energy of 954 kJ mol^{-1}. For a cation such as N_5^+ discovered by Karl Christe, the presence of a formal positive charge in one of the starting materials is very important in view of the high first ionization potential of N_2 (1503 kJ mol^{-1}).[149]

N_2F^+ cation and HN_3 have been proposed to be the ideal starting materials for the synthesis of N_5^+ because they already possess the desired types of bonds; N_2F^+ provides

[147] B. Zemva, K. Lutar, A. Jesih, W. J. Casteel, Jr., N. Bartlett, "A General Method for the Synthesis of Polymeric Binary Fluorides Exemplified by AgF_3, NiF_4, RuF_4 and OsF_4, J. Chem. Soc. Chem. Comm. (1989) 346–347.

[148] J. F. Lehmann, G. Schrobilgen, Synthesis and Characterization of Salts Containing the $BrO_3F_2^-$ Anion; A Rare Example of a Bromine(VII) Species, *J. Am. Chem. Soc.*, 127 (26) (2005) pp. 9416–9427.

[149] K. O. Christe, W. W. Wilson, J. A. Sheehy, J. A. Boatz, N_5^+: A Novel Homoleptic Polynitrogen Ion as a High Energy Density Material, Angew. Chem. Int. Ed., **38**, No. 13/14, (1999) 2004–2008.

the formal positive charge, and, in view of the weak N–F and strong H–F bonds, the HF elimination reaction is expected to be exothermic. Following the synthesis route, $N_2F^+AsF_6^- + HN_3 \rightarrow N_5^+AsF_6^- + HF$ (synthesis in aHF at 78°C), the $N_5^+AsF_6^-$ salt is a white solid that is sparingly soluble in anhydrous HF. It is poorly stable at 22°C and can be stored for weeks at −78°C without noticeable decomposition. Polynitrogen compounds are of significant interest as high energy density materials for propulsion or explosive applications.

2.1.3 Looking for the "naked" fluoride ion

Generally, the fluoride ion F^-, in polar protic solvents, is not a nucleophilic species. It is the smallest single anion with the largest negative charge, and therefore, it is a very strong H-bond acceptor. Because of these properties and its low polarizability, it is a weak nucleophilic species in protic solvents. However, in polar aprotic solvents where no H-bonds are present, and where the interactions with the counter-cations are limited, F^- behaves as a strong nucleophilic species and as a strong base.[150] In systems in which fluoride is associated with large organic cations and prepared in polar aprotic solvents such as acetonitrile CH_3CN, fluoride ion behaves as a "naked" fluoride ion, with a very strong basicity. The potential uses of such a naked fluoride ion are evident because—as H^+ or the magic acids are extremely acidic—, F^- would be a very strong Lewis base with huge catalytic possibilities in inorganic and organic synthesis. The first preparation of anhydrous tetramethylammonium fluoride, solvent-free prepared, with a large counter-cation as a source of naked fluoride anions with appreciable solubility in CH_3CN for example, gave rise to the synthesis of numerous new compounds.[151] The synthesis of truly anhydrous soluble naked fluoride ion has led to a renaissance in high coordination number chemistry.

Several scientists, G. Schrobilgen, K.O. Christe, K. Seppelt, and B. Zemva, among others, have investigated these naked fluorides in the form of truly anhydrous $N(CH_3)_4^+F^-$. This species has been used to synthesize fluoro- and oxofluoro-anions in which the central atom is frequently in its highest oxidation state and/or at the limits of its coordination, for example, PF_4^-; $PO_2F_2^-$; HPO_2F^-; XeF_5^-; $XeOF_5^-$; IOF_6^-; $IO_2F_5^{2-}$; $TeOF_6^{2-}$; TeF_7^-; TeF_8^{2-}; and IF_8^-. Several of these anions represent novel VSEPR (valence shell electron pair repulsion) geometries and, in the case of XeF_5^-, this pentagonal planar anion is, in VSEPR nomenclature, the first example of an AX_5E_2 arrangement which has a pentagonal planar geometry.[152] There have been numerous studies trying to discover the pure naked fluoride ion, but they showed that the counter-cation has some H-bonds connected with fluoride, and discussions still continue as to whether a pure naked fluoride ion really exists.[153]

[150] P. Kirsch, Modern Organofluorine Chemistry—Synthesis & Applications, Wiley-VCH (2004).

[151] R. Z. Gnann, R. I. Wagner, K. O. Christe, R. Bau, G. A. Olah, W. W. Wilson, Naked Fluoride Ion Sources: Synthesis, Characterization, and Coupling Reaction of 1-Methylhexamethylenetetramine Fluoride, J. Am. Chem. Soc. **119** (1997), 112–115.

[152] K. O. Christe, E. C. Curtis, H.P. Mercier, J. C. P. Sanders, G. J. Schrobilgen, D. Dixon, The Pentafluoroxenate (IV) Anion, XeF_5^-; the First Example of a Pentagonal Planar AX5 Species, J. Am. Chem. Soc., 113 (1991) 3351.

[153] K. Seppel, Does the Naked Fluoride Ion Exist? Angew. Chem. Int. Ed. Engl. **31** (1992) No. 3, 202–203.

2.2 Fluorolytic sol-gel method to produce high surface area nanofluorides[154]

The fluorolytic sol-gel process corresponds to a reaction of a metal alkoxide with hydrofluoric acid in alcooholic solution to form metal fluoride This can be schematically described as follows:

$$M(OR)_n + xHF \rightarrow M(OR)_{n-x}F_x + xROH\ (M = \text{metal ion})$$

Depending on the type and concentration of the reactants and solvents, reaction of metal fluorides with HF can yield either clear sols or gels. Aluminium isopropoxide is often used as a starting product in an alcoholic solution. After reaction with ethereal solution of hydrogen fluoride, a gel is formed due to the replacement of alkoxy groups by the fluoride groups. After drying, an additional fluorination of the dried gel under smooth conditions allows the removal of the attached organic components and leads to strong Lewis acidic aluminium fluoride with particularly high specific surface area, named HS-AlF_3. The most remarkable property of HS-AlF_3 is its outstanding Lewis acidity, which is far higher than that of $AlCl_3$. On a theoretical basis, it had been predicted that the Lewis acidity for an isolated "AlF_3-molecule" ranks among the highest ones at all.[155]

2.2.1 Aluminium chlorofluoride as catalyst in CFC and HCFC test-reactions

HS-AlF_3 is used as catalyst in reactions which must be catalyzed by a Lewis acid to proceed as below:

$$5CCl_2F_2 \rightarrow CCl_3F + 3CCl_3F + CCl_4 \quad (1)$$

$$3CHClF_2 \rightarrow CHCl_3 + 2CHF_3 \quad (2)$$

$$CBrF_2CBrFCF_3 \rightarrow CF_3CBr_2CF_3 \quad (3)$$

$$CCl_2FCClF_2 \rightarrow CCl_3CF_3 \quad (4)$$

The reactions (1) and (2) were used as post-fluorination agents in the course of HS metal fluoride preparation, especially that of HS-AlF_3. In Eqs. (3) and (4), as soon as first parts of the dry Al-F-gel are converted into HS-AlF_3, a dismutation reaction starts, which can be detected by GC. Thus, these fluorinating agents act conveniently as detector for Lewis acidity. However, reactions (1) and (2) proceed already in the presence of other weaker Lewis acids, and therefore, they are not pertinent for assessing high acidity. The isomerization of $CBrF_2CBrFCF_3$ (Eq. 3), which proceeds only under the catalytic action of the strongest known Lewis acids, that is, antimony pentafluoride SbF_5 or aluminium chlorofluoride (AlCF), is also catalyzed by HS-AlF_3, but not by $AlCl_3$. Thus, the isomerization reaction gives convincing evidence for the exceptionally high Lewis acidity of HS-AlF_3. Studying the isomerization of CCl_2FCClF_2 (Eq. 4) as a test reaction, $AlCl_3$, HS-AlF_3, ACF, α-AlF_3, and β-AlF_3 were compared regarding their respective catalytic activity. ACF and HS-AlF_3 were catalytically very active, whereas α- and β-AlF_3 were not at all, as expected. Nanoscopic ACF also catalyses functionalization reactions of

[154] St. Rüdiger, G. Eltanany, U. Gross, E. Kemnitz, Real sol-gel synthesis of catalytically active aluminium fluoride, J. Sol-Gel Sci. Techn., **41** (2007) 299–311.

[155] K.O. Christe, D.A. Dixon, D. McLemore, W.·W. Wilson, J.A. Sheehy and D.A. Boatz, On a quantitative scale for Lewis acidity and recent progress in polynitrogen chemistry, J. Fluorine Chem., **101** (2000)151–153.

refrigerants HFO-1234yf (2,3,3,3-tetrafluoropropene) and HFO-1234ze (1,3,3,3-tetrafluoropropene). These unprecedented transformation of an olefinic C—F bond into a C—H bond can be achieved by heterogeneous catalysis.[156] Based on theoretical investigations, the high Lewis acidity of HS-AlF_3 was assumed to be due to under-coordinated Al atoms, resulting from the high degree of disorder of this material.[156a]

2.2.2 Use of nanofluorides for various types of coatings

Although different types of techniques can be used to coat various materials with fluoride layers, such as CVD, PECVD, and fluorinated rf plasmas, the sol-gel fluorination process is very useful, because the particles consist of nanosized, colloidal metal fluoride homogeneously dispersed in a non-aqueous liquid. Thus, an anhydrous metal fluoride layer can be easily obtained by applying the sol on any solid material followed by a subsequent drying. Different methods can be used for the application of the sol, that is, soaking, dipping, spin coating, etc. The most current uses of fluorinated coatings include:

Catalysis: For technical reasons, it can be necessary to coat supported catalytic active solid material, for instance, for increasing mechanical stability or reducing flow resistance when used in a flow-reactor. The sol-gel fluorination synthesis provides a convenient way for depositing high surface area metal fluorides on active supports. When high-surface HS-AlF_3 is used as catalyst in a flow system, for example, the problems relative to the powdered nature of the substance can be solved as soon as the catalyst is supported on γ-Al_2O_3, Because its Lewis acidity, and consequently its catalytic activity, remains almost unchanged.[156a] This type of deposition of nanocatalysts is also of great interest for the microreactor technique.

Optics: The coating of metal fluoride layers in optics is probably one of the most important application fields. A reduction of light's reflection on the surface of glass and optical devices such as lenses for glasses, telescopes, and cameras, is of drastic importance. Anti-reflective systems are formed of alternating layers of transparent low and high refractive index materials. For a wave-length λ and an anti-reflective layer of thickness $\lambda/4$, the reflexions at the air/layer surface and at the layer/glass interface have a phase-shift to each other of $\lambda/2$; that is the condition for extinction. Total extinction is only possible when the two reflections are of the same intensity. This can be reached when the refraction index of the anti-reflective layer is equal to the geometric mean of air ($n = 1.0$) and glass ($n = 1.5$), that is, for $n = 1.225$. Whereas the refractive indices of metal oxides are higher than that of glass, metal fluorides such as MgF_2 ($n_{500} = 1.38$), AlF_3 ($n_{500} = 1.35$) and Na_3AlF_6 ($n_{500} = 1.33$) have lower indices of refraction, although not as low as 1.225, together with an excellent transparency within a broad range of wavelengths. MgF_2 thin layers have already found much interest.[157,158] MgF_2 sols in methanol are deposited on

[156] G. Eltanany, St. Rudiger and E. Kemnitz, Supported high surface AlF_3: a very strong solid Lewis acid for catalytic applications, J. Mater. Chem., **18** (2008) 2268–2275.

[156a] G. Meissner, K. Kretschmar, T. Braun, E. Kemnitz, Catalytic C—F Activation Steps at a Lewis Acidic Aluminum Fluoride, Angewandte Chem. Intern. Ed., **56** (2017) 16338–16341.

[157] T. Murata, H.·Ishizawa, I. Motoyama and A. Tanaka, Preparation of high-performance optical coatings with fluoride nanoparticle films made from autoclaved sols, Applied Optics **45** (2006) 1465–1468.

[158] J.K. Murthy, U. Gross, St. Rüdiger, E. Kemnitz and J.M. Winfield, Sol-gel-fluorination synthesis of amorphous magnesium fluoride, *J. Solid State Chem.*, **179** (2006) 739–746.

planar surfaces by spin-coating.[154] Microscopic investigations showed that the layers consist of densely packed particles of 10–20 nm in diameter.

Protective coating: Homogeneous, mechanically stable metal fluoride layers can protect against UV radiation and chemical and mechanical impact, and can form a barrier against microbial attack. A CaF_2 addition to lithium grease has already proven to be useful for friction reduction and stabilization under extreme pressure.

In conclusion, the access toward nanoscopic metal fluorides via fluorolytic sol-gel synthesis route opens a wide range of applications for metal fluorides due to the adaptative properties of these nanomaterials.

2.3 Nanoporous solids and organo-inorganic hybrid compounds

Beautiful examples of these are the nanoporous MIL solids for potential application in the storage or separation of gases and in catalysis (for Materials Institute Lavoisier), which were investigated by G. Ferey. They can store at room temperature 400 volumes of CO_2 per volume of product. These performances are due to the existence in the network of huge cages (20,000 Å^3) that can accommodate and retain the CO_2 molecules, thus contributing to the fight against pollution[159].

Metal-organic frameworks (MOFs) with porous coordination polymers (PCPs) composed of metal ions and organic bridging ligands as main building units and, in some cases, inorganic ligands, are involved in a variety of domains such as storage and separation, catalysis, drug delivery, and sensing ability. Thanks to the highest electronegativity and low electric polarizability of fluorine, the introduction of fluorine-functionalized in MOF-PCPs brings to the network new structural transformations, and improved gas and vapor adsorption/separation properties. Fluorine-functionalized MOFs/PCPs are summarized in Table 7.[160]

The hydrophobic character of the fluorinated pore surface provides a preferable adsorption and a high capacity of hydrophobic guests. The polar C–F bond may be useful for a preferable gas adsorption site. A. Cheetham et al. found that the enthalpy of H_2 adsorption in the 3D fluorinated MOF, {[Zn_5(triazole)$_6$(tetrafluoroterephthalate)$_2$(H_2O)$_2$]•4H_2O}, was enhanced using commercially available fluorinated tetrafluoroterephthalic acid, as shown in Fig. 41.[161,162] If pores can be filled with an appropriate number of fluorine-containing molecules, the composite may maintain sufficient accessibility for other guests with a fluorine-modified pore surface. Partial

[159] G. Férey, Some suggested perspectives for multifunctional hybrid porous solids, **Dalton Trans.**, pp. 4400–4415 (2009); Angewandte Chemie, 48 (2009) pp. 9398–9400.

[160] S. Noro, T. Nakamura, Fluorine-functionalized metal–organic frameworks and porous coordination polymers, *NPG Asia Materials* (2017) 9, e433.

[161] Hulvey, Z., Falcao, E. H. L., Eckert, J. & Cheetham, A. K. Enhanced H_2 adsorption enthalpy in the low-surface area, partially fluorinated coordination polymer Zn_5(triazole)$_6$ (tetrafluoroterephthalate)$_2$ (H_2O)$_2$•4H_2O. J. Mater. Chem. **19** (2009) 4307–4309.

[162] S. K. Mandal, H. W. Roesky, Designing Molecular Catalysts with Enhanced Lewis Acidity, pp. 1–61, Advances in Catalysis, Volume 54, B. C. Gates, H. Knözinger Eds. (2011); S. S. Sen, H. W. Roesky, Silicon-fluorine chemistry: from the preparation of SiF_2 to C–F bond activation using silylenes and its heavier congeners, *Chem. Commun.*, **54**, 5046–5057 (2018). It could be added that besides his numerous scientific interests, Herbert Roesky has been widely popularizing science (in particular chemistry) to general public, thus improving chemistry education and promoting the fascination of chemical experiments to younger audiences.

Table 7 **Summary of fluorine-functionalized MOFs/PCPs (from 160)**

MOFs/PCPs	Type of fluorinated ligands	Type of structures	Properties
$\{[Cd(NO_3)_2(2,6\text{-bpfn})_2]\cdot 2(p\text{-dimethoxybenzene})\}$	2,6-bpfn, organic bridging ligand	2D	Perfluoroarene-arene interaction
$[Zn_5(triazole)_6(tetrafluoroterephthalate)_2(H_2O)_2]$	Tetrafluoroterephthalate, organic bridging ligand	3D	High enthalpy of H_2 adsorption
$[Ni_{0.5}(tpt)_{0.5}(R\text{-opa})_{0.5}(H_2O)_{0.5}]$	$R\text{-opa}^{2-}$, organic bridging ligand	3D	Enhanced H_2 and CO_2 adsorption capacity
$[Zn_2(L1)(4\text{-(trifluoromethyl)pyridine})_2]$	4-(Trifluoromethyl) pyridine, organic terminal ligand	3D	Enhanced CO_2/N_2 and CO_2/CH_4 selectivity
$[Zn(L2)]$	$L2^{2-}$, organic bridging ligand	3D	Selective adsorption of hydrophobic heptane
$[Cu_4(L3)_4]$	$L3^{2-}$, organic bridging ligand	2D	High hydrophobicity, high tolerance to water, oil/water separation
$[Ag_6(tz)_6]$	tz^-, organic bridging ligand	3D	High hydrophobicity, high tolerance to water, high C6–C8 hydrocarbon adsorption capacity
$[Zr_6O_6(OH)_2(tdc)_4(RCOO)_2]$ ($RCOO^-$ = trifluoroacetate, 4-(trifluoromethyl) benzoate and pentafluorobenzoate	$RCOO^-$, organic terminal ligand	3D	High tolerance to water
$[Zn(SiF_6)(4,4'\text{-bpy})_2]$	SiF_6^{2-}, inorganic bridging ligand	3D	–
$[Cu(SiF_6)(4,4'\text{-bpy})_2]$ (SIFSIX-1-Cu)	SiF_6^{2-}, inorganic bridging ligand	3D	High CH_4 adsorption capacity, high CO_2/CH_4 selectivity, C_2H_2/C_2H_4 separation
$[Cu(TiF_6)(4,4'\text{-bpy})_2]$ (TIFSIX-1-Cu)	TiF_6^{2-}, inorganic bridging ligand	3D	High CO_2/N_2 and CO_2/CH_4 selectivity
$[Cu(SnF_6)(4,4'\text{-bpy})_2]$ (SNIFSIX-1-Cu)	SnF_6^{2-}, inorganic bridging ligand	3D	High CO_2/N_2 and CO_2/CH_4 selectivity

Continued

Table 7 Continued

MOFs/PCPs	Type of fluorinated ligands	Type of structures	Properties
$[Cu(SiF_6)(1,2\text{-bis}(4\text{-pyridyl})ethene)_2]$	SiF_6^{2-}, inorganic bridging ligand	3D	High CO_2/CH_4 selectivity
$[Cu(SiF_6)(4,4'\text{-dipyridylacetylene})_2]$ (SIFSIX-2-Cu)	SiF_6^{2-}, inorganic bridging ligand	3D	C_2H_2/C_2H_4 separation
$[Cu(SiF_6)(4,4'\text{-dipyridylacetylene})_2]$ (SIFSIX-2-Cu-i, i = interpenetrated)	SiF_6^{2-}, inorganic bridging ligand	3D	CO_2/N_2, CO_2/CH_4, and C_2H_2/C_2H_4 separation
$[Cu(TiF_6)(4,4'\text{-dipyridylacetylene})_2]$ (TIFSIX-2-Cu-i)	TiF_6^{2-}, inorganic bridging ligand	3D	C_2H_2/CO_2 separation
$\{[Cu_2(PF_6)(NO_3)(4,4'\text{-bpy})_4]\cdot 2PF_6\}$	PF_6^-, inorganic bridging ligand and inorganic guest	3D	–
$[Zn(SiF_6)(pyz)_2]$ (SIFSIX-3-Zn)	SiF_6^{2-}, inorganic bridging ligand	3D	CO_2/N_2, CO_2/CH_4, and C_2H_2/C_2H_4 separation
$[Ni(SiF_6)(pyz)_2]$ (SIFSIX-3-Ni)	SiF_6^{2-}, inorganic bridging ligand	3D	CO_2/N_2 and C_2H_2/C_2H_4 and CO_2/C_2H_2 separation
$[Cu(SiF_6)(pyz)_2]$ (SIFSIX-3-Cu)	SiF_6^{2-}, inorganic bridging ligand	3D	CO_2/N_2 separation
$[Co(SiF_6)(pyz)_2]$ (SIFSIX-3-Co)	SiF_6^{2-}, inorganic bridging ligand	3D	CO_2/N_2 separation
$[Fe(SiF_6)(pyz)_2]$ (SIFSIX-3-Fe)	SiF_6^{2-}, inorganic bridging ligand	3D	–
$[Cu_3(4\text{-(pyridin-4-yl)acrylate})(SiF_6)]$ (fsc-2-SIFSIX)	SiF_6^{2-}, inorganic bridging ligand	3D	High CO_2/N_2 and CO_2/CH_4 selectivity
$[Ni(NbOF_5)(pyz)_2]$	$NbOF_5^{2-}$, inorganic bridging ligand	3D	CO_2/N_2 and propylene/propane separation
$[Cu(BF_4)_2(4,4'\text{-bpy})_2]$ (ELM-11)	BF_4^-, inorganic terminal ligand	2D	Gate sorption

$[Cu(CF_3SO_3)_2(4,4'\text{-bpy})_2]$	$CF_3SO_3^-$, inorganic terminal ligand	2D	Gate sorption
$[Cu(PF_6)_2(4,4'\text{-bpy})_2]$	PF_6^-, inorganic terminal ligand	2D	Gate sorption, high CO_2/X (X = N_2, O_2, CH_4, and Ar) selectivity
$[Co(CF_3SO_3)_2(4,4'\text{-bpy})_2]$	$CF_3SO_3^-$, inorganic terminal ligand	2D	Gate sorption
$[Cu(PF_6)_2(\text{bpetha})_2]$	PF_6^-, inorganic terminal ligand	1D	Gate sorption, 2-butanone/EtOH and 2-butanone/MeOH separation
$[Cu(BF_4)_2(\text{bpp})_2]$	BF_4^-, inorganic terminal ligand	1D	Gate sorption, CO_2/CH_4 and CH_4/C_2H_6 separation
$[Cu(PF_6)_2(\text{bpp})_2]$	PF_6^-, inorganic terminal ligand	1D	Gate sorption, CO_2/CH_4 separation
$[Cu(CF_3SO_3)_2(\text{bpp})_2]$	$CF_3SO_3^-$, inorganic terminal ligand	1D	Gate sorption
$\{[Cu(CF_3SO_3)(\text{bpp})_2]\cdot PF_6\}$	$CF_3SO_3^-$, inorganic bridging ligand; PF_6^-, inorganic guest	2D	–
EMI-TSFA@$[Zn(\text{MeIM})_2]$ (EMI-TSFA@ZIF-8)	$TSFA^-$, inorganic guest	3D	High ionic conductivity
$[Co(NCS)_2(L4)_2]$	L4, organic bridging ligand	2D	High fluorophilicity
[Cu(bpbtp)(L5)]	bpbtp and L5, organic bridging ligand	2D	Preferential CO_2 and O_2 adsorption over N_2, Ar, and CO
$[Zr_6(\mu_3\text{-}OH)_8(OH)_{0\text{–}0.6}(\text{TBAPy})_2(C_xF_{2x+1}COO)_{4\text{–}3.4}]$ (x = 3, 7 and 9)	$C_xF_{2x+1}COO^-$, organic terminal ligand	3D	Enhanced enthalpy of CO_2 adsorption

1D, one-dimensional; *2D*, two-dimensional; *3D*, three-dimensional; *MOF*, metal–organic framework; *PCP*, porous coordination polymer.
2,6-bpfn, 2,6-bis(4-pyridylmethyl)hexafluoronaphthalene; *tpt*, 2,4,6-tri(4-pyridyl)-1,3,5-triazine; *H_2-R-opa*, phthalic acid with different functional groups; *H_4L1*, 4,4′,4″,4‴-benzene-1,2,4,5-tetrayltetrabenzoic acid; *H_2L2*, 4,4′-(hexafluoroisopropylidene)bis(benzoic acid); *H_2L3*, 4,4′-{[3,5-bis(trifluoromethyl)phenyl]azanediyl}dibenzoic acid; *tz*, 3,5-bis(trifluoromethyl)-1,2,4-triazolate; *H_2tdc*, 2,5-thiophenedicarboxylic acid; *4,4′-bpy*, 4,4′-bipyridine; *pyz*, pyrazine; *bpetha*, 1,2-bis(4-pyridyl)ethane; *bpp*, 1,3-bis(4-pyridyl)propane; *HMeIM*, 2-methylimidazole; *EMI-TSFA*, 1-ethyl-3-methylimidazolium bis(trifluoromethylsulfonyl)amide); *L4*, 1,3-bis(4-pyridylethynyl)-2-(3,3,4,4,5,5,6,6,7,7,8,8,8-tridecafluorooctyloxy)benzene; *H_2bpbtp*, 2,5-bis(perfluorobutyl)terephthalic acid; *L5*, 2,5-bis(perfluorobutyl)-1,4-bis(4-pyridyl)benzene; *TBAPy*, 1,3,6,8-tetrakis(*p*-benzoic acid)pyrene.

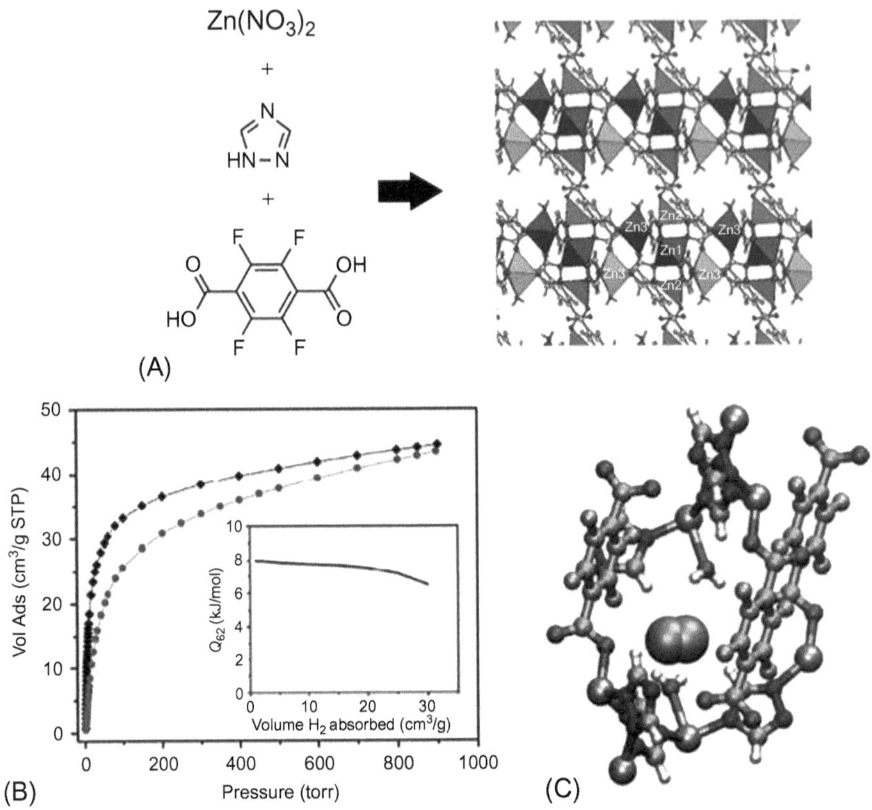

Fig. 41 (A) Crystal structure of $\{[Zn_5(triazole)_6(tetrafluoroterephthalate)_2(H_2O)_2]\cdot 4H_2O\}$ viewed down the *b* axis. (B) H_2 adsorption isotherms at 77 K (*black lozenges*) and 87 K (*red circles*), and Q_{st} absorption plots. (C) Adsorbed H_2 molecule (*center orange*) on the most favorable adsorption site as determined from simulation. Carbon atoms are depicted in cyan, hydrogen in white, nitrogen in blue, oxygen in red, fluorine in pink, and zinc in silver [from 161].

incorporation of ionic liquids with inorganic fluorinated anions into MOFs/PCPs is also effective for creating a fluorine-modified pore surface.[163]

Numerous metal-organic fluorides, lying at the border of inorganic and organic chemistries, have been investigated by H. Roesky, combining peculiar preparative chemistry and structural investigations. As an example, SiF_2 could be stabilized via the isolation of cyclic alkyl amino carbene (cAAC) ligand. The discovery of Me_3SnF as one of the most useful fluorinating agent for preparing organometallic fluorides of Group-4 metals overcomes the difficulty generally associated with the synthesis of

[163] Fujie, K., Otsubo, K., Ikeda, R., Yamada, T. & Kitagawa, H. Low temperature ionic conductor: ionic liquid incorporated within a metal–organic framework. Chem. Sci. **6** (2015) 4306–4310.

pure organometallic fluorides. This organometallic compound can even outperform its chloride analog (Me_3SnCl) as catalyst. The enhanced Lewis acidity attributed to the attachment of highly electronegative fluoride ion near the active site makes the metal center more electrophilic and therefore more catalytically active. ($CpMF_3$) compounds with M = Ti, Zr, Hf, could be obtained in high yield, following the reaction:

$$C_pMCl_3 + 3\,Me_3SnF \rightarrow C_pMF_3 + 3\,Me_3SnCl$$

with Cp = C_5H_5, C_5H_4Me, C_5Me_5 for M = Ti, and Cp = C_5Me_5 for M = Zr.

It can be added that $(R_2TiF_2)_3Al$ and $(R_2TiF_2)_3Ti$ compounds were prepared by the reduction of R_2TiF_2 using Al and Ti metals. This class of compounds combines impressively Werner coordination chemistry with metal-organic compounds by stabilizing molecular solids that are closely bound to metal-organic fluorides.[162]

2.4 Functionalization of materials via surface fluorination

Numerous improvements can be obtained through the functionalization of the materials at the nanometric scale. The substrates to be functionalized belong to many classes of materials, including oxides such as silica, TiO_2, and alumina, oxide-fluorides, hybrids, mesoporous materials. The concerned morphologies of the substrate materials range from powders, glass-ceramics, thin layers, and coatings and the applications concern catalysts, inorganic charges, superconductors, ionic conductors, UV absorbers, phosphors, materials for integrated optics, etc.[164]

Many physical and chemical methods are used, such as condensation from the vapor phase, mechanical milling, laser dispersion, or molecular-beam epitaxy, pyrolysis of fluorine containing compounds such as metal trifluoroacetates, sol-gel method yielding nanosized coatings as shown above. Thanks to their extreme reactivity, fluorine and fluorinated gases are currently used to efficiently modify the surface properties of numerous materials. Among these different techniques, radio-frequency (rf) plasma fluorination, PEF (for plasma-enhanced fluorination), allows drastic modification of the surface of various types of materials at low (room) temperature, including metals, ceramics, carbons, and polymers.[165] One of the major advantages of this technique is that the reactions are limited to the utmost surface and do not modify the bulk properties of the materials. In addition, they are carried out in most cases at room temperature, avoiding thermal degradation of the material. Depending on the experimental conditions and the fluorinated gas used, for example, CF_4 or c-C_4F_8, either a reactive etching or a deposition process may occur. The versatility of the technique can be illustrated using various examples: carbon-based materials, silica, silicates, alumina, and polymers. In plastic-based DSSC cells (see Section 4.4), CF_4/O_2 plasma treatments may increase the size of nanopores and nanochannels in the TiO_2

[164] A.Tressaud, Plasma-enhanced fluorination in materials science, Chimica Oggi/Chemistry Today, **29** (2011) 36–38.

[165] G. Castner, D.W. Grainger (eds.), "*Fluorinated surfaces, Coatings, and Films*", ACS Symposium series 787, ACS, Washington (2001).

nanoparticle electrode and, at the same time, passivate the TiO_2 nanoparticle surfaces. The electrolyte filling process allows the overall cell conversion efficiency to be increased up to 25%. The reaction yielding the fluorinated TiO_2 surface can be written as below:

$$CF_4 + O \rightarrow COF_2 + 2F$$

$$TiO_2\,(s) + F\,(g) \rightarrow TiO_xF_y\,(s) + TiF_4\,(s) + O_2\,(g)$$

With this process, the overall cell efficiency can reach about 9% using a masked frame measurement technique. The plastic-based electrolyte cells have been found to be much more stable compared with the liquid electrolyte cells.[166]

As far as the surface properties are concerned, X-ray photoelectron spectroscopy (XPS) is a particularly suitable technique for investigating changes in binding energies (BE) that are present within 1–100 nm of the surface.[167]

2.4.1 *Surface modification of carbon-based materials for Li-battery electrodes*

In electrochemical processes, a large number of electrodes are made of carbon-based materials. Their characteristics are strongly related to their graphitization level and thermal history; graphite and other carbon-based materials used as anode in Li-ion battery, exhibit higher capacities after a fluorination treatment, as shown by T. Nakajima.[168] Because PEF only affects the surface, the F-content is low; in addition, the treatment highly reduces the electrochemical decomposition of the propylene carbonate-based electrolyte onto the anode surface. Both effects explain the high first coulombic efficiency of PEF-treated samples, for example, electrodes formed of high-temperature treated petroleum cokes which are formed of bundles of carbon nanotubes (CNT) with hemispherical terminations. During the CF_4 rf plasma treatments, F· radicals, which are the most active species formed in the plasma, react with the external hemispheres of the CNT and destroy the most brittle part of the tips, yielding an opening of these tips. This process results in a very regular etching of the surface. In particular, the closed tips of CNT-like structures can be uniformly opened by plasma fluorination, even at room temperature, increasing the retention capacity of Li into the host structure. By the opening of these hemispherical terminations, the insertion of Li in the bulk of the electrode is increased, with a subsequent increase of the reversible capacity. After 10 cycles, $Q_{rev.}$ is equal to 244 mAh/g when compared to 192 mAh/g for the untreated petroleum coke.

[166] B. Lee, D. B. Buchholz, P. Guo, D-K. Hwang, R. P. H. Chang, Optimizing the Performance of a Plastic Dye-Sensitized Solar Cell, J. Phys. Chem. C, **115** (19) (2011) pp. 9787–9796.

[167] D. Briggs, M.P. Seah, Practical Surface Analysis, Vol. 1: Auger and XPS, John Wiley, New York (1990).

[168] K. Naga, T.Nakajima, Y. Ohzawa, Z. Mazej, B. Žemva, H. Groult, Surface-Structure Change and Charge/Discharge Behavior of Petroleum Cokes Surface-Modified by Thermally Activated ClF_3 and NF_3, J. Electrochem. Soc., **154** (2007) A347.

2.4.2 Functionalized silica and silicate minerals

Besides silica, which is used as substrate in many electronic applications, various classes of silicates, such as the phyllosilicate group, are of particular interest because the network offers numerous possibilities of substitutions and/or intercalations. In addition, these compounds have industrial uses, for example, in electronics, cosmetics, paints, plastics, etc. Using c-C_4F_8 as the fluorinated gas in PEF treatments, a PTFE-like layer is deposited onto the mineral or SiO_2 substrate, protecting it from further reaction or creating an antibacterial barrier using superhydrophobic xerogel coating.[169]

The surface of SiO_2 and silicate minerals can be functionalized in a very versatile way using either fluorinated gas (c-C_4F_8) or (oxy)fluorinated gas mixtures (O_2)/CF_4 rf plasmas and F_2-gas treatments as shown by L.P. Demyanova.[170] It is thus possible to either coat flaky or powdered samples, with a protective layer of carbon fluoride, 50–100 nm thick, acting as a surface "teflonization," or to create at the surface Al–F bonds by a reactive etching process caused by the gas mixture (O_2)/CF_4. Hydrophilic/hydrophobic or reactivity/passivity balances can be thus efficiently tuned.[171]

2.4.3 Switchable hydrophobic-hydrophilic layer for lithographic printing

Conventional lithographic printing processes using porous alumina for offset applications generally use "wet" routes. Recently, "dry" processes have been developed which are based on a heat-induced hydrophilic/oleophilic conversion of one or more layers of the coating.

The outmost surface of the hydrophilic anodized aluminium support can be hydrophobized by c-C_4F_8 rf-plasma PEF, by formation of PTFE-like external layer. This conversion from a hydrophilic to a hydrophobic state is characterized by an increase of the contact angle measured on the surface. The contact angles range from 20 degrees for untreated alumina to values higher than 130 degrees for c-C_4F_8 treatment. The formed hydrophobic layer can be further switched back to a hydrophilic one through the vaporization of the formed fluorocarbon using a laser beam.

[169] B. J. Privett, J. Youn, S. A Hong, J. Lee, J. Han, J.H. Shin, M. H. Schoenfisch, Antibacterial Fluorinated Silica Colloid Superhydrophobic Surfaces, Langmuir **27**(15) (2011) 9597–9601.

[170] P. Demyanova, A. S. Zaeva, Ratio of Starting Raw Material and Fluorinating Agent in the Process of Purifying Silica-Containing Raw Material, Theoretical Foundations of Chemical Engineering, **46**, No. 4 (2012) pp. 368–372. L. P. Demyanova, T. B. Makeeva, A. S. Buinovskii, Structural-phase changes of natural crystalline quartz during fluorination processing, Glass and Ceramics, 68 (2012) 11–12. L.P. Dem'yanova, Fluorolytic Treatment of Quartz-based materials leading to high quality products, Lambert Academic Publ. (Germany) 2011.

[171] L. P. Demyanova, A. Tressaud, Fluorination of alumino-silicate minerals: the example of lepidolite, J. Fluor. Chem. 130 (2009) 799–805. L.P. Demyanova, A. Tressaud, Labrugere C., Durand E., Majimel J., Petsyk P.A., Kozlov K.G., Fluorination of nanosized amorphous silica derived from silicate raw materials, Fluorine notes, Russia, Vol. 3(82) (2012) pp. 1–6.

Because the fluorinated layer is totally evaporated during the heating process, no ablation particles are retained on the material, leading to a method without solid residues.[172]

2.5 Improvements of polymer properties through fluorination

A. Kharitonov has shown that various properties of polymers are improved via direct fluorination.[173] This process is able to drastically reduce the evaporation of liquids which are stored inside polymer containers for packaging of industrial and consumer chemicals in particular fuel in automotive gas tanks. Rubber sheets used in sealing electrolytic capacitors become impermeable to paste and gas upon fluorination. The effect of the permeability reduction can be enhanced by incorporation of oxygen into the fluorinating mixture. Fluorination of ultrafiltration and microfiltration porous membranes results in an increase of the flux, decrease of fouling rate, and durability of membranes. Among other properties that can be improved, one can quote adhesion and printability of polymer items, thermal and mechanical properties of polymer composites. Friction coefficient of polymers also strongly decreased by direct fluorination.

Low density polyethylene (LDPE) films are treated by a combination of plasma treatment followed by direct fluorination.[174] Plasma treatment is generally carried out with ion bombardment. Combined treatment results for example, ion bombardment+fluorination, and reactive plasma +fluorination allow an enhancement in printability and in the barrier properties toward O_2 together with the maintenance of barrier properties toward water. Contrary to direct fluorination only, with these two dual fluorinations, a stabilized surface is achieved. Modification of the chemical-physical properties of ultra-high-molecular weight polyethylene (UHMWPE) were conducted by direct fluorination, and direct fluorination accompanied with UV irradiation, by XeF_2, and by TbF_4. The fluorinations result mainly in a formation of $-CF_2-$ groups. On the contrary, $-CHF-$ groups are prevailing in UHMWPE treated with XeF_2 and TbF_4. Surface texture of UHMWPE is quite different depending of the type of treatment, either with F_2 or with XeF_2. Direct fluorination results in a higher polarity of the polymer surface when compared with treatment with XeF_2. In the case of direct fluorination, both long-lived peroxy and fluoroalkylradicals are formed. In the case of treatment with XeF_2, only fluoroalkylradicals were detected.[175]

[172] H. Andriessen et al., Processless lithographic printing plate, European Patent by Agfa-Gevaert Co. # EP 1640175 A1(29/03/2006).

[173] A.P. Kharitonov, Direct fluorination of polymers—From fundamental research to industrial applications, Progress in Organic Coatings **61** (2008) 192–204.

[174] J. Peyroux, M. Dubois, E. Tomasella, N. Batisse, L. Frezet, E. Petit, A.P. Kharitonov, D. Flahaut, Plasma and fluorination combination for stable multifunctionality of LDPE packaging films, Plasma Processes and Polymers, Wiley-VCH Verlag, 2017, 14, https://doi.org/10.1002/ppap.201600066.

[175] P. Kharitonov, G.V. Simbirsteva, V. M. Bouznik, M. G. Chepezubov, M. Dubois, K. Guerin, A. Hamwi, H. Kharbache, F. Masin, Modification of ultra-high-molecular weight polyethylene by various fluorinating routes, J Polym Sci Part A: Polym Chem **49** (2011) 3559–3573.

2.6 Fluoropolymers for the protection of our cultural heritage

2.6.1 Protection of the surface of the stones

Due to their outstanding physical and chemical properties, especially their high hydrophobic properties, products and emulsions of acrylic (co-)fluoropolymers[176] polyurethanes or co-elastomers (e.g., hexafluoropropylene/fluoride vynilidène) are used for the treatment and protection of stone.[177] The advances obtained from these various products are regularly presented and discussed at symposia organized by ICOM. Very encouraging results were obtained, especially with the use of polyfluoroalkylméthacrylates compared to conventional non-fluorinated polymers such as Paraloid B72, which is a copolymer of ethyl (EMA) and methyl (MMA) acrylates.[178,179] In addition, these fluoropolymers are very resistant to microorganisms. The introduction of fluorine atoms, even in limited quantities in the side ester groups, increases the protective effect of partially fluorinated acrylic polymers, durability and potential for reversibility whatsoever on very porous stones (calcarenite) or very dense (marbles).[180] The performance of these materials is tested by capillary water absorption, color changes, water vapor permeability, and contact angles of the surface of the stone before and after accelerated aging after 1500h under UV.

2.6.2 Anti-graffiti systems

In addition to the deterioration of buildings caused by the ravages of time, graffiti raises increasing concern for the facades of both old and new buildings. While many painted walls exhibit an obvious artistic and justified sociological interest, a lot of "tags" are relatively primitive and feel like an assault. The cost for the removal

[176]New class of fluorinated acrylic polymers: materials for protective stone, ICOM-CC 18th Triennial Meeting, Copenhagen (2017). Note: With more than 1400 members, the Committee for Conservation of the International Council of Museums (ICOM-CC) aims to promote the conservation, research and analysis works with cultural and artistic interest and assist professional goals conservation (http://www.icom-cc.org/).

[177]Price CA, Stone Conservation: an Overview of Current Research, The Getty Conservation Institute, Santa Monica, USA (1996).

[178]Piacenti F., M. Camaiti, Manganelli Del Fa C., A. Scala, Fluorinated Aggregating materials for stone, in Conservation of Stone and Materials: Research-Industry-Media, International RILEM/UNESCO Congress, UNESCO headquarters, Paris (1993) MJ. Ed Thiel, London. E & F Spon (1993) pp. 740–747.

[179]V. Fassina, Arbissani R., C. Botteghi, Matteoli U., E. Passaglia, Agglietto M. Behaviour of 2,2,2-trifluoroethyl-methacrylate polymers as protective stone materials, in The Monuments conservazione dei nel bacino Mediterraneo, V. Fassina, H. Ott, F. Zezza Eds., Venice (1994) pp. 911–923. Botteghi C., Matteoli U., S. Paganelli, Arbizzani R., F. Ciardelli, Aglietto M., Taburoni E., V. Fassina, Polyfluoroalkylmethacrylates as materials for the protection of stones, Sci Technol. Cultural heritage, **1** (1992) 111–122.

[180]Alessandrini G., L. Toniolo, Peruzzi R. Poli T., F. Ciardelli, Aglietto M., V. Castelvetro, Chiantore O., M. Lazzari, New polymeric materials for the protection of stone surfaces, International Congress "Art & Chemistry," CNRS Editions, Paris, pp. 53–57 (2003); Alessandrini, G.; Toniolo, L.; and Colombo, C., Partially fluorinated acrylic Copolymers as coatings for calcareous stone materials, in Tradition and Innovation: advances in conservation: contributions to the Melbourne Congress, Roy A. Smith P., Eds (2000), pp. 1–6.

of these entries is very high; it was estimated at well over $600 million in the United States in 1996 and more than €600 million in the EU.[131] The graffiti-affected areas are experiencing very high increases: 170,000 m^2 for Paris in 1996, compared to 33,000 m^2 in 1985. There are three classes of anti-graffiti products: low cost reversible coatings (which must be renewed after each cleaning), semi-permanent coatings (usually acrylic resins which allow a limited number of cleanings), and permanent coatings (which especially use polysiloxane resins or fluorine resins). In the latter case, the procedure takes place in two stages: a preliminary coating should help limiting the adhesion of the paint on the surface; this step allows removal of the paint without damaging the protective layer. Among the various types of products tested, fluorinated polymers provide the most satisfactory answers to problems: improved water and oil repellent characteristics, impermeability, inertia with respect to solvents, and aging behavior.[181]

Copolymer composed of perfluoropolyether (PFPE) blocks with carboxylate functional groups investigated in water microemulsion is able to interact with polyfunctional reactive molecules, giving chain extension and crosslinking.[182] This "in-situ" reaction takes place directly on the stone surface. In addition to increased hydrophobicity of the substrate, the resistance to the dirtying caused by mural writings is improved (Fig. 42).[183] A protective layer is generated that is persistent on the stone surface, even after several staining and cleaning cycles. Monomolecular film of PFPE can also provide protection against aggressive acid rains.

The adhesion properties and surface conductivity of various non-fluorinated polymers (e.g., high-density polyethylene (HDPE), low density polyethylene (LDPE), ultrahigh molecular weight polyethylene (UHMWPE), polypropylene (PP), polyethylene terephthalate (PET), and polyvinylchloride (PVC)) can be improved by direct fluorination by F_2-gas, followed by a grafting of other molecules (e.g., acrylic acid, styrene, acrylonitrile, vinylidene chloride, aniline, and thiophene) from the gas phase. Graffiti and staining can be easily removed from the polymer surface in the

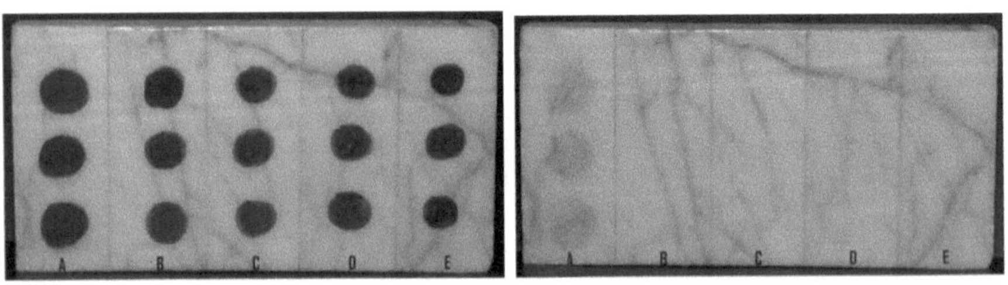

Fig. 42 Carrara marble sample taken after staining (left) and cleaning procedure (right) (A: Untreated zone; B–D: treated zones with PFPE-based copolymer) [from 183].

[181]R. Wood, The writing on the walls, Materials World, August 2004, pp. 30–32.

[182]Saidi S., F. Guittard, Géribaldi S., Polymers against graffiti, The Chemical News, March 2002, 3–14.

[183]M. Licchelli, Jacopo S. Marzolla, F. Carò, G. Moggi, Stone Surface Protection by Fluoropolymers from Decay Caused by Mural Writings, in Structural Analysis of Historical Constructions, New Delhi, P.B. Lourenço, P. Roca, C. Modena, S. Agrawal (Eds.) pp. 1405–1412 (2006).

case of PP and UHMWPE grafted with styrene and acrylonitrile, whereas, it is the printability that is improved when HDPE and LDPE are grafted with styrene and acrylonitrile.[184]

3 Industrial applications of fluoride minerals, gases, and inorganic fluorinated products

3.1 Applications of inorganic fluorinated products

It is the same peculiar properties of fluorine, that is, exceptional reactivity, high polarizing power, generally high binding energy, that have led to many applications of fluoride minerals, gases, and fluorinated products, as shown in Table 8. Concerning the most used mineral, fluorspar (CaF_2), a production rate of more than 6 MT per year (Table 9), will lead to a critical depletion of resources within a century. Fluorspar is a high criticality product for the European Union; the situation of the market in 2016 was marked by an increasing tension on supplies, due to strong Chinese, Indian, and European demands (see also the inset on the Ivigtut mine).

3.2 Contributions of fluorinated products to energy problems

Besides the fact that uranium hexafluoride is an unavoidable component for the functioning of the nuclear power plants, fluorine and fluoride compounds are required in various other energetic issues, in particular, for energy storage/production via lithium-ion batteries and for polymeric fuel cells' membranes, as illustrated in the book of T. Nakajima and H. Groult.[185]

3.2.1 Diversity of fluorinated components in lithium-ion batteries

Fluorinated materials intervene in the energy production via Li-ion batteries (reversible or not) and fuel cells. The two main types of reversible lithium batteries are polymer lithium-ion batteries with lithium metal negative electrode and Li-ion batteries that use a Li-insertion compound as a negative electrode. Whereas in the first type, the fluorinated molecules are limited to Li salt, the spectrum of fluorinated compounds used in Li-ion batteries is much wider and includes polymers, salts, solvents, additives, and possibly electrodes.

3.2.1.1 Primary Li-ion batteries

The first Li-ion primary batteries using graphite fluorides (CF_x, with $x \leq 1.3$) as positive electrodes were developed in the 1970s by N. Watanabe's group at Kyoto University and marketed by Matsushita (CF_x batteries/Li).[186]

[184] A. P. Kharitonov, G. V. Simbirtseva, V. G. Nazarov, V. P. Stolyarov, M. Dubois, J. Peyroux, Enhanced anti-graffiti or adhesion properties of polymers using versatile combination of fluorination and polymer grafting, Progress in Organic Coatings **88** (2015) 127–136.

[185] "*Fluorinated Materials for Energy Conversion*, T. Nakajima, H. Groult (Eds.), Elsevier, Oxford (2005).

[186] Watanabe N., Nakajima T., Touhara H., Graphite Fluorides, Elsevier, 1988.

Table 8 Main industrial applications of minerals, inorganic fluorocompounds, and fluorinated gases

Minerals, inorganic fluorocompounds and fluorinated gases	Uses
Fluorite, fluorspar: CaF_2	Metallurgic grade (60%–85%): anti-foaming agent, Al, Fe, Mg, and non-ferrous industries Ceramic grade (85%–96%): glass and enamel industries Acidic grade (>97%): HF production (3Mm^3 per year) Optical grade crystals (99%): optoelectronics, IR windows and UV laser at 2980 nm
Fluorocarbonates of bastnaesite type	Extraction and rare earth industries: electronics, opto-electronics, televisions, etc.
Fluorinated aluminosilicates: fluormica, muscovite	Coatings, wallpaper, rubber, plastics, adhesives industries cosmetics, glass-micas composites, glass-ceramics, ceramics for electronics, dentistry, compomers, composite resins
Fluoroaluminates minerals	Al industry, fluxes
Fluoroaluminates doped crystals	Optoelectronics, laser
Synthetic fluorides	
HF hydrofluoric acid	Aluminium industry, steels, foundries, alkylation catalysts, Ta and Nb purification, fluoro-organics synthesis
Alkaline fluorometallates	Various fluxes
Beryllium fluoroborates crystals ($KBe_2BO_3F_2$)	Non-linear optics
Borate fluorides, fluoro-oxoborates	Frequency-doubling crystals
F-doped zinc oxides, zinc hydroxifluorides	Components for QDSC and DSSC
Lithiated fluorosels: $LiPF_6$, $LiAsF_6$	Electrochemistry, lithium-ion batteries
Fluorophophates	Toothpastes, fine chemistry, biomaterials
MgF_2	Optical grade optics for excimer lasers, anti-reflective deposits
NaF	Dental additives, toothpastes, bactericidal, adhesives, paper and construction industries
Alkaline fluoborates: KBF_4	Solder flux, Al, Mg welding, electrochemistry, fine chemistry
NH_4F	Polishing agent, dye fixatives for textiles, rare earth extraction, etching and glass etching
NH_4HF_2	Glass etching, disinfectant, flux in metallurgy, pottery
ZnF_2	Flux solder
Complex fluorometallates	Catalysts, components in optics, electronics, Li-ion batteries

Table 8 Continued

Minerals, inorganic fluorocompounds and fluorinated gases	Uses
Fluorinated gases	
F_2	Nuclear industry, automobiles, etching in electronics, fine chemistry, fluoro-organics synthesis (see ARC—F_2-gas production room Fig. 27)
NF_3	Electronics
SF_6	Electrical insulation, power distribution plants
WF_6	Electronics, tungsten deposition
F_2, CF_4, c-C_4F_8, CHF_3	Radiofrequency fluorination surface treatments

These compounds are prepared by fluorination of carbon at high temperature (>350°C). Although having a low electrical conductivity, they possess certain advantages such as a high open circuit voltage (OCV) >2.5 V, a flat discharge potential, a long service life without discharge (>10 years at ambient temperature) and a wide range of possible uses (between −60°C and +90°C). For example, CF_x fluorides used in commercial batteries develop a theoretical capacity of 870 Ah kg^{-1}. The reactions taking place at the electrodes are as follows:

Negative electrode: $Li + solv \rightarrow Li^+ + solv + e^-$

Positive electrode: $CF + Li + solv \rightarrow C - F - Li - solv$

where solv is a solvating compound of Li^+ with one or more molecules of the solvent. The intermediate C-F-Li-solv phase eventually breaks down into low crystallinity carbon and LiF.[187]

Many types of fluorinated carbons have been tested as electrode materials. The starting type of carbon is essential to good electrochemical performance. For example, the capacitances of the cathodes of fluorinated carbon nanodiscs reach 1180 mAh g^{-1}.[188] In order to increase the performance of these batteries, in particular, higher energy densities and power, CF_x nanoparticles (∅: 5–50 nm) were synthesized by fluorination of carbon nanoparticles (CNP) obtained by electrolysis of molten alkali carbonates. These fluoronanocarbons (F-CNP) of compositions CF_x, with $x \sim 0.2$ actually showed a higher potential (>3 V vs Li^+/Li), and an absence of delay generally observed during the first discharge.[189]

[187] Touhara H., Fujimoto H., Tressaud A., Watanabe N., On the discharge reaction mechanism in graphite fluoride-lithium batteries, Solid State Ionics, **14** (1984) 163–171.

[188] Ahmad Y. et al., Pushing the theoretical limit of Li-CF_x batteries using fluorinated nanostructured carbon nanodiscs, Carbon, 94 (2015) pp. 1061–1072.

[189] Groult H., et al., Reaction mechanisms, structural changes and applications of fluorinated carbonaceous materials, Proceedings 2nd IISFC, Hyderabad, November 3–7, 2014.

Table 9 2014 Production and world reserves of fluorspar ores (CaF_2)

Country	South Africa	China	Spain	United States	Kenya	Mexico	Mongolia	Other countries	Total
Production 2014 (kt)	175	4400	107	N.d.	90	1200	340	70	6850
Reserves (kt)	41,000	24,000	6000	4000	5000	32,000	22,000	110,000	240,000

Source: http://minerals.usgs.gov.

It should be additionally noted that, because of their high chemical and thermal stabilities, CF_x graphite fluorides are also used as lubricants.

3.2.1.2 Secondary/reversible Li-ion batteries

Secondary/reversible Li-ion batteries also contain a large number of fluorinated components:

- Positive fluorinated electrodes from 2D or 3D networks of inorganic compounds: fluorophosphates ($LiFePO_4F$, $LiTiPO_4F$, Na_2FePO_4F ...) or fluorosulphates ($LiMSO_4F$ and Li ($Fe_{1-x}M_x$) SO_4F (M: Co, Ni, Mn) derived from the "tavorite" mineral).[190] This type of electrode may also be suitable for Na^+ ion-conducting batteries, such as vanadium and sodium fluorophosphate $Na_{1,5}VOPO_4F_{0,5}$.
- Binders which increase the performance of $LiFePO_4$-type cathodes, such as polymers, copolymers, or composites based on PVDF[191].
- Fluorinated electrolytes, such as $LiBF_4$ and $LiPF_6$, which are among the best conducting salts of the Li^+ ion. However, other fluorinated electrolytes are being developed: fluoro-ethers, fluorinated ionic liquids consisting of perfluorinated ions such as BF_4^- (tetrafluoroborate), $TSFA^-$ (trifluoromethanesulfonyl $((CF_3SO_2)_2N)^-$) or based on bis(trifluoromethane) sulfonimide $(NTf2)^-$ lithium salt ($CF_3SO_2NLiSO_2CF_3$) which have the advantage of present lower flammability and volatility[192].
- Additives such as lithium tetrafluoro-oxalato-phosphate (WAC) which makes it possible to increase the life of batteries using the couple MCMB/$Li_{1,1}$ $[Ni_{1/3}Co_{1/3}Mn_{1/3}]_{0,9}O_2$, with MCMB: "mesocarbon microbeads" for high energy application.

3.2.1.3 Electrodes based on transition element fluorides

High oxidation states transition element fluorides or carbon nanocomposites/metal fluorides (C/MF_n) obtained by chemical-mechanical grinding, achieve discharge capacities of $500\,mAh\,g^{-1}$ for voltage scales of 1.5–4.5 V and a current density of $7.58\,mAh\,g^{-1}$ at 70°C.[193] Theoretically, for FeF_3 iron fluoride, the potential could reach $712\,mAh\,g^{-1}$ for a complete three-electron reaction, whereas in $LiFePO_4$, the Fe (III)/Fe (II) couple would give a theoretical maximum four times less, $170\,mAh\,g^{-1}$. Thus, in fluorinated or hydroxyfluorinated Fe^{3+}-based systems, the electrochemical performances are linked to the presence of anionic vacancies.[194,195] Mixed insertion and conversion mechanisms have been observed in fluorinated

[190]Recham N., et al., A 3.6 V lithium-based fluorosulphate insertion, Nat. Mater., 9 (2010) p. 68.

[191]*Améduri B.*, From vinylidene fluoride (VDF) to the applications of VDF-containing polymers and copolymers: recent developments and future trends, *Chem. Rev.*, **109** (2009) pp. 6632–6639.

[192]Hagiwara R., Ito Y., Room temperature ionic liquids of alkylimidazolium cations and fluoroanions, J. Fluorine Chem., **105** (2000) pp. 221–232.

[193]Badway F., Pereira N., Cosandey F., Amatucci G.G., Carbon-metal fluoride nanocomposites: structure and electrochemistry of FeF_3:C, J. Electrochem. Soc., **150** (2003) p. A1209.

[194]Duttine M. et al., Tailoring the composition of a mixed anion iron-based fluoride compound: evidence for anionic vacancy and electrochemical performance in lithium cells, Chem. Mater., **26** (2014) p. 4190.

[195]J. Hu, Y. Zhang, D. Cao, C. Li, Dehydrating bronze iron fluoride as a high capacity conversion cathode for lithium batteries, J. Mater. Chem. A, **4** (2016) 16166–16174.

nanocomposites based on titanium oxyhydroxyfluoride.[196] The use of defect chemistry can provide new advanced electrode materials, because multivalent Mg^{2+} and Al^{3+} ions have been shown to easily migrate in cation-deficient anatase TiO_2.[197]

3.2.2 *All-solid fluoride ion batteries*

These FIBs, "fluoride-ion batteries," are no longer based on Li^+-ion transport, but on fluoride ion F^-. High-performance solid electrolytes are derived from the fluorite structure (CaF_2 type) or tysonite (LnF_3, Ln: rare earth).[198] These are $PbSnF_4$ (1.5×10^{-3} S cm^{-1} at 27°C), $Sm_{0.94}Ca_{0.06}F_{2.94}$ ($\approx 10^{-3}$ S cm^{-1} at 40°C), $Ce_{0.97}Sr_{0.03}F_{2.97}$ (5×10^{-4} S cm^{-1} at 20°C), or alternatively $La_{0.9}Ba_{0.1}F_{2.9}$ ($\approx 2 \times 10^{-4}$ S cm^{-1} at 50°C). The thermal dependence of the ionic conductivities of the best fluorinated solid electrolytes is given in Fig. 43. These crystalline lattices have vacancies that allow F^- ions to migrate easily,

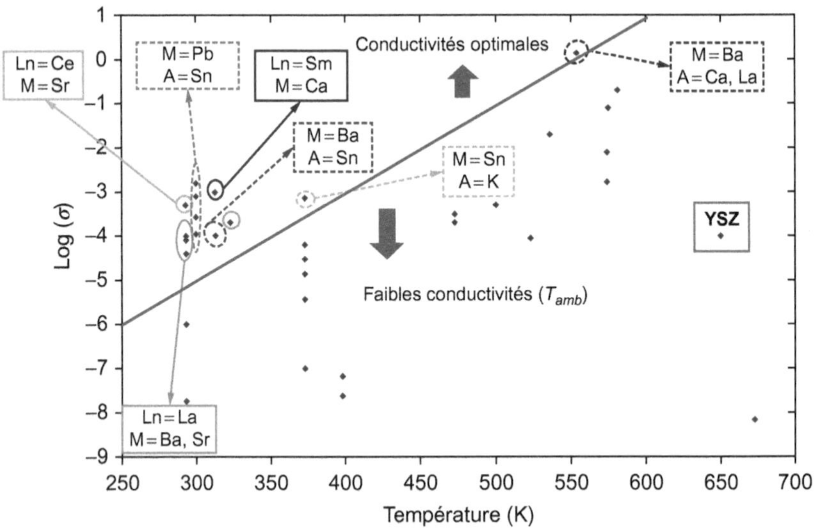

Fig. 43 Thermal dependence of the ionic conductivity of the best fluorinated solid electrolytes ($M_{1-x}A_xF_2$ fluorite structures (dashed lines), $Ln_{1-x}M_xF_{3-x}$ tysonite structure (solid lines). For comparison, the reference $Zr_{0.92}Y_{0.08}O_{1.96}$ is given (YSZ right frame). The central (*red*) line is an arbitrary limit set to identify most relevant electrolytes [from 198].

[196] Dambournet D. et al., Dual lithium insertion and conversion mechanisms in a titanium-based mixed-anion nanocomposites, J. Am. Chem. Soc., **133** (2011), pp. 13240–13251.

[197] T. Koketsu, J. Ma, B. J. Morgan, M. Body, C. Legein, W. Dachraoui, M. Giannini, A. Demortière, M. Salanne, F. Dardoize, H. Groult, O. J. Borkiewicz, K. W. Chapman, P. Strasser, D. Dambournet, Reversible magnesium and aluminium ions insertion in cation-deficient anatase TiO_2, Nature Mater., **16** (2017), pp. 1142–1148.

[198] A. Düvel, J. Bednarcik, V. Sepelàk, P. Heitjans, Mechanosynthesis of the Fast Fluoride Ion Conductor $Ba_{1-x}La_xF_{2+x}$: From the Fluorite to the Tysonite Structure, J. Phys. Chem. C, **118** (2014) 7117–7129.

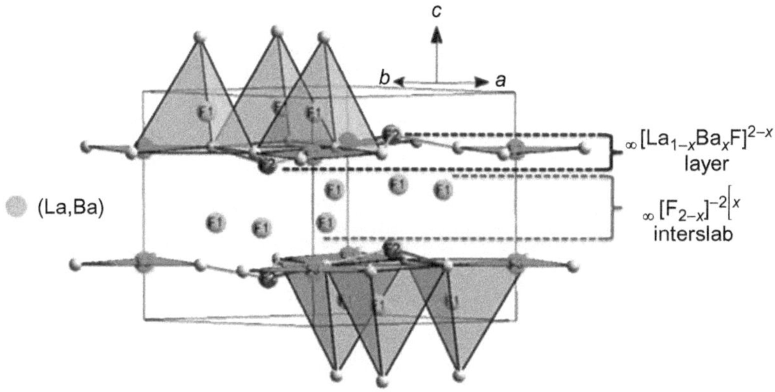

Fig. 44 Environments of fluorine atoms in $La_{1-x}Ba_xF_{3-x}$ tysonite ($x < 0.15$). F1 atoms (which are located in interslab)—and to a lesser extent F3 atoms—are responsible for a significant anionic mobility [from 199].

as shown in Fig. 44.[199,200] The relationships between structural and ionic conduction data that had been discussed some decades ago by J.M. Reau and P. Hagenmuller[201] have recently been clarified in the case of the $La_{1-x}Ba_xF_{3-x}$ tysonite solid solution ($x < 0.15$). Indeed, from a theoretical point of view, these phases are likely to deliver high potential (from 3.0 to 4.8 V) and high capacities (1000–1600 Wh kg^{-1} or 2000–5400 Wh L^{-1}). This concept based on fluoride shuttle was proposed by A. Reddy and M. Fichtner[202] has led to button cells that can work at 150°C.[203] The use of nanostructured fluorides has brought a renewed interest to this type of system. For example, the capacity of the system $BiF_3/La_{0,95}Ba_{0,05}F_{2,95}$ /Ce is 245 mAh g^{-1}, that is to say, that it reaches 80% of the theoretical capacity (302 mAh g^{-1}).

Solid polymer electrolytes would be a perfect solution to those safety issues, miniaturization and enhancement of energy density. A prototype battery was built based on a multifunctional single-ion polymer electrolyte based on polyanionic block copolymers comprising polystyrene segments. The *co*-polymer was obtained by grafting

[199] Chable J., Fluorinated solid electrolytes for FIB batteries, PhD thesis, Univ. Bordeaux/Univ. du Maine, 2015. J. Chable, B. Dieudonné, M. Body, C. Legein, M.P. Crosnier-Lopez, C. Galven, F. Mauvy, E. Durand, S. Fourcade, D. Sheptyakov, M. Leblanc, V. Maisonneuve, A. Demourgues, Fluoride solid electrolytes: investigation of the tysonite-type solid solutions $La_{1-x}Ba_xF_{3-x}$ ($x < 0.15$), Dalton Trans., **44** (2015) 19625–19635.

[200] Fedorov, P.P., Luginina, A.A., Kuznetsov, S.V., Osiko, V.V., Nanofluorides, J. Fluor. Chem.**132** (2011) pp. 1012–1039.

[201] Réau J.M., Hagenmuller P., Inorganic fluorides with high electrophysical properties, Rev. Inorg. Chem., **19** (1999) pp. 45–51.

[202] M. A. Reddy, M. Fichtner, Batteries based on fluoride shuttle, J. Mater. Chem. **21** (2011) pp. 17059–17062.

[203] Grenier A., Porras Gutierrez A.G., Groult H., Dambournet D., Modified coin cells to evaluate the electrochemical properties of solid-state fluoride-ion batteries at 150°C, J. Fluorine Chem., **191** (2016) pp. 23–29.

styrenesulphonyl(trifluoromethanesulphonyl) imide lithium salt (STFSILi) to neutral polystyrene-PEO-polystyrene (PS-PEO-PS) copolymer chain [P(STFSILi)-PEO-P (STFSILi)]. It gave excellent mechanical properties and an electrochemical stability window spanning 5 V versus Li(+)/Li couple.[203]

Fluoride-ion selective electrode: It should be noted that the conduction properties of F^- ions in tysonite-type phases have been used for a long time for the analysis of fluoride ions.[204] A LaF_3 single crystal doped with EuF_2 serves as a solid membrane between two solutions, one of which contains a reference sample of F^- ions (in general NaF), and the other, a quantity of F^- ions to be determined. This is the principle of the determination of fluorine by LaF_3 specific electrode.

3.2.3 Polymer fuel cells

Similarly, proton exchange membrane fuel cells (PEMFCs) have only been made possible by the discovery of the fully fluorinated ion exchange resin Nafion discovered in the 1960s at DuPont in the United States.[205] The membrane must conduct hydrogen ions (protons) but not electrons to avoid "short circuits" of the fuel cell. The membrane must also allow no gas crossover to the other side of the cell and should be resistant to the reducing environment at the cathode as well as the severe oxidative environment at the anode. The conduction principle of the Nafion membrane relies on liquid water humidification of the membrane to transport protons. This implies that it cannot be used at temperatures above 80–90°C, because the membrane would dry. The Nafion structure is shown in Fig. 45 together with the schematized fuel cell diagram. The exceptional properties of these perfluoro-sulfonyl membranes, high acidity, chemical, and thermal stability, led to exceptional products in many ways and are marketed under various names: Nafion, Flemion, 3M membrane, Hyflon Ion, Aciplex, etc. They are indeed used as a separator in fuel cells, but also as acidic catalysts, membranes in electrosynthesis or in gas drying systems. It should be pointed out that, due to the high cost of Nafion products, other perfluorinated ionomers in which perfluorosulfonate functions have been replaced by ionic sulphonimidide are currently being developed.[206,126] Recently, protic organic ionic plastic crystals have been shown as a promising alternative electrolyte materials for high temperature (100–200°C) PEMFCs.[207]

[204] Frant M.S., Ross J.W. Jr., Electrode for sensing fluoride ion activity in solution, Science, **154** (1966) p. 1553.

[205] P. R. Resnick, A Short History of Nafion®, Special issue devoted to Fluorine and Fluorinated Materials for the Centenary of H. Moissan Nobel Prize, SCF "Actualité chimique" – N° **301–302** (2006) pp. 144–147.

[206] Appleby A.J., Velev O.A., Lehelloco J-G., Parthasarthy A., Srinivasan S., Desmarteau D.D., Gillette M.S., Ghosh J.K., Polymeric perfluoro bis-sulfonimides as possible fuel cell electrolytes, Journal of the Electrochemical Society, **140** (1993) p. 109.

[207] J. Luo; A. H. Jensen; N. R. Brooks; J. Sniekers; M. Knipper; D. Aili; Q. Li; B. Vanroy; M. Wübbenhorst; F. Yan; L. Van Meervelt; Z. Shao; J. Fang; Z. H. Luo; D. E. De Vos; K. Binnemans; J. Fransaer, 1,2,4-Triazolium perfluorobutanesulfonate as an archetypal pure protic organic ionic plastic crystal electrolyte for all-solid-state fuel cells, Energy & Environmental Science **8** (2015) 1276–1291.

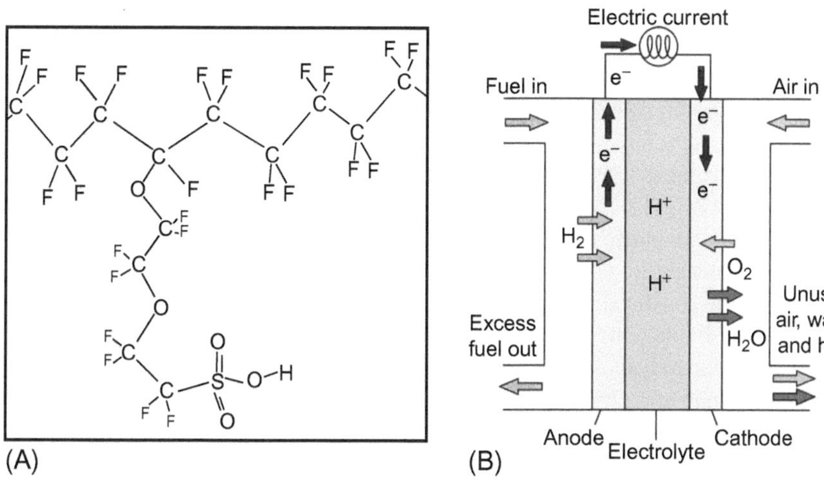

Fig. 45 (A) Structural network of Nafion, (B) schematized fuel cell diagram (by Mattuci—Own work, CC0, *https://commons.wikimedia.org/w/index.php?curid=42838482*).

4 Uses of fluorides in photonics

Although most of these studies were published only a few decades ago, the great breakthroughs achieved since the beginning of this century have raised researchers' hopes. The exceptional properties of the fluorides obtained in nanometric form may concern phosphors in solar batteries, LEDs, or laser materials.[208] The use of luminescent fluorocompounds in photodynamic therapy (PDT) will be illustrated below in Section 6.1.4.

4.1 Luminescence; up- and down-conversion

Due to the high ionicity that characterizes MF bonds and their low phonon energy, fluorinated inorganic compounds have generally a broad bandgap (>10 eV) and therefore high transparency in a wide spectral range: from near infrared (NIR: a few μm) up to the ultraviolet (VUV, "ultraviolet vacuum" 150–200 nm). When the host structure is doped with rare earth ions, $4f^n$ or $4f^{n-1}$ energetic states are located between the valence band and the conduction band (very often at the top of the forbidden band, i.e., close to the conduction band). The excitation of rare earth dopants often leads to

[208] Photonic and Electronic Properties of Fluoride Materials, A. Tressaud, K. Poeppelmeir (eds), Elsevier, 2016: (a) Gredin P., Mortier M., Optical properties of fluoride transparent ceramics, Chap. 4, pp. 65–87; (b) Heo J., Liu C., Fluorescence in nanostructured oxyfluoride glasses doped with RE ions, Chap. 7, pp. 139–157; (c) S. Pan, Y. Wang, K. R. Poeppelmeier, Frequency doubling oxide fluorides, borate fluorides and fluorooxoborates, Chapter 15, pp. 311–354, (d) Kirsch P., Fluorinated nematic liquid crystals: design, synthesis and properties, Chap. 8, pp. 159–176; (e) Calestani G., Mezzadri F., Multiferroism in fluorides, Chap. 14, pp. 285–307. (f) P.P. Fedorov, S.V. Kuznetsov, V.V. Osiko, Elaboration of Nanofluorides and Ceramics for Optical and Laser applications, Chap. 2, pp. 7–31.

luminescent properties (see Fig. 19) in a wide range of the optical spectrum (from NIR to VUV).[209] Such a luminescence that gives rise to a light emission at a lower energy than the exciting energy is called "down-conversion." On the other hand, when the excited species are in an energy state higher than that of the exciter photon, the mechanism is called "up-conversion." These phenomena were theorized in the 1960s by F. Auzel.[210] The most widely used host structures are the fluorite, $LiLnF_4$, $NaLnF_4$ or the tysonite LnF_3 (Ln: Y/Gd) type. Various synthetic methods are used to obtain "up-conversion nanoparticles" (UCNP), for example from trifluoroacetate precursors: $Ln\ (TFA)_3$ or oleates.

Thanks to the drastic development of nanomaterials, important breakthroughs in biology and medicine concern medical imaging, medical diagnostics, but also photovoltaics, or photocatalysis.[211] For example, semiconductor quantum dots (SCQD) is a promising alternative to dyes as markers in biology. The first doped $NaYF_4$ UCNP nanofluorides showed outstanding results as biomarkers.[212] Since then, thousands of publications have appeared on the subject.[213,208]

Methods for obtaining fluorinated transparent ceramics include self-fluorination using precursors such as $BaF_2{\cdot}HF$ hydrofluoride or $Sr_{1-x-y}R_x(NH_4)_yF_{2+x-y}$ solid solutions, as shown by P.P. Fedorov.[214] These ceramics are generally obtained by hot forging or SPS ("spark plasma sintering").[211] Thus, Yb-doped CaF_2 transparent ceramics made from nanopowders are one of the most promising routes for laser applications, with laser emission efficiencies similar to those obtained with single crystals.[208a] Fig. 46 illustrates "up- and down-conversion" luminescence in "core-shell" nano-crystals of $NaREF_4$:RE@$NaREF_4$ fluorides (RE = rare earth) in solution.[213]

Toward active nano-fluorides: Another way to obtain nanoscaled single-phase materials with luminescent properties is the fluorolytic sol-gel method using calcium lactate pentahydrate and RE acetate hydrate in methanolic water-free HF, as promoted by E. Kemnitz.[215] This room-temperature synthesis enables the preparation of clear colloidal solutions, from which nanosized particles can be precipitated. This one-pot-wet chemical route was applied for instance to fluorite-type CaF_2 particles doped with RE: $Ca_{1-x}RE_xF_{2+x}$. The particles are quasispherically shaped and formed by monodisperse nanocrystals of ca. 5 nm for CaF_2:Eu and CaF_2:Tb. Luminescence

[209] Fouassier C., Latourrette B., Portier J., Hagenmuller P., Nature de la fluorescence de l'europium divalent dans les fluorures, *Mater. Res. Bull.*, **1976**, *11*, p. 933.

[210] Auzel, F., Compteur quantique par transfert d'énergie entre deux ions de terres rares dans un tungstate mixte et dans un verre. Compt. Rend. Acad. Sci. Paris **262**, 1016–1019 (1966).

[211] Ang L.Y., Lim M.E., Ong L.C., Zhang Y., Application of upconversion nanoparticles in imaging, detection and therapy, Nanomedicine, **6** (2011) p. 1273.

[212] Heer S., Kompe K., Gudel H.U., Haase M., Highly efficient multicolour upconversion emission in transparent colloids of lanthanide-doped $NaYF_4$ nanocrystals, *Adv. Mater.*, 2004, **16**, p. 2102.

[213] S. Han, X. Qin, Z. An, Y. Zhu, L. Liang, Y. Han,W. Huang, X. Liu, Multicolour synthesis in lanthanide-doped nanocrystals through cation exchange in water, Nature Commun. (2016) **7**: p. 13059.

[214] Fedorov, P.P., et al. Nucleation and growth of fluoride crystals by agglomeration of nanoparticles. J. Crystal Growth **401** (2014) 63–66; Transparent Oxyfluoride Glass Ceramics, J. Fluorine Chem. 172, 22–50, 2015.

[215] Y B. Ritter, T. Krahl, K. Rurack, E. Kemnitz, Nanoscale CaF_2 doped with Eu^{3+} and Tb^{3+} through fluorolytic sol–gel synthesis, J. Mater. Chem. C, 2014, 2, 8607–8613.

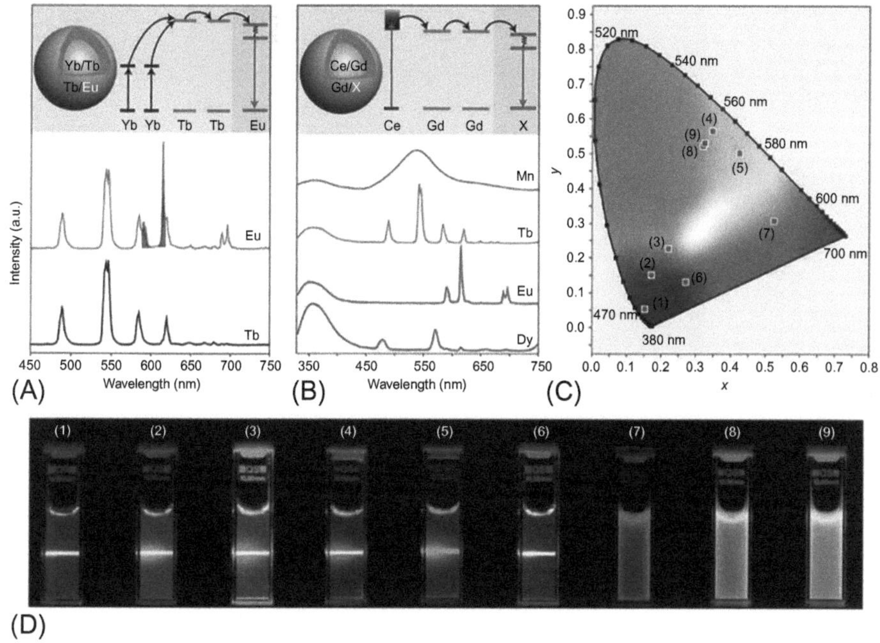

Fig. 46 "Up-conversion" luminescence spectra of (A) core-shell nano-crystals of $NaYbF_4$:Tb @$NaTbF_4$ before and after $EuCl_3$ exchange. The Eu^{3+} emission is underlined in red. Under a 980-nm excitation, a cooperative up-conversion process takes place first between the Yb^{3+}/Tb^{3+} pairs, which is followed by a migration in the Tb sub-network. The excitation energy is trapped by Eu^{3+} ions. (B) Down-conversion luminescence of core-shell nanocrystals of $NaYbF_4$: Tb@$NaTbF_4$ treated with $TbCl_3$, $EuCl_3$, $MnCl_2$, $DyCl_3$. The insert shows the energy transfer mechanism: under UV irradiation, the excited states of Ce^{3+} are populated in priority, followed by a migration in the Gd sub-network and a cross transfer between Ce^{3+} and Tb^{3+}, Eu^{3+}, Mn^{2+}, Dy^{3+} ions. (D) Luminescence of various colloidal solutions with (C) corresponding chromatic coordinates (IEC System) [from 213].
Reprinted with permission of Nature Commun.

properties revealed characteristic transitions $^5D_0/^7F_J$ for Eu^{3+} and $^5D_4/^7F_J$ for Tb^{3+}, with $^5D_0/^7F_2$ (611 nm) for Eu^{3+} and $^5D_4/^7F$ (581 nm_4) for Tb^{3+} as the most prominent transitions. The particles are highly luminescent with quantum efficiencies of 15.1% for $Ca_{0.9}Eu_{0.1}F_{2.1}$ and 17.3% for $Ca_{0.9}Tb_{0.1}F_{2.1}$ 0.2 M sols. These colloidal solutions exhibit improved quantum yields (factor of 3–4) when compared with oxide systems from the literature.

4.1.1 Optically active single-crystals of fluorides: the example of $LiCaAlF_6$

Experiments are also being pursued on the crystal growth of extended cm-sized high purity single crystals of optically active fluorides. As an example, we will consider $LiCaAlF_6$ single crystals with 3–4 in. diameters (Fig. 47) that were grown by

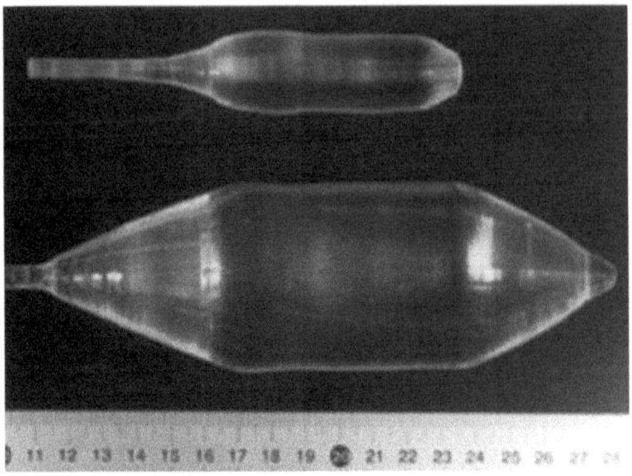

Fig. 47 Single crystals of $LiCaAlF_6$ and $LiSrAlF_6$ grown by the Czochralski technique under fluorinating CF_4 atmosphere [Credit Dr. K. Shimamura, NIMS, Tsukuba, Japan].

K. Shimamura using the Czochralski technique under a fluorinated CF_4 atmosphere. After considering reflection loss, the sample showed transmittance over 95% at 157 nm. A precise measurement of the refractive index showed that the birefringence vanished at around 152 nm. Thus, the use of $LiCaAlF_6$ as lens was demonstrated.[216] The scintillation properties of Ce-doped $LiCaAlF_6$ crystal, well known as an effective UV laser material, showed a Ce^{3+} emission at 286–305 nm with a single exponential decay time of 35 ns provides a scintillation pulse. Ce^{3+} is responsible for the ultradeep traps formation, revealing thermostimulated luminescence. Overlapping of color-center absorption and Ce^{3+} ion emission bands, however, limits the scintillation efficiency of $LiCaAlF_6$:Ce at high radiation doses.[217] Absorption, emission, and thermostimulated luminescence (TSL) of as-grown and X-ray irradiated pure and Ce-doped $LiCaAlF_6$ crystals showed that Ce-doped samples demonstrate intensive TSL in the UV region (280–310 nm) with a main peak at 350°C. In addition to its previously known UV laser and TSL properties, $LiCaAlF_6$:Ce crystal showed to be also a promising thermoluminescent dosimetric material.[218] Finally, the energy storage in a crystal is defined by the dissimilarity of basic matrix properties, intrinsic defect structure, preferred Ce^{3+} ion disposition and charge compensation defect type. These parameters accounted for a higher colorability and thermostimulated luminescence (TSL) efficiency of as-grown

[216] K. Shimamura' H. Sato, A. Bensalah, H. Machida, N. Sarukura, T. Fukuda, Growth of $LiCaAlF_6$ single crystals with an extended diameter and their optical characterizations, J. Alloys Comp. 343, Issues 1–2, (16) pp. 204–210, 2002.

[217] A. Gektin,N. Shiran, S. Neicheva, V. Gavrilyuk, A. Bensalah, T. Fukuda, K. Shimamura, $LiCaAlF_6$:Ce crystal: a new scintillator, Nucl. Instr. Meth. Phys. Res. Section A, **486**, (1–2) (2002), pp. 274–277.

[218] Gektin A, Neicheva S, Shiran N, Fukuda T, Shimamura K, A new effective thermoluminescent material: LiCaAlF6:Ce, Radiat Prot Dosimetry. **100** (1–4) (2002) 377–80.

Ce-doped $LiSrAlF_6$ than the Ca-based fluoride.[219] Such an example clearly illustrates the great potential of optically active fluorides.

4.2 Fluoro-glasses, (oxy)fluorinated glass-ceramics

Vitreous fluorinated materials, and (oxy)fluorinated glass-ceramics are also used in optics, thus continuing the tradition of fluoride glasses discovered by J. Lucas in 1975.[220] Their uses concern optical amplification on planar waveguides at telecom wavelengths (in particular, with the luminescence of Er^{3+} at 1.55 μm). The main advantage of the glass-ceramization process is to obtain a composite material that retains the macroscopic properties of a glass in which the rare earth is found mainly in the crystallized phase, thus increasing the luminescence yields. The oxyfluorinated glass-ceramics are thus composed of fluorinated nanocrystals embedded in the oxygenated vitreous matrix. They can be used as transparent fluoride ceramics for high power lasers.[221] These systems combine the advantages of fluoride crystals and oxygenated glasses. Note that the matrix retains its mechanical, chemical, and thermal properties and can be fiber-reinforced, making these doped oxyfluorinated glass-ceramics suitable for use as laser fibers.[205a] Thus, oxy-fluoride glass-ceramics offer an alternative for the hosting of RE ions, making them suitable for the fabrication of optical devices. Transparent oxy-fluoride glass-ceramics doped with RE^{3+} can be obtained from glasses present in the K_2O–Na_2O–Al_2O_3–SiO_2–LaF_3 system by heat treatments slightly above the glass transition (T_g) temperature. Following this process, $NaLaF_4$ nanocrystals with sizes up to 30 nm precipitate, with doping RE generally incorporated into the nano-crystallized phase.[222] Nanoparticles, 8 nm in diameter, of RE-doped fluorite-type (where RE stands for either single rare-earth ion, such as Er^{3+}, Yb^{3+}, Eu^{3+}, Dy^{3+}, Ho^{3+}, Tm^{3+} or combinations of those ions) were extracted by chemical etching from the oxyfluoride nanoglass-ceramics template and analyzed by transmission electron microscope with energy dispersion spectroscopy. These nanoparticles show durable up-conversion photoluminescence, which is neither concentration nor impurity quenched after 6 months of aging in ambient atmosphere. High doping levels in these nanoparticles ensure high (up to 15% quantum yield of up-conversion) luminescence.[223]

[219] A.Gektin, S.·Neicheva, V.Voronova, V.·Kornienko, K.·Shimamura, N.·Ichinose, Energy storage in Ce-doped $LiCaAlF_6$ and $LiSrAlF_6$ crystals, Radiation Measur., **38** (4–6) (2004) pp. 459–462.

[220] Poulain M., Poulain M., Lucas J., Brun P., Verres fluorés au tétrafluorure de zirconium: propriétés optiques d'un verre dopé au Nd^{3+}, *Mat. Res. Bull.*, **10** (1975), *10*, p. 243.

[221] A. J.Stevenson, H. Serier-Brault, P. Gredin, M. Mortier, Fluoride Materials for Optical Applications: Single Crystals, Ceramics, Glasses, and Glass-Ceramics J. Fluor. Chem. **132**, (12) (2011) pp. 1165–1173.

[222] A. de Pablos-Martín M.O.·Ramírez A. Durán[a], L.E. Bausá[b], M.J.Pascual, Tm^{3+} doped oxy-fluoride glass-ceramics containing $NaLaF_4$ nano-crystals, Optical Materials, 33(2) (2010) 180–185.

[223] Tikhomirov VK, Mortier M, Gredin P, Patriarche G, Görller-Walrand C, Moshchalkov VV., Preparation and up-conversion luminescence of 8 nm rare-earth doped fluoride nanoparticles, Opt Express. **16**(19) (2008) 14544–9.

4.3 Other uses in photonics: Non-linear optical fluorides, photochromic and holographic fluorocompounds, excimer KrF lasers

The $MBe_2BO_3F_2$ beryllium and alkaline fluoroborates, in particular those based on K and Rb, are the only non-linear optical crystals (NLO) capable of producing harmonic generation in the deep UV below 200 nm per second-harmonic generation. These crystals opened the way to deep UV laser technology.[208c]

4.3.1 Bistable centers in semi-conducting fluorite crystals: media for real-time holography

CdF_2 crystal is an ionic dielectric that can be converted into a semiconducting state via doping with column-III elements, Ga or In, acting as donors. During a subsequent annealing of the crystals in a reducing atmosphere (Cd vapor or H_2) an "additive coloration" of the crystal is observed due to interstitial fluorine ions, F^-, which compensate the charge excess of the dopants, diffuse out of the bulk of the crystal to its surface, where they recombine with the reducing agents (Cd^{2+} or H^+ ions). The charge neutrality of the crystal is maintained by an opposite current of electrons that are segregated off this agent and diffuse into the crystal bulk. These electrons are localized in the conduction band or at hydrogenic donor orbitals centered on the trivalent impurity ion, and the crystal is thus converted into a semiconductor.[224] Photoinduced deep-to-shallow state transition underlies the photochromism of CdF_2:Ga and CdF_2:In. Real-time phase holograms can be recorded in such crystals, allowing to follow up optical processes in a wide frequency range.[225] In the case of CaF_2 crystals with color center, transmittive and reflective holograms can be performed in the IR region ($<10\,\mu m$).[226]

Recently, in a CaF_2:Eu doped single crystal, clear 3D atomic images around trivalent Eu have been observed for the first time using a new neutron holographic method.[227]

4.3.2 Krypton fluoride lasers

Krypton fluoride lasers are one important particular type of excimer laser. This type of laser produces coherent electromagnetic radiation by dissociation of an excited KrF* complex formed by absorbing energy from an external source:

[224] Kingsley, J.D.; Prener, J.S. Free charge carrier effects in cadmium fluoride. *Phys. Rev. Lett.* 1962, *8*, 315–316. Prener, J.S.; Kingsley, J.D. Mechanism of the conversion of CdF2 from an isolator to a semiconductor. *J. Chem. Phys.* 1963, *38*, 667–671.

[225] A. I. Ryskin, A. S. Shcheulin and A. E. Angervaks, Semiconductor CdF2:Ga and CdF2:In Crystals as Media for Real-Time Holography, Materials **5** (2012) 784–817.

[226] K. A. Aksenova, A. E. Angervaks, A. S. Shcheulin, A. I. Ryskin, Holographic characteristics of calcium fluoride crystals in the IR region, J. Opt. Technol. **82** (11) (2015) pp. 760–766.

[227] K. Hayashi, K. Ohoyama, N. Happo, T. Matsushita, S. Hosokawa, M; Harada, Y. Inamura, H. Nitani, T. Shishido, K. Yubuta, Multiple-wavelength neutron holography with pulsed neutrons, *Science Advances*, **3** (8) (2017) e1700294. https://doi.org/10.1126/sciadv.1700294.

$$2Kr + F_2 + energy \rightarrow 2KrF^*$$

$$KrF^* \rightarrow KrF + h\nu \text{ at } 248\,nm$$

$$2KrF \rightarrow 2Kr + F_2$$

The KrF^* complex is stable only in the excited state, and it dissociates immediately in krypton and fluorine as soon as it came back to its ground state after photon emission corresponding to the transition, that is, electromagnetic radiation at $\lambda = 248\,nm$, in the ultraviolet. The krypton fluoride laser is used in laser microlithography, because its rather short λ allows to work on very small patterns. The pattern size diminished from 800 nm in 1990 to 10 nm in 2018. However, it can be noted that KrF laser will certainly be replaced by the argon fluoride laser (ArF), with $\lambda = 193\,nm$ only.

The KrF laser has also been used to produce soft X-ray emissions from irradiated plasma by brief pulses of this laser radiation (commercial KrF lasers have typical pulses of 20–30 ns). The wavelength emitted by the KrF laser is strongly absorbed by organic tissues, including proteins, nucleic acids, and lipids, making it a potentially effective tool in medicine, especially in surgery.

4.4 Fluorinated components in photovoltaic systems

4.4.1 Transparent conducting films based on fluorine-doped tin oxide (FTO)

Transparent conducting films (TCFs) are thin films used as an important component in a number of electronic devices, including liquid-crystal displays, LEDs, OLEDs, touch screens and photovoltaics.[228] These materials possess wide bandgaps whose energy value is greater than those of visible light. In applications such as solar cells, a wider range of transparency beyond visible light is required to make efficient use of the full solar spectrum. While indium tin oxide (ITO) is the most known of transparent inorganic conducting oxides, fluorine-doped tin oxide (FTO), or doped zinc oxide are also largely used.[229] FTO is a material that displays competitive opto-electronic properties and offers higher chemical, mechanical and thermal resistance and effective deposition. FTO possesses a fundamental electronic direct band gap of $Eg = 3.6\,eV$, and an optical band gap that can often exceed 4 eV depending on the level of fluorine incorporation. Fluorine acts as a substitutional donor occupying an oxygen site of the rutile network. The resulting resistivity of FTO films is as low as $4 \times 10^{-4}\,\Omega cm$.

[228] A. Klein, C. Körber, A. Wachau, F. Säuberlich, Y. Gassenbauer, S. P. Harvey, D. E. Proffit, T. O. Mason, Transparent Conducting Oxides for Photovoltaics: Manipulation of Fermi Level,Work Function and Energy Band Alignment, Materials **3,** (2010) 4892–4914.

[229] J. E. N. Swallow, B. A. D. Williamson, T. J. Whittles, M. Birkett, T. J. Featherstone, N. Peng, A. Abbott, M. Farnworth, K. J. Cheetham, P. Warren, D. O. Scanlon, V. R. Dhanak, T. D. Veal (2017). Self-compensation in transparent conducting F-doped SnO_2. Advanced Functional Materials. *DOI:* 10.1002/adfm.201701900.

4.4.2 Fluorinated components in dye-sensitized solar cells (DSSC)

Dye-sensitized solar cells (DSSC, or Grätzel cell) are photo-electrochemical systems based on a semiconductor set between a photo-sensitized anode and an electrolyte. These systems were invented in the 1980s by M. Grätzel at UC Berkeley and later at EPF, Lausanne, with the publication of the first high efficiency DSSC in 1991.[230]

Among DSSC numerous attractive features, low-cost conventional roll-printing techniques with semi-flexible and semi-transparent electrodes can be noted, a property that cannot be applicable to glass-based systems. The liquid electrolyte presents a serious challenge for making a cell suitable for uses in all weather.

A cell design is formed of three main parts. A transparent anode (F-TCO, or FTO), made of fluorine-doped tin dioxide (SnO_2:F) is deposited on the back of a glassy plate. On the back of this conductive plate, a thin layer of sintered TiO_2 nanoparticles with extremely high surface area forms into a highly porous structure TiO_2 only absorbs a small fraction of the solar photons (those in the UV). After immersion in molecular sensitizers, that is a mixture of a photosensitive ruthenium-polypyridine dye (black-dye) with a solvent, a thin layer of the dye is left covalently bonded to the surface of the TiO_2. Another separate plate made of a thin layer of the iodide electrolyte spread over a conductive sheet is then joined to the first plate and sealed together to prevent the electrolyte from leaking. So, as in a conventional battery, the titanium dioxide (anode), and the platinum layer (cathode) are placed on each side of a liquid conductor (electrolyte) (*see* "Processes involved in the energy conversion of dye-sensitized TiO_2 solar cell (DSSC)."

Processes involved in the energy conversion of dye-sensitized TiO_2 solar cell (DSSC)

1. The incident photon is absorbed by Ru complex photosensitizers adsorbed on the TiO_2 surface.
2. The photosensitizers are excited from the ground state (S) to the excited state (S^*). The excited electrons are injected into the conduction band of the TiO_2 electrode. This results in the oxidation of the photosensitizer (S^+).

$$S + h\nu \rightarrow S^* \quad (5)$$

$$S^* \rightarrow S^+ + e^- \, (TiO_2) \quad (6)$$

Black Dye, an anionic Ru-terpyridine sensitizer

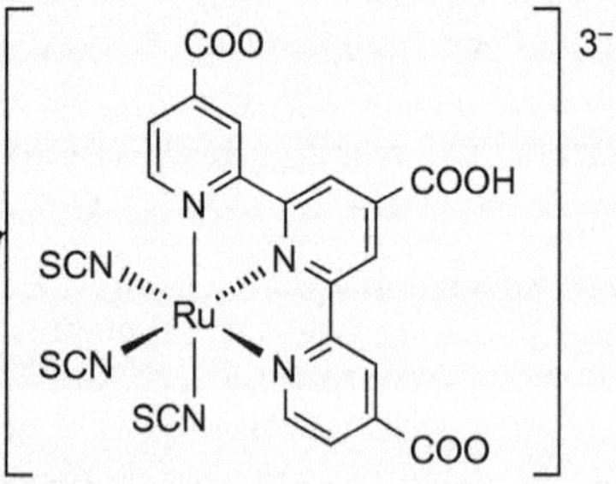

[230]B. O'Regan; M. Grätzel, A low-cost, high-efficiency solar cell based on dye-sensitized colloidal TiO_2 films. Nature **353** (6346) (1991) 737–74.

3. The electrons injected in the TiO_2 conduction band are transported toward the back contact (TCO) and finally reach the counter electrode through the circuit.
4. The oxidized photosensitizer (S^+) accepts electrons from the I^- ion redox complex leading to regeneration of the ground state (S), and two I^- ions are oxidized to elementary iodine which reacts with I^- to the oxidized state, I_3^-.

$$S^+ + e^- \rightarrow S \quad (7)$$

5. The oxidized redox mediator, I_3^-, diffuses toward the counter electrode and is then reduced back to I^- ions.

$$I_3^- + 2e^- \rightarrow 3I^- \quad (8)$$

The efficiency of a DSSC depends on four energy levels of the component: the excited state (approximately LUMO) and the ground state (HOMO) of the photosensitizer, the Fermi level of the TiO_2 electrode, and the redox potential of the mediator (I^-/I_3^-) in the electrolyte.[231]

Dye-sensitized TiO_2 solar cell (DSSC): processes involved in the energy conversion (S: dye sensitizer, I-/I-3: charge mediator) [from C. Longo; M-A. De Paoli, Dye-sensitized solar cells: a successful combination of materials, J. Braz. Chem. Soc. vol. 14 no. 6 São Paulo Nov./Dec. 2003, https://doi.org/10.1590/S0103-50532003000600005].

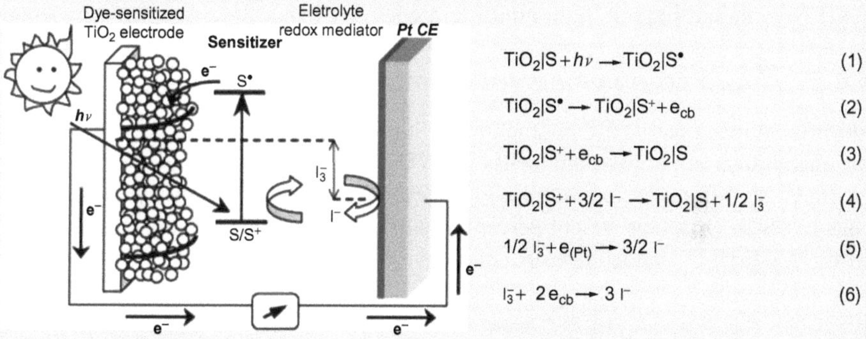

Although the photovoltaic conversion efficiency is less than the best thin-film cells, the performances are regularly improved. Commercial applications can significantly contribute to renewable electricity generation by 2020, and fluorinated components highly contribute to the improvement of these systems.

[231] Hara, K., Arakawa, H. (2005). "Chapter 15. Dye-Sensitized Solar Cells," in A. Luque; S. Hegedus. Handbook of Photovoltaic Science and Engineering.

The performance of DSSCs can also be improved through a fluorine-doped TiO_2 dense layer for use as a blocking layer,[232] or even through a N/F doped TiO_2.[233] The F-doped TiO_2 layer is deposited on commercial F-doped SnO_2 (FTO) substrate by the hydrolysis of a $TiCl_4$ aqueous solution with NH_4F. the performance of dye-sensitized solar cells (DSSCs) are improved through a fluorine (F)-doped TiO_2 dense layer as a blocking layer. The F-doped TiO_2 layer is deposited on commercial F-doped SnO_2 (FTCO) substrate by the hydrolysis of a $TiCl_4$ aqueous solution with NH_4F. The photocurrent density vs. voltage characteristics show that DSSCs fabricated with the F-doped TiO_2 layer exhibits higher conversion efficiency (~5.24%) than ones with the undoped TiO_2 layer (~4.76%), and the bare FTO substrate (~3.88%). This improvement is attributed to the combined effects of reduced interfacial resistance and effective blocking behavior by the F-doped TiO_2 dense layer.[234,235]

4.4.3 Fluorinated photoelectrode in QDSSC (quantum dot-sensitized solar cell)

Due to its large energy band of 3.60 eV, zinc hydroxyfluoride ZnOHF has been proposed as a photoelectrode in QDSSC (quantum dot-sensitized solar cell) type solar cells, but also in dye-sensitized water decomposition cells, perovskite-based solar cells, organic solar cells, etc.[236] In addition, it has been possible to deposit rare earth fluoride nanoparticles of <10 nm on multilayer carbon nanotubes via sodium dodecyl sulfate. These hybrid materials could be used as field emitters with brightness higher than conventional organic light emitting diodes (OLEDs).[237]

4.4.4 Fluoro co-polymers for membranes in DSSC systems

Multifunctional coating of photocurable fluorinated polymer have been proposed for membranes in photovoltaic cells.[238] PVDF-HFP nanocomposite membranes can be doped by different nanoparticles to increase the efficiency in dye-sensitized solar cells (DSSC). Different weight percentages of cobalt sulfide (CoS) were electrospun with PVDF-HFP nanocomposite membranes and activated by an ionic liquid

[232] S. I. Noh, K.N. Bae, H.J. Ahn, T.Y. Seong, Improved efficiency of dye-sensitized solar cells through fluorine-doped TiO_2 blocking layer, Ceramics Internat. **39**, (7) (2013), pp. 8097–8101.

[233] Park SK, Yun TK, Bae JY, Fabrication N, F, and N/F-Doped TiO_2 Photoelectrodes for Dye-Sensitized Solar Cells. J Nanosci. Nanotechnol. 2015 (8) pp. 5967–70.

[234] X Q Zhang, Y P Wu, Y Huang, Z H Zhou, S Shen, Effect of fluorine doped TiO_2 on the property of perovskite solar cell; IOP Conf. Ser.: Mater. Sci. Eng.**182** (2017) 012001.

[235] M. Che, Y. Fang,J. Yuan, Y. Zhu, Q. Liu, J. Song, F-doped TiO_2 Compact Film for High-Efficient Perovskite Solar Cells, Int. J. Electrochem. Sci. 12 (2017) 1064–1074.

[236] Chen H.N. L. Zhu, Q. Hou, W. Liang, H. Liu[a] W. Li, ZnOHF nanostructure-based quantum dots-sensitized solar cells, J. Mater. Chem., 22 (2012) p. 23344.

[237] X.W. Wei, J. Xu, X.J. Song, Y.H. Ni, P. Zhang, X. J. Xia, G. C. Zhao, Z. S. Yang, Multi-walled carbon nanotubes coated with rare earth fluoride EuF_3 and TbF_3 nanoparticles, Mater. Res. Bull., **41** (2006) pp. 92–98.

[238] M. Grätzel, Dye-sensitized solar cells, A review, J. Photochem. Photobiol. C: Photochem. Reviews **4** (2003) 145–153.

electrolyte to form electrospun PVDF-HFP/CoS nanocomposite membrane electrolytes (esCPMEs). The photovoltaic performance of the dye-sensitized solar cell assembled using the esCPME with 1 wt% of CoS was measured to be 7.34% higher than that of the dye-sensitized solar cell assembled using the esPME without CoS (6.42%).[239] In the case of addition of LiCl, a higher overall cell efficiency is obtained, as the cells having 1.0% LiCl in PVdF showed 8.73% power conversion efficiency compared to 7.87% of the cells without LiCl. Moreover, compared to the normal DSSC with liquid electrolyte, the proposed nanofiber separator-made DSSC only lost 1.9% efficiency, showing its promise to be used in the next generation of flexible solar cells and textile-structured solar cells.[240]

A luminescent europium complex can also be embedded in the fluoropolymer that acts as luminescent down-shifting material converting UV photons into visible light. The combination of this system with ruthenium-free organic dye-sensitized solar cells gives a 70% relative increase in power conversion efficiency as compared with control uncoated devices, which is the highest efficiency enhancement reported on organic DSSC systems by means of a fluoropolymeric layer. Such systems exhibit excellent outdoor stability that can be attributed to the combined action of the luminescent material that acts as UV-screen and the highly photostable, hydrophobic, fluoropolymeric carrier that further prevents photochemical and physical degradation of the solar cell components.[241]

4.4.5 Fluorinated components in perovskite solar cells

Perovskite solar cells based on organometal halides derive from DSSC technology. In a liquid-based DSSC structure, the adsorption of methylammonium lead halide perovskite ($CH_3NH_3PbX_3$ (X = Br, I)) on a nanocrystalline TiO_2 surface produced a photocurrent with a power conversion efficiency of around 3%–4%, as first discovered in 2009 by Miyasaka.[242] The perrovskite optical band gap is between 1.5 and 2.3 eV. The first use of perovskite in a solid state solar cell was in a dye-sensitized cell using $CsSnI_3$ as a p-type hole transport layer and absorber. The efficiency was doubled after 2 years by optimizing the perovskite coating conditions. A long-term, stable, and ~10% efficiency perovskite solar cell was developed in 2012 by substituting the solid hole conductor by a liquid electrolyte. Perovskite solar cells

[239] E. Vijayakumar, A. Subramania, Z. Fei, P J. Dyson, High-performance dye-sensitized solar cell based on an electrospun poly(vinylidene fluoride-*co*-hexafluoropropylene)/cobalt sulfide nanocomposite membrane electrolyte, **RSC Adv.**, **5** (2015) 52026–52032.

[240] I. A. Sahito, F. Ahmed, Z. Khatri K.C. Sun, S.H. Jeong, Enhanced ionic mobility and increased efficiency of dye-sensitized solar cell by adding lithium chloride in poly(vinylidene fluoride) nanofiber as electrolyte medium, J. Mater. Sci. **52**, (24) (2017) pp. 13920–13929.

[241] G. Griffini F. Bella, F. Nisic, C. Dragonetti, D. Roberto, M. Levi, R. Bongiovanni, S. Turri, Multifunctional Luminescent Down-Shifting Fluoropolymer Coatings: A Straightforward Strategy to Improve the UV-Light Harvesting Ability and Long-Term Outdoor Stability of Organic Dye-Sensitized Solar Cells, Adv. Energy Mater., **5** (2015) 1401312–1401321.

[242] A. Kojima, K. Teshima, Y. Shirai, T. Miyasaka, Organometal Halide Perovskites as Visible-Light Sensitizers for Photovoltaic Cells, *J. Am. Chem. Soc.*, **131** (17), (2009) pp. 6050–6051.

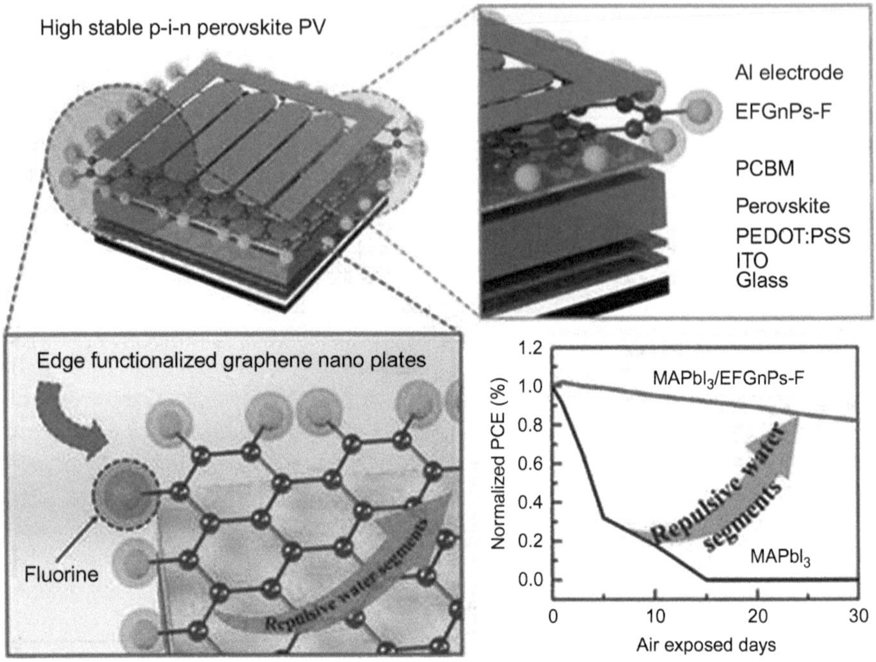

Fig. 48 Experimental set-up of a p-i-n perovskite solar cell made of methyl ammonium lead iodide ($MAPbI_3$), protected by a thin layer of an edge-fluorinated graphene nanoplatelets (EFGnPs-F). (PCBM is the fullerene derivative [6,6]-phenyl-C_{61}-butyric acid methyl ester). The improvement of hydrophobic properties of EFGnPs-F versus pristine MAPbI3 have been characterized by contact angle measurement (from Kim et al. 244).

are a highly promising photovoltaic technology, because the latest world efficiency record reaches 22.7% (fall 2017 https://www.nrel.gov/pv/ and [243]) with the use of cheap organometal halide perovskite materials.

Edge fluorine-functionalized graphene nanoplatelets (EFGnPs-F) protecting a p-i-n structure of perovskite solar cells (i.e., a wide, undoped intrinsic semiconductor region between a p-type semiconductor and an n-type semiconductor region) allows the achievement of 82% stability relative to initial performance over 30 days of air exposure without encapsulation (Fig. 48). The enhanced stability of the methyl ammonium lead iodide ($MAPbI_3$) perovskite-based devices arises from F-substitution on EFGnPs. Indeed, fluorocarbons, such as PTFE and others, are well-known for their superhydrophobic properties and for limiting chemical degradation. These hydrophobic moieties tightly protect perovskite layers from air degradation.[244]

[243] W.S. Yang, B. W. Park, E. H. Jung, N. J. Jeon, Y.C. Kim, D.U. Lee, S.S. Shin, J. Seo, E.K. Kim, J.H. Noh, S.I. Seok. Iodide management in formamidinium-lead-halide–based perovskite layers for efficient solar cells. *Science*, **356** (2017) (6345): 1376 DOI: 10.1126/science.aan2301.

[244] Kim GH, Jang H, Yoon YJ, Jeong J, Park SY, Walker B, Jeon IY, Jo Y, Yoon H, Kim M, Baek JB, Kim DS, Kim JY., Fluorine Functionalized Graphene Nano Platelets for Highly Stable Inverted Perovskite Solar Cells. Nano Lett., **17**(10) (2017) pp. 6385–6390.

p-type inorganic fluorine-doped $CsInI_3$ perovskite, $CsInI_{3-x}F_x$ with $x = 0.05$, exhibits high hole-mobility semiconductor abilities. The use of a solid component allows to suppress the problem of liquid electrolytes in DSSC. The resulting solid-state dye-sensitized solar cells consist of $CsSnI_{2.95}F_{0.05}$ doped with SnF_2, nanoporous TiO_2 and the dye Ru^{II} N719 complex, and show conversion efficiencies of up to 10.2%. With a bandgap of 1.3 eV, this enhances visible light absorption on the red side of the spectrum to outperform the typical dye-sensitized solar cells in this spectral region.[245]

5 Uses of fluorides in electronics and other modern technologies

5.1 New fluorinated carbon materials for electronics

It has long been known that, once intercalated by fluorinated species, graphitized carbon microfibers may reach an electrical conductivity equivalent to or even greater than that of copper: $\sigma_{25°C} = 1.6 \times 10^5\ cm^{-1}$.[246] More recently, new forms of carbon such as graphene have appeared (a few layers of graphite with a thickness of a few nm that can reach several hundreds or thousands of nm along the graphitic planes). By adjusting the fluorine content onto the surface of graphene, it is possible to modulate the conductivity of the material from semimetal to insulator. In field effect transistors (FET) with planar heterostructure (HFET), based on graphene/fluorographene/graphene, the on/off switching frequency is 10^5 and can hold up to 100°C.[247] A modulation of the width of the band gap can also lead to interesting optical properties, as in the graphene/fluorographene diodes with a switching frequency $>10^5$. Such intraplanar heterojunctions are created for the design of nanoribbons. These highly fluorinated graphenes can be used as effective dielectrics 5 nm wide in nanoelectronics, with particular mention for flexible electronics applications.[248,249]

In Fig. 49, the proof of concept of an FET system with fluorographene as dielectric and pristine graphene as channel was demonstrated. The transfer characteristics indicate that the electrical properties of pristine graphene were maintained, yielding a high FET mobility of ~760 cm^2/V, which is compatible with typical mobility of CVD

[245] Chung I., Lee B, He J, Chang RP, Kanatzidis MG, All-solid-state dye-sensitized solar cells with high efficiency, Nature 2012 May 23; 485(7399):486–9. https://doi.org/10.1038/nature11067.

[246] Tressaud A.; Dordor P.; Marquestaut E.; Hagenmuller P.; Fujimoto H.; Flandrois S. Electrical properties of room-temperature fluorine intercalated graphite fibers. *Synth. Metals*, **40** (1991) 17–185.

[247] D.Chronopoulos, A. Bakandritsos, M. Pykal, R. Zbořil, M. Otyepka, Chemistry, properties, and applications of fluorographene, Applied Materials Today **9** (2017) 60–70.

[248] J. S. Moon H.-c. Seo, F. Stratan, M. Antcliffe, A. Schmitz, R.S. Ross, A. Kiselev, V.D. Wheeler, L.O. Nyakiti, D.K. Gaskill, K.U. Lee, P.M. Asbeck, IEEE Electron Device Lett. **34** (2013) 1190–1192.

[249] K.I. Ho, C.S. Lai, C.Y. Su, Nanoelectronics Based on Fluorinated Graphene, Chapter 16 in New Fluorinated Carbons: Fundamentals and Applications, O. Boltalina, T. Nakajima, Eds., Elsevier 2016. Ho K.I., Huang C.H., Liao J.H., Zhang W., Li L.J., Lai C.S., Su C.Y., Fluorinated graphene as high performance dielectric materials and the applications for graphene nanoelectronics, Sci Rep. 2014 Jul 31;4:5893. https://doi.org/10.1038/srep05893.

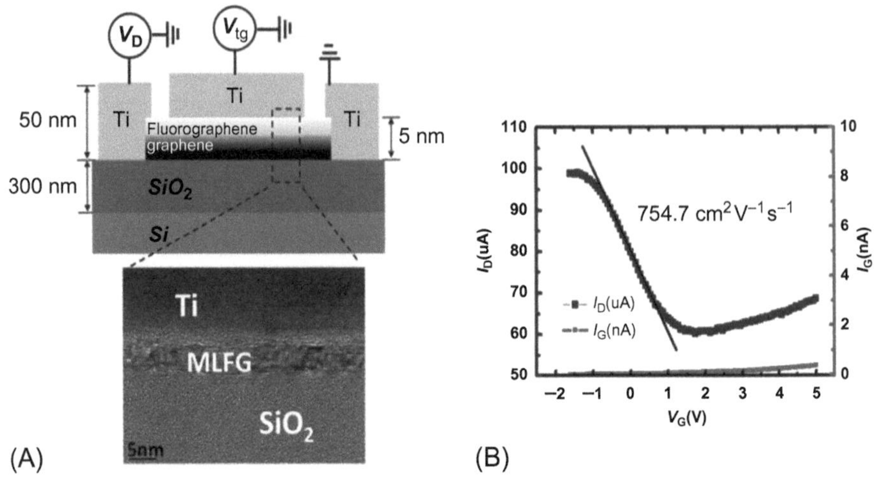

Fig. 49 Graphene-based FET composed of a fluorographene gate dielectric and the corresponding TEM cross-sectional image. (B) The transfer characteristics and the monitored gate current (I_G) of these transistors, which demonstrate hole mobility of 760 cm^2/Vs at $V_D = 0.5$ [from 249].

graphene. Multi-layer fluorographene (MLFG) is a new way to prepare top-gated graphene-based transistors compatible with conventional IC technology.[249] In addition, fluorographene could also act as a passivation layer for screening charged impurity on the surface of substrate (e.g., SiO_2). The graphene device supported by the fluorographene film, exhibits a significantly enhancement of electrical properties of graphene transistors in terms of carrier mobility and current swing.[250] Thus, fluorinated graphene could be an excellent supporting buffer layer for devices assembly of emerging 2D-layered materials, such as MoS_2-based electronics, in order to improve its carrier mobility.

5.2 Fluorinated organic compounds for molecular electronics

Fluorinated organic molecules are involved in molecular electronics as liquid crystals, transparent conductive films, phosphorescent emitters, dye-sensitized solar cells, etc.

- *Fluorinated nematic crystals* are used as liquid crystal displays (LCDs).[251] Their uses include light and unbreakable LCD systems, holography, microwave switching, the new generation of "smart window" based on liquid crystals doped with dichroic dyes.

[250] K.-I. Ho, M. Boutchich, C.-Y. Su, R. Moreddu, E.S.R. Marianathan, L. Montes, C.-S. Lai, Self-Aligned High-Mobility Graphene Transistor: Decoupling the Channel with Fluorographene to Reduce Scattering, Adv. Mater., **27** (2015) 6519–6525.

[251] P. Kirsch, Fluorinated nematic liquid crystals: Design, Synthesis and Properties, Chapter 8, pp. 159–176, Photonic and Electronic Properties of Fluoride Materials, A. Tressaud, K. Poeppelmeir, Eds, Elsevier (2016).

Fluorinated nematic crystals are the most common components in flat panel displays. In 2014, 160 km^2 of these phases were produced worldwide.

The “nematic fluorine” effect discovered in 1950 is used to improve the polarity and mesophasic characteristics of positive dielectric compounds. This is the case for example of the 2,3-difluorophenylene group. One of the most widely used methods for accessing active phases is orthometallation (generally a lithiation). The safety problems that arise during these syntheses require reactions at low temperature because of the risk of decomposition at higher temperatures.

- *Iridium fluoride complexes as phosphorescent emitter (OLED)*. The three primary colors, red, green, and blue, are obtained in organic light-emitting diodes (OLED), especially from iridium-based complexes. However, in the complex case of Ir^{3+} that emits in blue, it is the fluorine element which has made it possible to obtain the maximum of quantum yield (around 20%–25%), that is to say, significantly higher than the yields of other emitters.[252] The stability of the two most efficient blue-emitting phosphorescent materials, iridium(III) bis(4′,6′-difluorophenylpyridinato)tetrakis(1-pyrazolyl) borate (FIr6), and bis(2-(4,6-difluorophenyl) pyridyl-N,C2')iridium(III)picolinate (FIrpic) has been recently improved.[253]

5.3 Recent uses of fluoro-copolymers

For backsheet protection layers: System lifetime is key to investment returns. Because investment returns for solar systems can increase by more than 30% if the lifetime of the solar system can be increased from 10 to 25 years. PV backsheets play a critical role, providing long-life protection to the solar panel. PVF film-based backsheets (Tedlar®) have been tested for more than years in all kinds of climates (desert, tropics, temperate, marine, mountain).

For Li-ion batteries: Composite gel polymer electrolyte (CGPE) based on poly(vinylidene fluoride-hexafluoropropylene) (PVDF-HFP) *co*-polymer with Al-doped $Li_{0.33}La_{0.56}TiO_3$ (A-LLTO) particles covered with a modified SiO_2 layer possess high ionic conductivity, a large electrochemical stability window, and interfacial stability—all superior to that of the pure gel polymer electrolyte (GPE). The cyclability of Li-ion batteries is improved, because the CGPE effectively suppressed the growth of Li dendrites.[254]

For membranes in supercritical systems: Catalyst-loaded co-polymeric PVDF-HFP membranes offer a larger flexibility over conventional reactors, due to nanoporous homogeneous networks with palladium nanoparticles homogeneous

[252] Y. Takahira, E. Murotani, K. Fukuda, V. Vohra, H. Murata, Design, synthesis, and properties of a series of charged iridium(III) complexes with a neutral bidentate ligand for deep-blue phosphorescent emitter, J. Fluorine Chem. **181** (2016), 56–60.

[253] R. Seifert, I. Rabelo de Moraes, S. Scholz, M. C. Gather, B. Luessem, K. Leo, Chemical degradation mechanisms of highly efficient blue phosphorescent emitters used for organic light emitting diodes, Organic Electronics, **14** (2013) pp. 115–123,.

[254] H. T. T. Le, D. T. Ngo, R. S. Kalubarme, G. Cao, C. N. Park, C. J. Park, Composite Gel Polymer Electrolyte Based on Poly(vinylidene fluoride-hexafluoropropylene) (PVDF-HFP) with Modified Aluminum-Doped Lithium Lanthanum Titanate (A-LLTO) for High-Performance Lithium Rechargeable Batteries, *ACS Appl. Mater. Interfaces*, **8** (2016) (32), pp. 20710–20719.

distributed. Their potential applications in catalysis include supercritical assisted phase inversion and supercritical assisted gel drying. These co-polymeric fluoromembranes are characterized by very high porosity (higher than 90%) and a very uniform catalyst distribution.[255]

5.4 Fluorinated colored pigments and UV absorbers for solar protection

Rare earth sesquisulfides RE_2S_3 and ARE_2S_4 (A = Ca, Sr), with structure deriving from γ-Th_3P_4, have been used for long time as inorganic pigments because they exhibit pronounced color. Deriving from these 2D structural networks, RE fluorosulfides RESF and $ARE_2F_4S_2$ (A = Ca, Sr) prepared via solid state or gas-solid routes by A. Demourgues show improved colors[256] (Fig. 50). The color of these

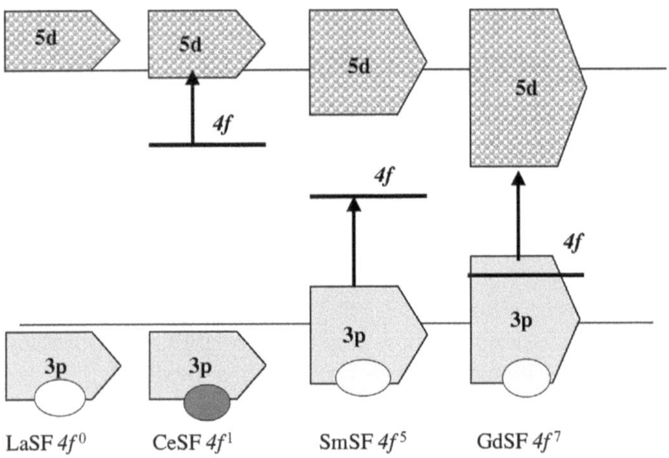

Fig. 50 Schematic band diagram of LnSF fluorosulfides (color of the materials are shown in ellipses: *light yellow* for La, *pronounced yellow* for Sm and Gd, and *deep red* for Ce) (from 256). [Credit Dr. A. Demourgues, ICMCB-CNRS, Pessac, France]. [Recent calculations revealed that the optical transitions in CeSF should be rather due S-3***p*** to Ce-4***f*** interatomic transitions. J.M. Tomczak, L.V. Pourovskii, L. Vaugier, A. Georges, S. Biermann, Rare-earth vs. heavy metal pigments and their colors from first principles, PNAS 2013 January, 110 (3) 904–907. https://doi.org/10.1073/pnas.1215066110].

[255] S.-Cardea, E. Reverchon, Nanostructured PVDF-HFP membranes loaded with catalyst obtained by supercritical CO_2 assisted techniques, Chem. Engin. Process. Process Intensif. **50**, (7) (2011) pp. 630–636.

[256] A. Demourgues, A. Tressaud, Preparation of Rare Earth Fluorosulfides and Oxyfluorosulfides, in "Efficient Preparations of Fluorine Compounds", H. Roesky Ed., Wiley, Chapter 64 (2012) pp. 415–424. A. Demourgues, A. Tressaud, H. Laronze, P. Macaudière, Rare earth fluorosulfides LnSF and $Ln_2AF_4S_2$ as new colour pigments, J. Alloys Comp. **323–324**, 223 (2001).

compounds can be related to various energetic parameters such as ionization energies, crystal field splitting, or nephelauxetic effect. In these RE fluorosulfides, the origin of the color is related to the position of the energy levels where electronic transitions with respect to the valence and conduction bands and closely depends on the crystal structure, as shown in Fig. 50:

- The conduction band is characterized by the RE 5d states and its splitting, governed by the crystal field, increases as the rare earth size decreases (from La to Gd).
- The valence band is characterized by the 3p states of the sulfur and its bandwith increases when the Ln-S distance decreases, that is, when covalence increases.
- The RE 4f levels, related to the RE ionization energy, increases as the number of f electrons decreases.

  Because the nephelauxetic effect and the crystal field are stronger with sulfur that fluorine, interactions between RE and sulfur are predominant in the mechanism of color. Most of transitions, except Ce (4f $\rightarrow$ 5d), are the result of charge transfer mechanism from 3p(S) band to rare earth (4f, 5d) bands. The absorption edge energies of rare earth fluorosulfides vary in a first approximation as the difference between the fourth and the third ionization energies of rare earth (Fig. 51A). The presence of fluorine participates in the mechanism of color because the rather ionic RE-F bonds reinforce the covalence of the RE-S interactions. Recent results revealed that the red coloration of CeSF is caused by the large density of weakly dispersing states (resulting from both the 2D dimensionality of the crystal structure and the localized character of Ce-4*f* states). The relevant optical transitions in CeSF should be S-3*p* $\rightarrow$ Ce-4*f* interatomic transitions, rather than Ce (4f $\rightarrow$ 5d).

Titanium oxides are known as inorganic UV absorber materials; they are characterized by an absorption edge at a wavelength in the 380–400 nm range corresponding to the charge transfer O(2p)-Ti(4 s/3d). However, Ti oxide varieties exhibit high refractive indices (rutile form: $n = 2.7$; anatase form: $n = 2.5$), thus lowering the diffusion of visible light at $\lambda = 500$ nm. As a consequence of these high refractive indices, a whitening of the medium is generally observed, limiting the use of such materials as UV absorbers in the case of wood protection. With a lower refractive index due to the presence of fluorine, Ti oxy-fluorides and hydroxy-oxyfluorides ($n = 1.9$ in the visible range for $Ti_{0.75}(OH)_{1.5}F_{1.5}$), exhibit improved UV-shielding properties with absorption edges around 3.2 eV close to the one of anatase TiO_2 Fig. 51B. In addition, these fluorinated compounds do not exhibit photocatalytic activities because of both the stabilization of F^- ions in the vicinity of Ti^{4+} cations and the limitation of charge carriers in such ionic compounds, which is an important feature for applications in UV protection.[257]

[257] X. Rocquefelte, F. Goubin, Y. Montardi, N. Viadere, A. Demourgues, A. Tressaud, M.H. Whangbo, S. Jobic, Analysis of the refractive index of TiO_2, $TiOF_2$, and TiF_4: Concept of Optical Channel as a guide for understanding and designing Optical Materials, Inorg. Chem. **44,** 3589–3593 (2005). Demourgues A., Penin N., Durand E., Weill F., Dambournet D., Viadere N., Tressaud A. New titanium hydroxyfluoride $Ti_{0.75}(OH)_{1.5}F_{1.5}$ as a UV absorber, Chem. Mater. 2009, **21**, n° 7, 1275–1283.

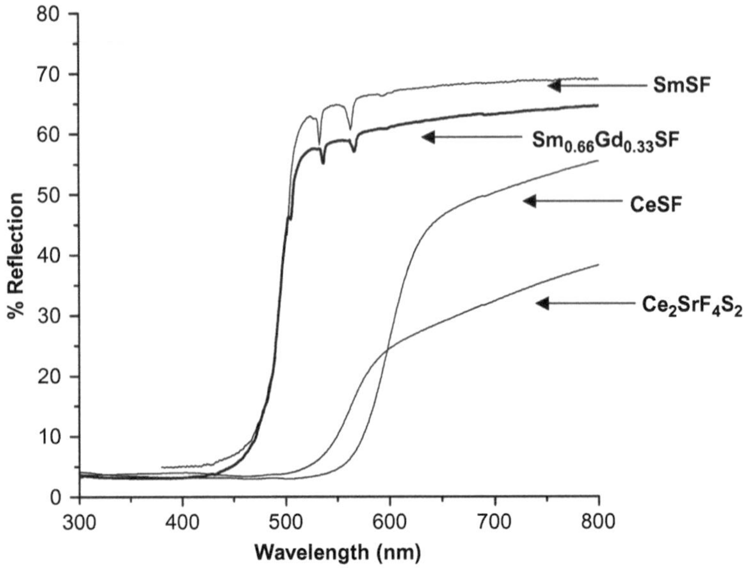

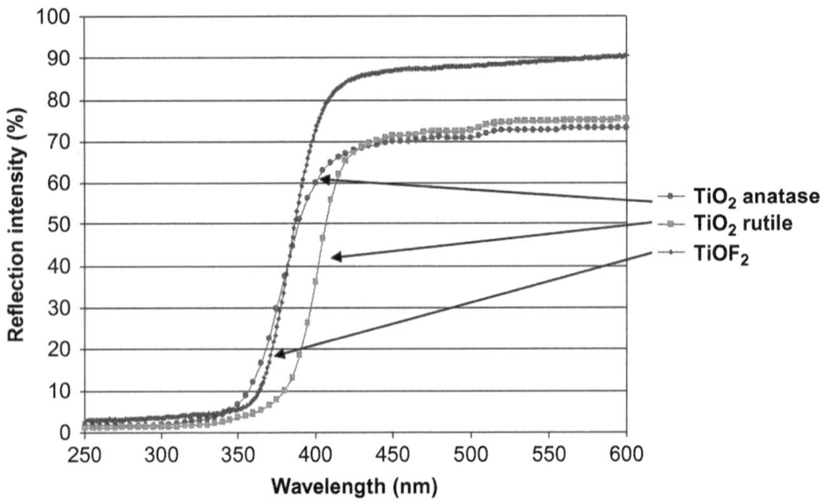

Fig. 51 (A) Absorption edges in diffuse reflectance spectra of various fluorosulfides colored pigments (CeSF, SmSF, $Sm_{0.66}Gd_{0.33}SF$ and $Ce_2SrF_4S_2$) (from 256), and (B) UV-visible reflectance in UV absorbers: Ti oxyfluoride.and anatase and rutile TiO_2 (from 257). [Credit Dr. A. Demourgues, ICMCB-CNRS, Pessac, France].

5.5 High T_C superconductors obtained by insertion of fluorine into oxides and oxypnictides (La_2CuO_4, LaFeAsO, BiS_2 layer structures)[258,259]

Since the discovery in 1986 of new high T_c superconductors, many investigations have tried to increase to room temperature the critical temperature at which resistance drops abruptly to zero. In the superconducting state, there is no thermal loss, and the electric current through a loop of a superconducting material can persist indefinitely without any power source. There is an immense field of applications related to this property such as fast digital circuits, including those based on Josephson junctions and rapid single flux quantum technology, powerful superconducting electromagnets used in magnetic levitation (Maglev) trains, magnetic resonance imaging (MRI), magnetic confinement fusion reactors (e.g., tokamaks), etc.

When semiconducting La_2CuO_4-type compounds are submitted to fluorination with F_2 gas, an intercalation of fluorine occurs in between the 2D layers of the cuprate, forming a superconducting oxide-fluoride $La_2Cu(O, F)_{4+x}$ with $T_c = 40\,K$.[260] Similar behavior is obtained for the YBCO phases $YBa_2Cu_3O_{7-\delta}$.[261]

Recently studied, novel, layered superconductors have a structure similar to those of Cu-oxide: the Fe-based superconductors (SC), with 2D networks consisting of conductive layers formed of Fe squares. As in the case of Cu oxides, the structure is composed of alternate stacks of the Fe-based superconducting layers and various types of blocking layers. The notable point of the Fe-based SC family is that there are several types of superconducting layers such as Fe_2Pn_2 (Pn: P and As) or Fe_2Ch_2 (Ch: S, Se, Te). In LaOFeP deriving from ZrCuSiAs structure, the superconducting layers are (Fe_2P_2) layers.

In the oxypnictide series REOM*Pn* (with RE = rare earth, M = Mn, Fe, Co, Ni and *Pn* = P, As) by reaction with REF_3, superconducting phases are formed. The schematic reaction can be written: LaOFeAs → $LaO_{1-x}F_xFeAs$, with $T_c = 43\,K$.[258] Electron carriers generated by F-doping into oxygen sites are injected into FeAs metallic layers. Note that the carrier doping layer is spatially separated from the conduction layer

[258] H. Takahashi, K. Igawa, K. Arii, Y. Kamihara, M. Hirano, H. Hosono, Superconductivity at 43 K in an iron-based layered compound $LaFeAsO_{1-x}F_x$, Nature Letter **453**, (2008) 376–378.

[259] Y. Mizuguchi, Review of superconductivity in BiS2-based layered materials, J. Phys. Chem. Solids, **84** (2015) 34–48.

[260] B. Chevalier, A. Tressaud, B. Lepine, K. Amine, J.M. Dance, L. Lozano, E. Hickey, J. Etourneau, Stabilization of a new superconducting phase by low-temperature fluorination of La_2CuO_4, Physica C, **167** (1990) 97–101.

[261] A. Tressaud, B. Chevalier, B. Lepine, K. Amine, J.M. Dance, L. Lozano, J. Darriet, J. Etourneau, R. Tournier, A. Sulpice, P. Lejay, J.L. Diot, P. Maestro Influence of low-temperature fluorination process on the characteristics of superconductors $YBa_2Cu_3O_{7-\delta}$ and $Bi_4(Ca, Sr)_6Cu_4O_{16+\delta}$, Solid State Ionics, **32/33** (1989) 1109–1113.

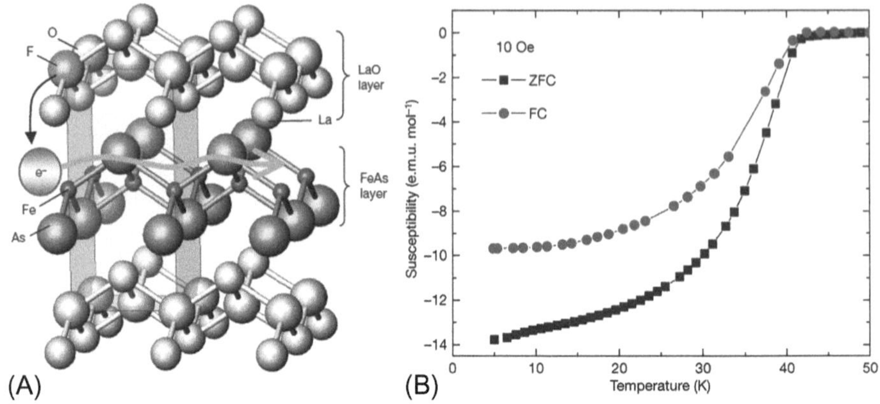

Fig. 52 : (A) Superconducting properties via fluorine insertion in LaOFeAs of tetragonal ZrCuSiAs-type structure. Note that electron carriers generated by F^- doping into oxygen sites are injected into FeAs metallic layers. (B) Temperature dependence of magnetic susceptibility for a sample annealed at 116°C. Data are shown for zero-field cooled (ZFC) and field-cooled (FC) measurements at 10Oe [from 258].

(Fig. 52A). So far, the highest record of T_c is 58K observed in $SmFeAsO_{1-x}F_x$.[262,263] Fig. 52B shows the appearance of an SC state at low temperature.

Superconductivity (SC) and charge-density wave (CDW) are two contrasting yet relevant collective electronic states. In the layered europium bismuth sulfofluoride, $EuBiS_2F$, a CDW-like transition occurs at 280K, below which SC emerges at very low temperature (T_c below 1 K), without any extrinsic doping.[264] Thus, the fluorinated SC $EuBiS_2F$ exhibits an unprecedented behavior that simultaneously accommodates SC, CDW, and *f*-electron valence instability.

5.6 Magnetism and multiferroism in fluorides

5.6.1 Ferro- and ferrimagnetic fluorides

Ionic ferromagnets are rather peculiar objects, and so are the ferromagnetic fluorides.[265] However, the strong ionic character of these compounds allows simplifications in the interpretation of their magnetic coupling, based on the J.B. Goodenough superexchange rules.[266]

[262] X.H. Chen, T. Wu, G. Wu, R.H. Liu, H. Chen, D.F. Fang, Superconductivity at 43 K in $SmFeAs_{O1-x}F_x$, Nature **453** (2008) 761.

[263] M. Fujioka, S.J. Denholme, T. Ozaki, H. Okazaki, K. Deguchi, S. Demura, H. Hara, T. Watanabe, H. Takeya, T. Yamaguchi, H. Kumakura, Y. Takano, Phase diagram and superconductivity at 58.1 K in a-FeAs-free $SmFeAsO_{1-x}F_x$ Supercond. Sci. Technol. **26** (2013) 085023.

[264] H. F. Zhai, Z.T. Tang, H. Jiang, K. Xu, K. Zhang, P. Zhang, J.K. Bao, Y.L. Sun, W.H. Jiao, I. Nowik, I. Felner, Y. K. Li, X.F. Xu, Q. Tao, C.M. Feng, Z.A. Xu, G.H. Cao, Possible charge-density wave, superconductivity, and *f*-electron valence instability in $EuBiS_2F$, Phys. Rev. B **90** (2014) 064518.

[265] J.M. Dance, A. Tressaud, Ferro- and ferrimagnetism in fluorides, Chap. 9, in Inorganic Solid Fluorides, P. Hagenmuller Ed., Academic Press (1985).

[266] J. B. Goodenough, Magnetism and the Chemical Bond, R.E. Krieger Publ. Co. (1963).

In fluorides, ferromagnetism has been observed in several structural types:

- In the $LiREF_4$ type (RE = rare earth), deriving from the scheelite structure, the exchange couplings are small with respect to the magnetic dipole interactions and this point affects the short-range order. Ferromagnets $LiTbF_4$ and $LiHoF_4$ have a very low ordering Curie temperatures $T_C < 4\,K$ (Table 10).
- In layer-type compounds A_2CuF_4 (A = alkaline ion), the network consists of $(CuF_4)_n$ slabs separated by alkaline ions. The ferromagnetism present in these layers is due to an ordering of the d electronic orbitals of Jahn-Teller type Cu^{2+} ions, characterized by the alternation of long and short Cu^{2+}-F bonds. The resulting antiferrodistortive arrangement gives rise to ferromagnetic electron transfers from filled $d(x^2 - z^2)$ orbitals on their empty homologues $d(x^2 - y^2)$.
- A large class of $A^{II}B^{IV}F_6$ fluorides derive from the ReO_3 (and perovskite) structure by a cationic ordering. When both A and B cations are d transition elements, as in $Pd^{II}Pd^{IV}F_6$, the magnetic moments of e_g electrons of the divalent palladium ($t_{2g}^6\,e_g^2$ configuration) can be ferromagnetically coupled via superexchange interactions with the empty e_g orbitals (t_{2g}^6 configuration) of tetravalent palladium. Ordering Curie temperatures and low-temperature magnetization of ferromagnetic fluorides are given in Table 10.

Table 10 **Magnetic data of ferro- and ferrimagnetic fluorides**

Ferromagnetic fluorides		
Compound	**T_C (K)**	**M_{sat} (BM) at 4.2 K**
$LiREF_4$-type		
$LiTbF_4$	2.86	8.9
$LiHoF_4$	1.30	6.98
TbF_3	3.95	8.09
A_2CuF_4-type		
K_2CuF_4	6.25	J/k = 11.2 K
Rb_2CuF_4	17.5	J/k = 8.8 K
Cs_2CuF_4	9 0.8	J/k = 8.3 K
$A^{II}B^{IV}F_6$-type		
$NiPdF_6$	6.0	1.42
$NiPtF_6$	8.0	1.10
Pd_2F_6	10.0	1.80
$PdPtF_6$	25.0	1.70
Pt_2F_6	16.0	1.60
$NiMnF_6$	39	4.55
$ZnMnF_6$	9.5	2.87
$CdMnF_6$	8	2.34

Ferrimagnetic fluorides		
Compound	**T_C (K)**	**M_{sat} (BM) at 4.2 K**
Hexag. polytypes perovskite (6H)		
$CsFeF_3$	60	1.31
$RbNiF_3$	140	0.76
$TlNiF_3$	150	0.42 (at 77 K)
Chiolite		
$Na_5V_3F_{14}$	21	1.94
$Na_5Cr_3F_{14}$	18	2.97
$Na_5Fe_3F_{14}$	90	4.98
$Na_5Fe_2CoF_{14}$	94	4.33
$Na_5Co_3F_{14}$	108	3.10
Weberites		
Na_2FeFeF_7	84	0.75
Na_2CoFeF_7	80	1.2
Na_2NiFeF_7	90	1.5
Na_2NiCoF_7	126	0.9
Other types		
$NaMnCrF_6$	21	6
$LiMnCrF_6$	23	4.1
$CrTiF_5$	26	1.7
$CrVF_5$	40	0.8

Ordering (Curie or Néel) temperatures; low-temperature saturation magnetization (BM); exchange constants (K).

Besides ferromagnetism, numerous other types of magnetic behavior have been observed. When there is no compensation between the different magnetic subnetworks, ferrimagnetism can be generated. Several fluoride series of ferrimagnets include:

- Chiolite type ($Na_5Fe_3F_{14}$);
- Weberite-type ($Na_2\ M^{II}M^{III}F_7$), M being a 3d transition element;
- Hexagonal polytypes of perovskite (AMF_3), such as the so-called 6H type of $RbNiF_3$, $TlNiF_3$, $CsFeF_3$; the 12R-type of $CsM_{1-x}Mn_xF_3$, with M = Ni or Co and $0.5 < x < 0.75$;
- $AMnCrF_6$, with A = Na or Li;
- $M^{II}M^{III}F_5$: $MnAl_{1-x}FexF_5$ with $0 < x < 0.58$ and also $Cr^{II}M^{III}F_5$, with M = Ti, V, Cr. The hydrated form of $Fe^{II}M^{III}F_5{\cdot}2H_2O$, with M = Fe or Co.
- FeF_3 is a particular case, because it is an antiferromagnet with a 3D Néel temperature of 362 K that is above room temperature. Because the compound exhibits also a weak ferromagnetism due to a canting of the magnetic moments, a moment is observed up to the Néel temperature that makes FeF_3 one of the few transparent magnetic materials with a ferromagnetic component above room temperature.

5.6.2 Other types of magnetic behaviors

Many other observed magnetic behaviors of inorganic fluorides include antiferromagnetism, idle-spin magnetism, spin-flop, frustration effects, and low-dimensional magnetism (clusters, 1D, or 2D types).[267] Moreover, as illustrated by the one-dimensional $Zn(viz)_4[IrF_6]$ system, with viz. = 1-vinylimidazole, Ir^{IV}-based $[IrF_6]^{2-}$ species appears to be a robust building-block for the construction of molecule-based architectures. This type of compounds may open perspectives for using 4d and 5d transition metal fluorides for designing $[ReF_6]^-$ molecule-based magnetic materials, or for creating $[IrF_6]^{2-}$ or $[OsF_6]^{n-}$ ($n = 1, 2$) building-blocks for electronic and magnetic quantum materials synthesized by soft chemistry routes.[268]

5.6.2.1 Pd_2F_6 and $M^{II}M^{IV}F_6$ compounds as piezo-conductive species

Although, the insulating properties of Pd_2F_6 and similar compounds have been confirmed at ambient pressure with resistivity and activation energies are about $10^{10}\ \Omega$ cm and 1.5 eV, respectively, the pressure dependence of the conductivity of Pd_2F_6 shows an increase by six orders of magnitude up to $P = 60$ kbar, with resistivity about $10^4\ \Omega$ cm and activation energies as low as 0.07 eV. When pressure is unloaded, the initial resistivity is recovered after a hysteresis process. The pressure dependence of the energy diagram can be explained by a decrease of the gap between Pd^{II} and Pd^{IV} e_g levels with new e_g^1 Pd^{III} levels that would appear. A hopping mechanism would arise between Pd^{II} and Pd^{III} species and between Pd^{III} and Pd^{IV}. The values of the activation energies observed at high pressures are similar with those of semiconductor oxides showing hopping properties. However, the still-important value of

[267] Ferey, G., Leblanc, M., de Pape, R., Pannetier, J. Competing spn interactions and frustration effects in fluorides, Chapter 10, Inorganic Solid Fluorides, Hagenmuller, P. (Ed.), Academic Press Inc. (1985).

[268] K. S. Pedersen et al., Iridates from the molecular side, Nature Commun. 7:12195 | https://doi.org/10.1038/ncomms12195, (2016).

resistivity (10^4 Ω cm) could be explained by a lower mobility of the localized electrons. The material could be proposed as a pressure sensor because of its linear response of resistivity variation in large pressure ranges.[269]

5.6.3 Multiferroic fluoride materials

In multiferroic materials, two so-called "ferroic" orders coexist, that is, ferromagnetism, ferroelectricity, and/or ferroelasticity. Among these phases, magnetoelectric materials, that is, magnetically and electrically polarizable, constitute the most promising subclass, allowing the realization of multifunctional devices. Among these, electric-field–controlled magnetic memories and frequency-dependent devices, such as filters, oscillators, phase shifters and transducers, as well magnetic (and electric) field sensors, are notable.

Fluorides represent an interesting group of potential multiferroic compounds because on one hand, the high electronegativity of fluorine with respect to oxygen gives rise to different hybridization properties and a stronger tendency to induce polar properties. On the other hand, the replacement of fluorine for oxygen lowers the negative charges and allows the replacement of high valency, d^0 transition metals by lower valency transition metals carrying unpaired electrons.[270]

Two major classes of multiferroic transition metal fluorides have been discovered: the $BaMF_4$ series (M = Mn, Fe, Co, Ni) and the tetragonal tungsten bronze structure (TTB) fluorides of formula $K_xM^{II}_xM^{III}_{5-x}F_{15}$. Based on symmetry analysis, the presence of a large magnetoelectric coupling in $Ba_2Ni_7F_{18}$ was predicted, but no direct observation could confirm this hypothesis.[271]

5.6.3.1 The $BaMF_4$ series

The structure can be schematized as being composed of diagonal slabs of a perovskite-type lattice. The Mn^{2+} compound, $BaMnF_4$ forms a modulated structure, whereas the other transition metal compounds do not, this result being accounted for on the basis of simple geometrical considerations. The polar distortion in $BaNiF_4$, chosen as representative of this series of fluorides, gives rise to a Dzyaloshinskii-Moriya interaction, and the sign of this interaction is determined by the orientation of the ferroelectric polarization, which can be inverted by an external electric field. This ferroelectricity-induced magnetic ordering that is intimately coupled to the electric polarization, could induce a controlled switching of the magnetic order parameter by switching the polarization using an electric field. This mechanism that would be indeed a very promising route toward magnetoelectric applications has not been unfortunately confirmed by any direct observation. It should be noted

[269] F. Langlais, G. Demazeau, J. Portier, A. Tressaud, P·Hagenmuller, Mise en évidence sous hautes pressions d'une transition électronique dans le fluorure Pd_2F_6, Solid State Commun., **29** (1979) 473.

[270] Scott, J.F., Blinc, R. Multiferroic magnetoelectric fluorides: why are there so many magnetic ferroelectrics? J. Phys.: Condens. Matter **23** (2011) 113202.

[271] Nénert, G., Palstra, T. T. M., Prediction for new magnetoelectric fluorides. J. Phys.: Condens. Matter **19** (2007) 406213.

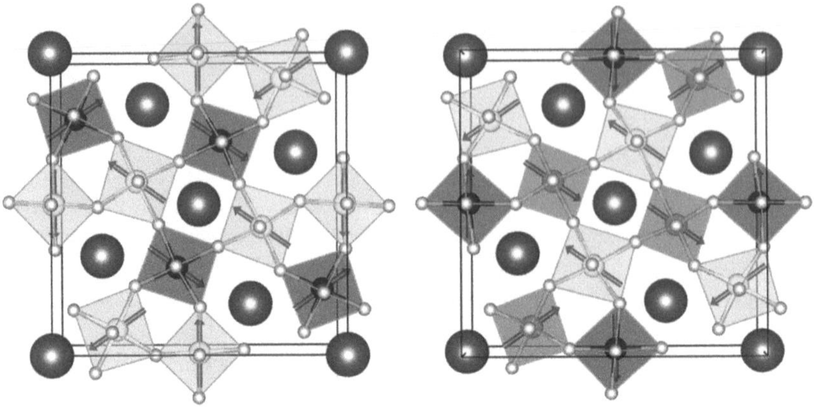

Fig. 53 Magnetic structure of $K_3Fe_5F_{15}$. Dark and light coordination octahedra indicate Fe^{2+} and Fe^{3+}, respectively, intermediate color displays mixed occupancy. The unbalanced charge distribution between the two layers gives the major contribution to the magnetic resultant [from 279].

that, because their optical transparency is much higher than that of oxides, these compounds have been recently used as optical materials, such as $BaZnF_4$.[272]

5.6.3.2 The TTB fluorides series $K_xM^{II}_xM^{III}_{5-x}F_{15}$

The TTB structure is well known in oxide materials. Its general formula can be written as $(A_1)(A_2)_2C_2B_5O_{15}$ and can give rise, as perovskite, to very large possibilities of chemical substitutions. The B octahedral site can be occupied by a wide number of transition metals, whereas alkaline elements, lead, or rare earths are located in the A1 and A2 sites, which constitute along *c* axis channels of square or pentagonal sections (see Fig. 53). On the contrary, the C site (with triangular section) is generally empty and can be occupied only by the smaller Li^+ ion. Because of many possible cationic substitutions, today, the TTB series comprises a large number of functional materials exhibiting electro-optic, ferroelectric, ferroelastic, pyroelectric, and piezoelectric properties.

Some cases of "complete" multiferroics are observed in the TTB fluorides family, displaying the coexistence of ferroelectric, ferromagnetic, and ferroelastic properties, in addition to charge order related to the presence of transition metal cations in different oxidation states.

The corresponding fluoride-bronze compounds have a general formula A_xMF_3 ($0.0 \leq x \leq 1$), (A = alkaline; M = d transition metals). These series were discovered by R. De Pape.[273] Several structural types occur within the composition domain: TTB, HTB (hexagonal tungsten bronze), and perovskite- and ReO_3-related types.[274]

[272] Villora, E.G., Shimamura, K., Jing, F., Medvedev, A., Tekekawa, S., Kitamura, K., Ferroelectric and optical properties of single crystal $BaZnF_4$. Appl. Phys. Lett. **90** (2007) 192909.

[273] R. De Pape, Une série de composés fluorés de type bronze, C. R. Acad. Sci. **260** (1965) pp. 4527–4530.

[274] A. Tressaud, F. Menil, R. Georges, J. Portier, P. Hagenmuller, Propriétés magnétiques des composés de type "Bronzes fluorés" M_xFeF_3 (M = K, Rb, Cs), Mat. Res. Bull., Linus Pauling Issue, **7** (1972) 1339–1345.

The TTB-type fluoride series can be written as the oxide series, $K_xM_x^{II}M_{5-x}^{III}F_{15}$ ($2.0 \leq x \leq 3.0$), pointing out the mixed valence character of the system. In addition, several structural phase transitions were evidenced by I.N. Flerov.[275] Magnetism is introduced in TTB structure by the substitution of fluorine for oxygen, lowering the oxidation state of the d elements with the consecutive presence of unpaired electrons. The substitutions in the M sites are very large: M = Fe, V, Cu, Co/Fe, Mn/Fe, Mn/Cr, and Fe/Cr.[276]

The structure of TTB has been accurately determined on untwinned tetragonal $K_{2.65}Fe_5F_{15}$ single crystal.[277] Two distinct modulations, charge ordering and cooperative tilting, involving the $[MF_6]$ octahedra coexist within the structure. While the charge order is peculiar to these mixed-valency fluorides, the nature of the tilt modulation involving the $[MF_6]$ octahedra is clearly related to those of corresponding TTB Nb-based oxyniobates. This gives rise to the doubling of the *c* axis leading to a charge order superstructure.

Analogous to the TTB oxides, $K_xM_x^{II}M_{5-x}^{III}F_{15}$ compounds crystallize in non-polar space groups, related to the inversion symmetry breaking due to the off-centering of the M ions within their coordination octahedra. The switching of the electrical polarization is expected to be possible in these systems. Ferroelasticity origins from the cooperative tilting of the MF_6 octahedra within two possible variants giving rise to domains, which are suggested to couple not only to polar distortions but also to the magnetic degree of freedom.[278]

The appropriate choice of the M^{II}/M^{III} couples and specific ratios can allow tuning of the magnetic behavior, ranging from AFM to strongly frustrated ferromagnetism and modulating the ordering temperatures, but also yielding more effective magnetoelectric coupling coefficients by adapting the engineering concept of the systems.[279]

5.6.3.3 Other multiferroic ABF_3 fluoroperovskites

Fluoro-perovskite ABF_3 have also generated great interest for their coupling properties between magnetism and ferroelectricity. $NaMnF_3$ exhibits a orthorhombically distorted perovskite (of the $GdFeO_3$ type) with the presence of a spin-canting in the ground antiferromagnetic phase. Multiferroic signatures had been predicted under stress. Recently, this fluoroperovskite has been found to be ferroelectric under epitaxial strain becoming a promising alternative to conventional oxides for multiferroic

[275] M.V. Gorev, I.N. Flerov, A. Tressaud, M.S. Molokeev, A.V. Kartashev, E.I. Pogoreltsev, O.A. Bayukov, Phase transitions in fluoride KFe_2F_6 with TTB structure, J. Fluor. Chem. **168** (2014) 204–211.

[276] Blinc, R., Cevc, P., Potočnik, A., Žemva, B., Goreshnik, E., Hanžel, D., Gregorovič, A., Trontelj, Z., Jagličić, Z., Laguta, V., Perović, M., Dalal, N.S., Scott, J.F., Magnetic properties of multiferroic $K_3Cr_2Fe_3F_{15}$. J. Appl. Phys., **107** (2010) 043511.

[277] S. Fabbrici, E. Montanari, L. Righi, G. Calestani, A. Migliori, Charge Order and Tilt Modulation in Multiferroic $K_xM_x^{II}M_{1-x}^{III}F_3$ ($0.4 < x < 0.6$) Transition Metal Fluorides with Tetragonal Tungsten Bronze Structure, Chem. Mater., **16** (16) (2004) pp. 3007–3019.

[278] D. Pajić, Z. Jagličić, Z. Trontel, Slow magnetic dynamics in the $K_3M_3^{II}M_2^{III}F_{15}$ multiferroic system, J. Appl. Phys. **112** (2012) 073908; https://doi.org/10.1063/1.4757006.

[279] G. Calestani, F. Mezzadri, Multiferroism in fluorides, Photonic and Electronic Properties of Fluoride Materials, Chapter 14, pp. Tressaud, K. Poeppelmeir, Eds, Elsevier, 2016.

applications.[280] Quasi-epitaxial thin films of $NaMnF_3$ were grown directly onto $SrTiO_3$ substrates. Piezoresponse force microscopy has shown a preferred polarization pointing out of the plane was found in the as-grown state. This compound, which exhibits distorted geometric effects due to the displacements of Na cations from the center of the XII coordination number site (A site of the perovskite), is expected to generate at room temperature a ferroelectric instability under strain leading to a stable polar ground state where the latter instability is condensed. Such types of result, together with the magnetic order previously reported, open new ways to future multiferroic and magnetoelectric investigations in fluoroperovskites.

Another fluoroperovskite, $CsPbF_3$, is the only experimentally synthesized ABF_3 fluoride with a polar ground state. The lone pair cation Pb^{2+} on the B-site serves to stabilize a polar ground state, analogous to the role of lone pair cations on the A-site of oxide perovskites. A generalization of these results, might be extended to the oxide perovskites, as well as to the organic-inorganic hybrid halide perovskite photovoltaics.[281,282]

6 Contributions of fluorinated products in medicine, pharmacy, and biotechnologies

If we ask the question today, "Where do we find fluoride around us?" The first answer is obviously, "in the toothpaste." But do we know that fluoride and fluoride products are involved in topics of primary importance to humans, in many aspects of our everyday lives, and in our health and well-being? Indeed, in addition to the "teflonned" utensils found in kitchens, fluorinated products are used as substitutes for cardiovascular surgery, bioceramic prostheses, anti-caries products, artificial blood, molecules for the treatment of cancer, malaria, and many other diseases, early detection of Alzheimer's disease, etc.

6.1 *Fluorinated products in biochemistry and medicinal chemistry*

Given the importance of the subject, many recent books on the biomedical applications of fluoride molecules are available (*see titles of more specialized books at the end of the book*). Beside the biomaterials that will be discussed later, we can distinguish two main categories of fluorinated molecules for biomedical applications: molecules containing one or more fluorine atoms and used as drugs or as phytosanitary products and those that greatly assist with medical diagnosis, using advanced biomedical techniques.

[280] M. Yang, Amit KC, A. C. Garcia-Castro, P. Borisov,E. Bousquet,D. Lederman, A. H. Romero, C. Cen, Room temperature ferroelectricity in fluoroperovskite thin films, *Scientific Reports* **7** (2017) Article number: 7182, doi:https://doi.org/10.1038/s41598-017-07834-0.

[281] Smith EH, Benedek NA, Fennie CJ. Interplay of Octahedral Rotations and Lone Pair Ferroelectricity in $CsPbF_3$, Inorg Chem. **54**(17) (2015) 8536–43.

[282] R. Xi Yang, **J.** M. Skelton, E. Lora da Silva, **J.** M. Frost, A. Walsh, Spontaneous Octahedral Tilting in the Cubic Inorganic Cesium Halide Perovskites $CsSnX_3$ and $CsPbX_3$ ($X = F$, Cl, Br, I)**,** J. Phys. Chem. Lett. ***8*** (19) (**2017**) 4720–4726.

6.1.1 New drugs

Since the beginning of the last second half of the 20th century, there has been a real explosion in the use of fluoride molecules for applications in medicine. In the 1950s, only a few examples of fluoride products were used: fluoridated anesthetics, fluorouracil with antitumor properties, and especially fluorocorticoids discovered by J. Fried.[283] The study of these fluorocorticoids showed that the presence of a single fluorine atom in a molecule could profoundly modify the behavior of this molecule vis-à-vis its biological environment (see above, Chapter 1: "Advances in organic fluorine chemistry"). The rapid growth of the medicinal chemistry of fluorinated molecules can be largely illustrated by the exceptional number of fluorine-containing drugs currently on the pharmaceutical market. Thus, the percentages of fluorinated compounds rose from 2% in 1970 to 8% in 1980, 13% in 1990, and 18% at the beginning of the 21st century, with 6 products in the "top-12", in a wide variety of medical fields (anticancer, anti-inflammatory, analgesic, or antidepressant). It should be emphasized that fluorinated molecules are also very present in the field of biocompatible materials, and in the phytosanitary field. Thus, in agrochemicals, they represented 3% in 1970 to about 50% of the market today.

Fig. 54 shows some examples of the most used molecules in medicinal chemistry.

- 5-Fluorouracil is an inhibitor of cell division (anticancer): it inhibits the biosynthesis of DNA and RNA.
- Antibiotics also include Cipro, which is part of the family of fluoroquinolones, which exhibit a better tissue and intracellular diffusion with respect to quinolones.
- In the field of anesthetics, fluorinated products such as halothane ($CF_3CHClBr$), enflurane ($CHFClCF_2OCF_2H$), isoflurane ($CF_3CHClOCF_2H$), sevoflurane ($HC(CF_3)_2OCH_2F$), and desflurane ($CF_3CHFOCF_2H$) have replaced ether or chloroform because of their greater safety. The benefits of these agents are multiple; they are absorbed and excreted by the lungs, their action is brief, and their metabolization has minimal effects. They are, therefore, specific and very easy to use.
- Vinflurine is a molecule for anti-cancer treatment (Pierre Fabre Co.). It derives from navelbine originally extracted from Madagascar periwinkle (*Catharanthus roseus*) by P. Potier in 1995. The alkaloids contained in the plant are particularly effective for breast cancer.
- An example of an enzyme inhibitor is difluoromethylornithine (eflornitine). It is an irreversible inhibitor of ornithine decarboxylase, a key enzyme for the biosynthesis of polyamines. Initially studied for the treatment of African trypanosomiasis (Trypanosoma brucei gambiense) at the origin of sleeping sickness, new pharmacological applications have emerged during its clinical use. Eflornithine is thus used as a cream in the treatment of facial hirsutism in women. By blocking ornithine decarboxylase in the bulb of the hair follicle, it slows hair growth.
- Prozac, fluoxetine hydrochloride of formula $C_{17}H_{18}F_3NO{\cdot}HCl$, is one of the best-selling antidepressants on the American and European markets.

[283] Fried J., E.F. Sabo J. Am. Chem. Soc., **76** (1954) p. 1455.

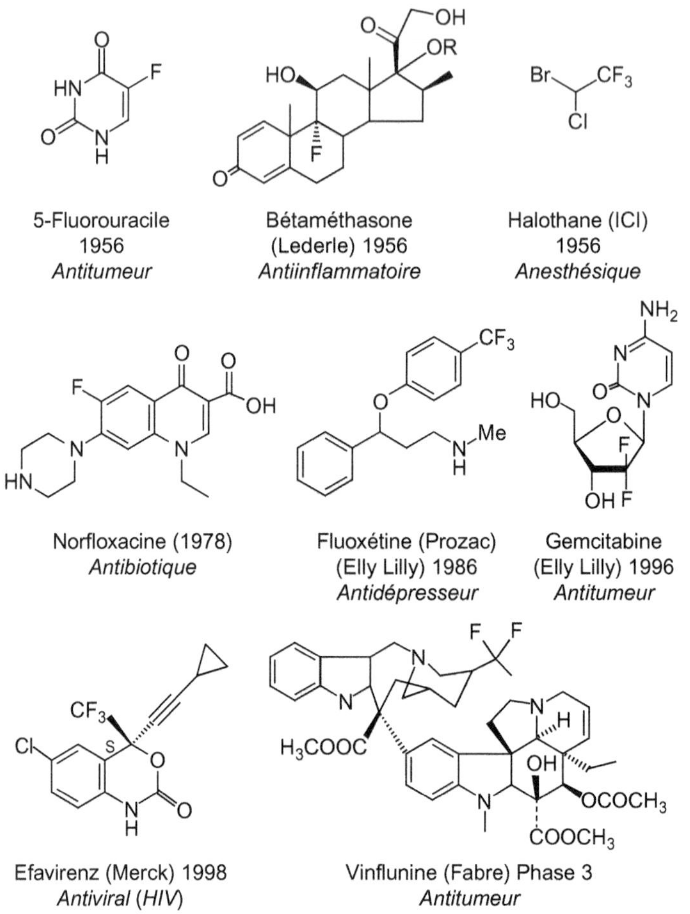

Fig. 54 Some fluorinated molecules important in medicinal chemistry [from 140].

- Flutamide is one of the nonsteroidal antiandrogens. These are drugs that oppose the effect of testosterone, a male hormone. The indication for flutamide (2-methyl-*N*-[4-nitro-3-(trifluoromethyl) phenyl]-propanamide) is the treatment of cancer. In cancer of the prostate, cancer cells need testosterone to grow and multiply, and flutamide inhibits testosterone binding to cytoplasmic or even nuclear receptors.
- These molecules and fluorinated drugs include Mefloquine (anti-malaria), Efavirenz (anti-AIDS treatment), Tafluprost (anti-glaucoma), and Celebrex (anti-osteoarthritis).

The concept of "the fluorine effect" in medicinal chemistry has been developed in France by various groups, particularly by D. Bonnet-Delpon and J.-P. Bégué. Most strategies exploit the physicochemical characteristics of the fluorine atom, that is, its small size and very high electronegativity. Its small size allows a poorly fluorinated molecule to be recognized by the target macromolecule of the natural substrate. On the other hand, its electronegativity gives the C—F bond a very great strength, which results in a chemical inertia greater than that of the C—H bond. Thus fluorine, once substituted to a hydrogen of a bioactive molecule, blocks the metabolically sensitive site. In other strategies, the characteristics of the fluorine atom or fluoroalkyl groups, for example the CF_3 trifluoromethyl group, are used to modulate the physicochemical and structural properties of the molecule. Thus, the electronic distribution, the acidity, the hydrogen bonds, the hydrophobicity, the conformation of the molecule are affected.[284] These modulations are used to increase the affinity of the active ingredient with the target protein and to promote the transport phenomena.

Among the decisive contributions of fluorinated molecules, it can be noted:

- The metabolic stability that determines the bioavailability of the molecule, that is to say, its ability to reach its biological target. Indeed, as soon as it enters the body, the molecule must resist to processes that tend to its elimination, including the various barriers (intestinal, blood brain, cell, etc.). Thus, the replacement of the CH_3 group by a CF_3 group of a muscarinic receptor allowing the parasympathetic regulation of the heart, lungs and digestive system, does not alter the affinity for the receptor, but protects this site from hydroxylation by Cytochrome P450 enzymes.[285] In the case of artemisinin, a potent antimalarial, the hydrolysis processes are slowed down, greatly improving its bioavailability.[286,287] This results in an extraordinary increase in antimalarial activity in vivo.

  The increase in the acidity relative to the presence of fluorine atoms results in the reinforcement of the hydrogen bonds, which can contribute to increasing the affinity of an active substance for its biological target. This is the case, for example, of α-difluoro- or trifluoromethylated alcohols.[283]
- Lipophilicity is a factor that also plays a very important role for the absorption and transport of active ingredients; however, the fluorinated molecules are above all hydrophobic. The introduction of fluorine atoms is, therefore, rather intended to modulate lipophilicity, because too lipophilic molecules are most often excluded from pharmaceutical products because of their poor solubility in water and possible toxicity.
- Specific conformational effects are created by the presence of fluorine atoms inducing multiple factors: sterical effects, hydrogen bonds, hydrophobic effects, dipole-dipole interactions, electronic repulsion. The affinity of an active ingredient for the target protein should be indeed modified. One can cite the increase of the cytotoxicity of taxoids (anticancer), when the phenyl of the isoserine chain is replaced by a CF_3 group.[288]

[284] Böhm H.J., Banner D., Bendels S., Kansy M., Kuhn B., Müller K., Obst-Sander U., Stahl M, Chem.Bio. Chem. **2004**, *5*, p. 637.

[285] Mitch C.H., Brown T.J., Bymaster F.P., Calligaro D.O., Dieckman D., Merrit L., Peters S.C., Quimby S.J., Shannon H.E., Shipley L.A., Ward J.S., Hansen K., Olesen P.H., Sauerberg P., Sheardown M.J., Sweedberg M.D.B., Suzdak P., Greenwood B., J. Med. Chem., **40** (1997) p. 538.

[286] (a) Ford Jr H., Siddiqui M.A., Driscoll J.S., Marquez V.E., Kelley J.A., Mitsuya H., Shirasaka T., J. Med. Chem., **38** (1995) p. 1189. (b) Jeannot F., Mathé C., Gosselin G., Nucleosides, Nucleotides & Nucleic Acids, **20** (2001) p. 755.

[287] Chorki F., Grellepois F., Ourévitch M., Crousse, B., Charneau S., Grellier P., Charman W.N., McIntosh K.A., Pradines B., Bonnet-Delpon D., Bégué J.P., J. Med. Chem., **47** (2004) p. 1423.

[288] Ojima I., Kuduk S.D., Slater J.C., Gimi R.H., Sun C.M., Chakravarty S., Ourévitch M., Abouabdellah A., Bonnet-Delpon D., Bégué J.P., Veith J.M., Pera P., Bernacki R.J., Biomedical Frontiers of Fluorine Chemistry, I. Ojima, J.R. Mc Carthy, J.T. Welch Eds, ACS Symp. Ser. 639, Washington DC (1996), pp. 158–181.

Table 11 **Properties of the usual positron-emitting atoms**

Isotope	Half-life (min) ($t1/2$)	Specific radioactivity (GBq/mol)	Emission β^+ (%)	Maximum kinetic energy β^+ (keV)	Average distance in water (mm)
^{15}O	2	3.4×10^{12}	99.9	1723	8.2
^{13}N	10	6.9×10^{11}	100	1190	5.4
^{11}C	20	3.4×10^{11}	99.8	961	4.1
^{18}F	110	6.3×10^{10}	96.9	635	2.4

6.1.2 New ^{18}F fluoride-based radiotracers for positron emission tomography (PET imaging)

Positron emission tomography (PET) is certainly one of the most powerful medical imaging techniques currently available. Its development is linked to molecules marked by a positron-emitting atom (β^+). This is the case of the radioisotope of fluorine, ^{18}F. The nuclear reaction that is most widely used to obtain, in a cyclotron, the artificial isotope of fluorine is ^{18}O (p, n) ^{18}F. The target consists of water enriched in ^{18}O, a stable isotope (0.1% isotopic abundance). The fluorine is obtained in the form of fluoride ion $^{18}F^-$ in aqueous solution. It is the radionuclide of choice because of its relatively long lifetime (109.8 min) and the relatively low energy of the emitted positron. Table 11 compares the advantages of ^{18}F with respect to various conventional positron emitters. When it disintegrates, ^{18}F emits a positron. It travels in the body to a distance depending on its initial kinetic energy and during the final shock with an electron there is an annihilation reaction where the two particles, the positron and the electron, disappear to give birth of two photons of equal energy (511 keV), emitted at 180 degrees from each other. It is the detection of these two radiations that forms the basis of PET (Fig. 55). The retroprojection of all the coincidence lines, resulting from the radioactive events, leads to tomographic sections of the whole body or organs studied (brain or heart, for example).

Fluor-18 has been used for the labeling of many relatively simple or complex molecules of biological interest,[289] including high molecular weight macromolecules such as peptides, proteins or oligonucleides. This chemistry, however, presents several difficulties. First, the isotopic labeling is theoretically limited to molecules that have the possibility of accepting a fluorine atom in their structure (but this is the case of nearly one molecule of therapeutic interest out of three). If so, it is then possible to substitute one of the atoms of the pristine non-fluorinated molecule. The biological properties and the safety of this new molecule must be checked because we have seen that, generally, the introduction of a fluorine atom could lead to drastic modifications of physical properties. The synthesis, final purification and formulation of the radiopharmaceutical,

[289] F. Dollé, C. Perrio, L. Barré, M.C. Lasne et D. Le Bars, Fluorine 18 labelled molecules: Synthesis and application in medical imaging, L'actualité chimique – N° **301–302**, pp. 93–98 (2006). F. Dollé, D Roeda, B. Kuhnast and M.C. Lasne, Fluorine-18 Chemistry for Molecular Imaging with Positron Emission Tomography, pp. 3–66, Fluorine and Health, A.Tressaud, G. Haufe Eds., Elsevier B.V. (2008).

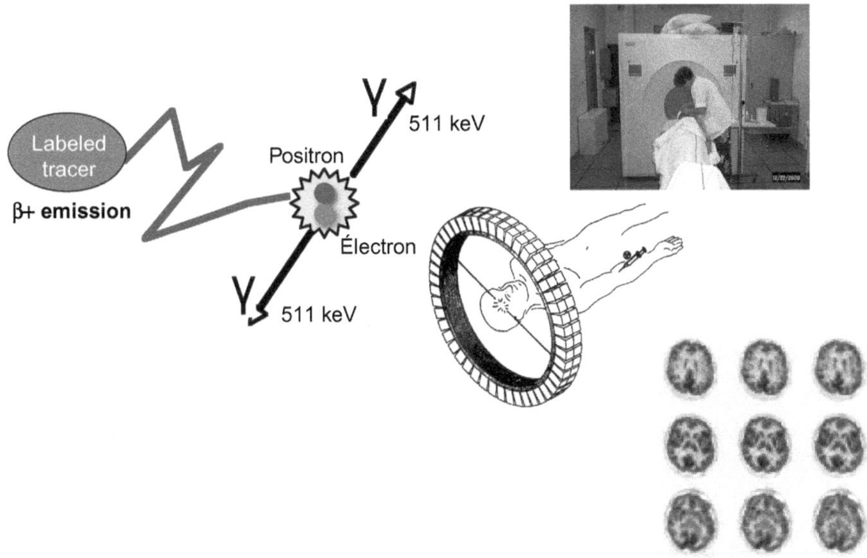

Fig. 55 Principle of positron emission tomography (PET).
From Dollé, F., Perrio, C., Barré, L., Lasne M.C., Le Bars, D., Fluorine 18 labelled molecules: Synthesis and application in medical imaging, L'actualité chimique – N° **301-302** (2006) pp. 93–98.

which is generally administered intravenously, must be carried out in the shortest possible time, with the incorporation of the radioisotope as late as possible. The introduction of ^{18}F into a molecule can be carried out either by electrophilic reaction from molecular fluorine $[^{18}F]\ F_2$ or by nucleophilic reactions from fluoride ion $[^{18}F]\ F^-$.[290,291]

In its electrophilic form ($[^{18}F]\ F_2$ or derived reagents), fluorine-18 generally substitutes a hydrogen or a metal carried by a vinyl or aromatic carbon. The process for obtaining fluoro-L-DOPA is given in Fig. 56. This molecule makes it possible to measure the storage, degradation and turnover of intracerebral dopamine. This dopamine plays an important role in the control of movement (Parkinson's disease), in psychosis and schizophrenia.

The nucleophilic form, generally an alkaline fluoride, for example, $K[^{18}F]F$ is by far the most common in aliphatic or (hetero) aromatic substitution reactions (see Fig. 57). The most commonly used molecule is 2-$[^{18}F]$ fluoro-2-deoxy-D-glucose (denoted $[^{18}F]$ FDG),[292] which was marketed in 1998 and is used in tens of thousands

[290] Welch M.J., Redvanly C.S., Handbook of Radiopharmaceuticals - Radiochemistry and Applications. Welch MJ, Redvanly CS (ed). Wiley-Interscience Pub, New York (2003), p. 1.

[291] Okarvi S.M., Recent progress in fluorine-18 labelled peptide radiopharmaceuticals. Eur. J. Nucl. Med., *28* (2001) p. 929.

[292] Hamacher K., Coenen, H.H., Stoecklin G., Efficient stereospecific synthesis of NCA 2-$[^{18}F]$fluoro-2-deoxy-D-glucose using aminopolyether supported nucleophilic substitution. *J. Nucl. Med.*, **27** (1986) p. 235.

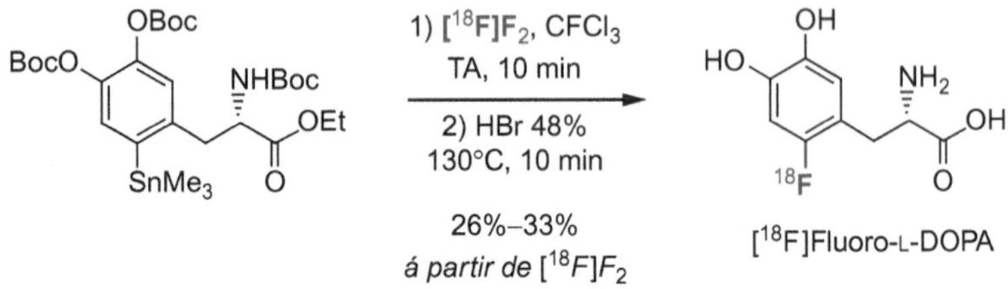

Fig. 56 [^{18}F] electrophilic fluorination: the example of fluoro-L-DOPA (3,4-dihydroxyphenylamine).

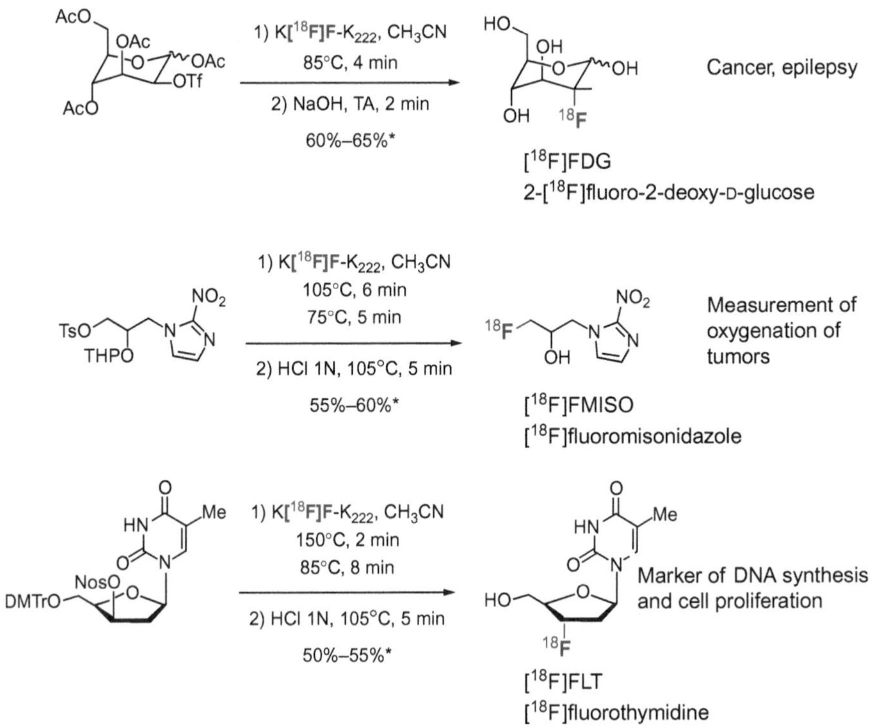

Fig. 57 Aliphatic nucleophilic radiofluorination via K [^{18}F] F.

of exams per year in France. This glucose (sugar) is transformed in the cell into [^{18}F] FDG-6-phosphate. However, due to the presence of fluorine at the 2-position, [^{18}F] FDG-6-phosphate is not metabolized and accumulates in the cell at a rate proportional to glucose uptake, which directly reflects the need energy of the concerned cell. All areas in which a significant variation in carbohydrate metabolism occurs, as is the case in tumors whose energy consumption is exacerbated by cell proliferation, are clearly highlighted. [^{18}F]FDG is thus used for the diagnosis of cancers (lung, intestine, upper aerodigestive tracts, lymphoma, melanoma) and more recently, of drug-resistant epilepsy.

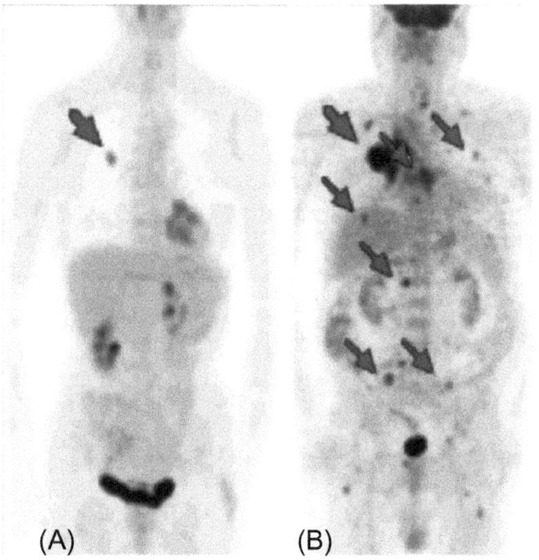

Fig. 58 Effect of [^{18}F] FDG on tumor cells (A) small isolated tumor, (B) larger metastasized tumor. Note the physiological fixation of [^{18}F] FDG in the brain, heart, kidney, liver, bladder [from 289].

Other molecules are efficient, for example [^{18}F] DCFBC for the detection of prostate cancer.[293,294] In Fig. 58 is a comparison of two PET images: either an isolated tumor or a tumor giving rise to metastases. This type of information makes it possible to refer the patient to the most appropriate therapy: either surgery or radiotherapy coupled with chemotherapy. Myocardial dysfunction can also be followed by this technique. Alzheimer's disease is a neurodegenerative disease of the brain tissue that causes a progressive and irreversible loss of mental function. Today, it affects many elderly people. Although there is no cure for the disease, some medicines such as anticholinesterase inhibit the progression of the disease by reducing memory loss, language, and reasoning. PET scans, combined with MRI techniques and single photon emission computed tomography (SPECT)using ^{99m}Tc, allow early detection of the disease, which results in decreased metabolism of several parts of the brain (temporal, parietal and posterior lobe, especially the hippocampus) and also to monitor in vivo the effectiveness of new treatments. In various chapters of the book, "Fluorine and Health" (Elsevier, 2008), data on the use of [^{18}F] FDG and other currently tested labeled molecules can be found.

6.1.3 Fluorinated products for magnetic resonance imaging (MRI)

Because of its high sensitivity in nuclear magnetic resonance, the ^{19}F nucleus provides a valuable tool for nuclear magnetic resonance imaging (MRI).[295] The localization and concentration in the body of fluoride drugs such as the anti-tumor 5-fluorouracil,

[293] Gouverneur V., Proceedings ICFC Congress, May 2009, Kyoto, Japan.

[294] Mease R. C. et al., Clinical Cancer Research 14 (2008) 3036–3045.

[295] J.X. Yu, W. Cui, D. Zhao, and R. P. Mason, Non-Invasive Physiology and Pharmacology Using ^{19}F Magnetic Resonance, pp. 197–278, Fluorine and Health, A.Tressaud, G. Haufe Eds., Elsevier B.V. (2008).

the anesthetic gemcitabine, or the antidepressant fluoxetine could be followed, and so the retention in the body of fluorinated products from pollution, poisoning, or the possible presence of products used in agrochemicals. In the human body, fluorine is present in small amounts and in the state of mineral fluoride only; it gives a very wide NMR signal. Thus, any molecule containing fluorine (drug, probe molecule, external pollution) and present in the body or in the excretions can be detected. It is obviously a matter of being very attentive to the toxicological risks, and of preventing degradation products such as mono- or difluoroacetate to be formed. The trifluoromethyl groups CF_3 are perfectly suitable because they are better resistant to degradation, for example, the mobility in the liver and bladder of the anti-cancer Gemcitabine (Gemzar) which is used in addition to radiotherapy, the location of perfluorocarbons used to oxygenate tissues, or fluorinated derivatives of vitamin B6 which can serve as pH probe in vivo.

6.1.4 Photodynamic therapy (PDT)

A new treatment for cancer and tumors has recently appeared: photodynamic therapy (PDT). This method involves a photosensitizer and a light source leading to the production of oxygen-fluorinated derivatives such as 4,4-difluoro-4-bora-3a,4a-diaza-s-indacene (BODIPY) act as photosensitizers by direct contact at the level of the tissues to be treated, making it possible to generate, under a suitable radiation, active species: hydroxyl radicals ($^{\bullet}OH$), superoxides (O_2^-) and more especially, the singlet form of $^1O_2$ dioxygen ($^1\Delta_g$ configuration). These species react very locally with the components of the cells to be treated such as unsaturated lipids, amino-acid residues, and nucleic acids. Among the different fluorescent dyes, boron-fluorine derivatives (typically 4,4-difluoro-4-bora-3a,4a-diaza-s-indacene, BODIPY) have a strong ability to generate singlet oxygen.[296]

LaF_3:Ce,Tb nanoparticles have been biofunctionalized, with a novel surface coating that makes them stable in cell culture media. Once introduced into cells, these nanoparticles are expose to therapeutic doses of x-rays, and colony-forming assays are used to quantify inhibition of cell reproduction resulting from the radiation.[297]

6.2 Perfluorocarbons in medicinal technologies: Contrast agents, artificial blood, retinal surgery

Perfluorocarbons (PFCs), that is, compounds wherein all C–H bonds have been replaced by C–F bonds, include perfluoroalkanes, fluoroalkenes, fluoroalkynes, and perfluoroaromatic compounds. They are characterized by an extreme contrast

[296] L. Yao, S. Xiao, F. Dan, Boron-Fluorine Photosensitizers for Photodynamic Therapy, J. Chem. 2013, Article ID 697850.

[297] J. Nadeau, K. Kudinov, D. Bekah, D. Cooper, S. Bradforth, Lanthanum fluoride particles to enhance radiation therapy, SPIE, Newsroom, https://doi.org/10.1117/2.1201604.006430, (2016).

between the strength of their *intra*molecular C–C and C–F bonds and the weakness of their *inter*molecular interactions. The perfluoroalkylated compounds combine a range of exceptional properties.[298] We have seen that perfluorocarbons and their derivatives are used as fluoropolymers, refrigerants, and solvents. Because of their outstanding chemical and biological inertia, associated with a high gas-dissolution capacity, extreme hydrophobicity and marked lipophobicity they are also used in medicinal technologies. The low polarizability of the fluorine atom results in low van der Waals forces. Consequently, liquid PFCs display very low intermolecular cohesiveness, a feature they share with gases. Low cohesiveness is reflected by the highest vapor pressures (relative to molecular weight), lowest surface tension and lowest water solubility of all liquids, yielding an exceptional property for liquid PFCs to dissolve gases.[298] Because of these behaviors, PFCs have been the basis for an abiotic way for delivering intravascular oxygen.

6.2.1 Contrast agents in ultrasound or magnetic resonance technologies

Their low solubility in the water, combined with a high vapor pressure, makes it possible to stabilize injectable microbubbles which serve as contrast agents in ultrasound or magnetic resonance technologies. These elastic microbubbles act as reflectors and resonators when subjected to an ultrasonic wave field, as experienced by M. P. Krafft (289). While microbubbles of air injected into the blood dissolve very rapidly under the joint action of arterial pressure and Laplace pressure, the introduction of fluorocarbon stabilizes osmotically the microbubbles and significantly prolongs their intravascular lifespan.[299,300] In addition, microbubbles interact with incoming ultrasonic waves and generate harmonics. The manipulation of the incident waves and the filtering of the emission frequency make it possible to retain the only harmonics and to thus release a specific image of the microbubbles and, consequently, of the tissues in which they are located, that is, the cardiac cavities, blood vessels, and capillaries. This results in a considerable amplification of the contrast between healthy and pathological tissues. Thus, the use of such microbubbles in echocardiography makes it possible to achieve a more effective diagnosis of structural or functional cardiac abnormalities (see Fig. 59).

[298] M. P. Krafft, J. G. Riess, Perfluorochemical-Based Oxygen Therapeutics,Contrast Agents, and Beyond, pp. 447–486, in Fluorine and Health, A.Tressaud, G. Haufe Eds., Elsevier, Amsterdam. (2008). M. P. Krafft, J. G. Riess, Chemistry, Physical Chemistry, and Uses of Molecular Fluorocarbon-Hydrocarbon Diblocks, Triblocks, and Related Compounds, Chem. Rev., **109** (2009) 1714–1792.

[299] Nanda N.C., D.W. Kitzman, D.H. Crouse, H.C. Dittrich, G. Hall, Imagent improves endocardial border delineation, Echocardiography. **20** (2003) 151–161.

[300] Schutt E.S., Klein D.H., Mattrey R.M., Riess J.G., Injectable microbubbles as contrast agents for diagnostic ultrasound imaging: The key role of perfluorochemicals, Angew. Chem. Int. Ed., **42** (2003) 3218–35.

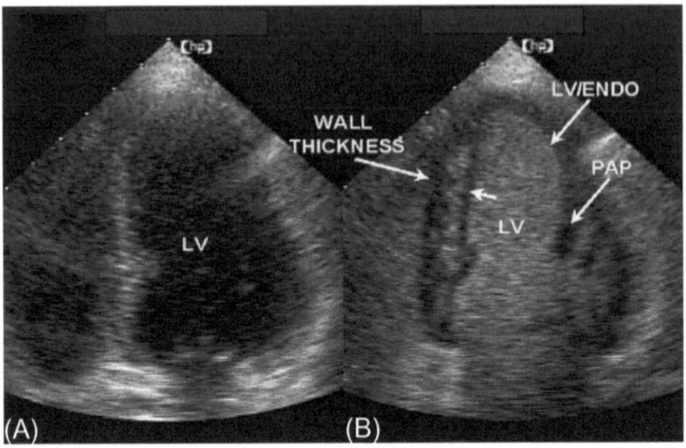

Fig. 59 Opacification effect of the left ventricular cavity: (A) without contrast agent, (B) after injection of microbubbles of a fluorinated contrast agent. The various constituents are perfectly visualized (LV: left ventricle, LV/ENDO: endocardial border, PAP: papillary muscle) allowing evaluation of the thickening of the myocardial heart muscle. Note: when watching the heart in motion, normal functioning heart muscle thickens as it contracts; abnormal functioning heart muscle moves less and does not thicken.
From Krafft, M.P., Riess, J.G., Perfluorochemical-Based Oxygen Therapeutics, Contrast Agents, and Beyond, pp. 447–486, in Fluorine and Health, A. Tressaud, G. Haufe Eds., Elsevier, Amsterdam (2008). Original from N.C. Nanda, D.W. Kitzman, L.J. Crouse, H.C. Dittrich, Imagent improves endocardial border delineation, inter-reader agreement, and the accuracy of segmental wall motion assessment Echocardiography, 20 (2003) pp. 151–161.

6.2.2 PFCs as blood substitutes

The research and development of "blood substitutes," based on sub-micron fluorocarbon emulsions, is continuing[301] that should allow the delivery of oxygen to the tissues and thus reduce the need for blood transfusion surgery. Stable, injectable PFC emulsion has reached Phase III clinical trials. Further clinical trials are being carried out to assess their practical use as blood substitutes. However, for the moment, no effective system which is able to oxygenate tissues and prevent possible ischemia is commercially available. Ongoing clinical research focuses on blood sparing and on the important issue of protecting the organs, especially the intestine, at risk of ischemia during surgery. PFC-based oxygen carriers could also help facing blood shortages.

6.2.3 PFCs as respiratory gas transport agent

These perfluorinated molecules are used as a respiratory gas transport agent in acute respiratory syndrome cases in adults and newborns. Among the emulsions currently undergoing clinical evaluation, a mention should be made of Oxygent, an emulsion

[301] Riess J.G., Injectable oxygen carriers (blood substitutes)—Raison d'être, chemistry, and some physiology, Chem. Rev., **101** (2001) 2797–920. Riess J.G., Krafft M.P., Fluorocarbon emulsions as in vivo oxygen delivery systems: background and chemistry. In: Blood Substitutes, R.M. Winslow (Ed.), Elsevier, Amsterdam (2006), 259–75.

containing 60% by weight by volume of fluorocarbon, mainly perfluorooctyl bromide, $C_8F_{17}Br$, emulsified using phospholipids. The ventilation of the lungs or bronchi by these emulsions allows a gas exchange facilitating the reopening of clogged alveoli and the recovery of respiratory mechanisms in the animal. It should be remembered that the fluorocarbons administered are not metabolized and are excreted with exhaled air. The choice of perfluorooctyl bromide is guided by its slight lipophilicity linked to its polarizable terminal bromine atom, allowing a compromise between the stability of the emulsion and the rate of excretion of the product. It should be noted that the experimentation of blood substitutes to the man raises ethical issues.

Perfluoroalkanes do not bioaccumulate and are rapidly excreted from the body, primarily via expiration, with the rate of excretion as a function of the vapor pressure; the half-life for octafluoropropane (C_3F_8) is $<2\,min$, compared to about a week for perfluorodecalin ($C_{10}F_{18}$).

6.2.4 PFCs in retinal surgery

This same type of highly fluorinated liquid is used in retinal surgery. Various diseases of the natural vitreous humor itself like opacification, hemorrhage, inflammation, as well as retinal detachment, may require its removal by so-called vitrectomy. Primary (idiopathic, rhegmatogeneous) retinal detachment is induced by a retinal tear, followed by penetration of vitreous fluid between the sensoric retina and the pigment epithelium. More frequently, this could be due to a shriveling of the membrane. The vitreo-retinopathies of diabetic origin are characterized by the growth of blood vessels and the vascularization of the retina. During operations to replace damaged parts, or to remove the vitreous when it has been invaded by blood from diabetic neovessels, for example, the vitreous humor must be evacuated (vitrectomy). Perfluorocarbon liquids such as perfluorooctane (C_8F_{18}) and perfluorodecalin ($C_{10}F_{18}$) allow, once vitrectomy is performed, to replace the vitreous humor and, by their high density, to stabilize the retina by pressing against the epithelial wall. Perfluorocarbon products can be used pure or blended with silicone oil, as was the case for the first product on the market: Oxane Hd (Bausch & Lomb Inc., Rochester, NY, United States) which is a mixture of 1perfluorooctyl-5-methylhex-2-en and silicone oil.[302] Denser than water, the specific gravity of this product was $1.02\,g\,mL^{-1}$. The question of the biocompatibility of these products is being investigated.[303] The physical-chemical data of perfluorocarbons used during vitrectomy as intra-ocular tamponade are grouped in Table 12.

6.2.5 PFCs in cosmetics

In cosmetics, because of PFCs' lipophobic, hydrophobic, and high stability properties, they can form a third phase, facilitating emulsion formulation. PFCs are used in color products for enhancing color endurance; they do not interfere with the normal

[302] Rico-Lattes I., J. C. Quintyn, V. Pagot-Mathis, X. Benouaich, and A. Mathis, Fluorinated Molecules in Eye Surgery, pp. 407–420, Fluorine and Health, A.Tressaud, G. Haufe Eds., Elsevier, Amsterdam (2008).

[303] Menz D. H., J. H. Dresp, Biocompatibility of Highly Fluorinated Liquids Used in Ophthalmic Surgery pp. 422–445, in "Fluorine and Health", A.Tressaud, G. Haufe Eds., Elsevier, Amsterdam (2008).

Table 12 **Physical chemical data of perfluorocarbons used during vitrectomy as intra-ocular tamponade**

	Perfluoro-1,3-dimethyl-cyclohexane 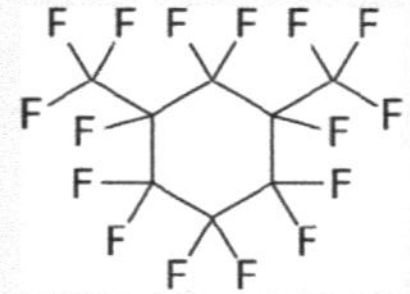	**Perfluoro-*n*-octane** 	**Perfluor-decaline (cis + trans Isomere)** 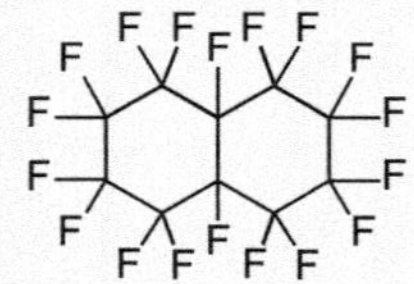	**Perfluoroperhydro-phenanthrene** 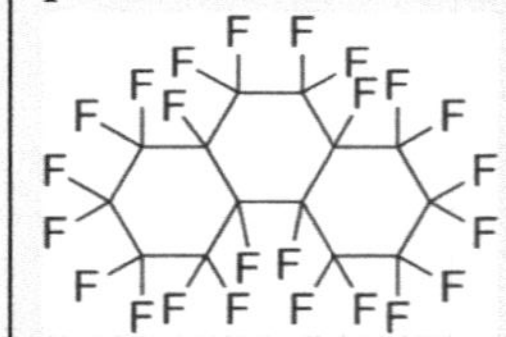
Formula	C_8F_{16}	C_8F_{18}	$C_{10}F_{18}$	$C_{14}F_{24}$
Molecular weight, g/mol	400	438	462	624
Boiling point, °C	102	104	142	215
Melting point, °C	−70		trans >10 cis <−10	−20
Density, g/mL	1.828	1.780	1.917	2.03
Refractive index (20°C)	1.2895	1.270	1.3130	1.3348
Viscosity mPas (25°C)	1.919	1.4	5.53	28.4
Surface tension, mN/m	16.6	14	17.6	19
Vapor pressure, mbar	48	43	8.8	<1
CTSH °C, lipophily	21	37	22	45
O_2-solubility, vol%	–	48–52	40–44	–

functions of the skin and they are non-greasy, non-irritating, non-sensitizing, and odorless. The appearance, smoothness, and softness of the skin is enhanced, and because of their high gas-dissolving ability, they can be loaded with oxygen. Suggested applications include sun creams, anti-aging creams, gel lipsticks, eye contouring products, oxygen masks, etc.

Because of all these exceptional properties, PFC provide new useful tags for molecular recognition, selective sorting and template (e.g., selective pairing of peptide and of nucleic acid). They also stabilize membranes by forming micro- and nanosized fluorous environments. In addition to their uses in biomedical applications as contrast agents, artificial blood, lung surfactant replacement and ophthalmologic aids, highly fluorinated PFCs are useful as a range of inert materials: for example, targeted drug delivery, fluorosilicons, expanded polytetrafluoroethylene, components of contact lenses, reconstructive surgery (e.g., vascular grafts) and other devices.

6.3 Fluorinated compounds as bio-materials

Fluorine-containing bioceramics and derived biomaterials have applications in clinical operations, mostly for bone and dental systems, as shown in Table 13.

6.3.1 Prophylactic effect of fluorine and fluorinated dental biomaterials

The first toothpaste containing fluorine was Fluocaril, created in 1946, containing both fluoride (NaF) and sodium monofluorophosphate (Na_2PO_3F) in its "bi-fluorinated" form. His commercial presentation is also reminiscent of "neon green."

Fluoride is an essential trace element for the mineralization of bone, dentin and tooth enamel, and may be considered the most effective element for the prophylaxis and treatment of dental caries. The major mineral component of bones and teeth is apatite, a calcium phosphate of the general formula $Ca_{10}(PO_4)_6(OH,Cl,F)_2$. As in the natural mineral of the same type discussed in next section, Section 3.1.2, the crystal lattice allows

Table 13 Applications of fluorinated bioceramics and biomaterials

Type of biomaterial	Type of applications	Main effects related to the presence of fluoride
Ceramics Bio-glasses	Bone substitute, bone replacement, bio-engineering of bone tissues	Improvement of the properties of surfaces and interfaces
Ca-P ionic cements	Bone substitute odontological applications	Improved cement strength; biological effects of the release of F-ions; anti-cariogenic
Glass ionomer cement, compomers and other derived materials	Dental filling	Biological effects of the release of F^- ions; anti-cariogenic
Deposits	Deposits on metal prostheses	Improvement of deposition process and surface properties

substitutions between OH^- hydroxyl and F^- fluoride anions and between PO_4^{2-} and CO_3^{2-} anions. Due to its mineralizing character, fluoride ion will lead to better crystallized compounds, and fluoro-hydroxyapatites will be more thermally and mechanically stable than hydroxyapatites. Over a span of 50 years, large-scale campaigns for the fluoridation of drinking water have allowed large populations of people to maintain good dental health. The fluoride ion, whose mineralizing properties are well recognized, has a positive effect on the balance between demineralization and remineralization that occurs on the surface of the tooth, displacing the balance in favor of remineralization. This mechanism intervenes by various processes: decrease in the dissolving properties of saliva vis-à-vis the dental mineral, reduction of the solubility of the apatite while improving its mineralization.[304] But it can also act on the bacterial population by decreasing enzymatic activities. Fluoride is thus distributed to very large populations around the world via international health agencies, and despite controversies led by anti-fluoride pressure groups, its use at the optimized rates does not pose major problems and is effective as for example, the results of the survey given in Table 14. The optimal amount of fluoride in water has been shown to be about 1 ppm.[305,306]

In addition to toothpastes and fluoride-based applications, bioactive ceramics and glasses containing fluoride ions can play an effective role in conservative dentistry. These products restore the physiological balance of the damaged oral cavity, releasing the active ingredients for long periods of time. For example, the ionomer glasses whose composition is given in Table 15, or the resin-modified ionomer glasses, "compomers," are used to seal cavities following caries, alternatively to composites. Whatever the compositions used, the effectiveness of fluoride diffusion has been demonstrated by an improvement of the interfaces, a decrease in the depths of the lesions and a reduction of mineral losses in the dental restorations performed.[307]

Table 14 Rates (%) of damaged, missing, or treated teeth in two regions of Ireland showing the beneficial effects of water fluoridation over a 30-year period [305]

Northern Ireland, Northern part (non-fluoridated water)			**Republic of Ireland, Eastern part (fluoridated water)**		
1963	1983	1996/7	1961–63	1984	1993
5.5	4.4	3.3	4.7	2.6	1.4

304 J. W. Nicholson, B. Czarnecka, Fluoride in dentistry and dental restoratives, pp. 333–378, in "Fluorine and Health", A. Tressaud, G. Haufe Eds., Elsevier, Amsterdam (2008).

305 H.T. Dean, F.A. Arnold Jr and E. Elvove, Additional studies of the relation of fluoride domestic waters to dental caries experience in 4,425 white children aged 12–14 years of 13 cities in 4 states, Public Health Rep., **65** (1942) 1403–1408.

306 H.V. Churchill, Occurrence of fluorides in some drinking waters of the United States, J. Ind. Eng. Chem., **23** (1931) 996–998.

307 J. Arends, G.E.H.M. Dijkmann and A.G. Dijkman, Review of fluoride release and secondary caries reduction by fluoridating composites, Adv. Dent. Res., **9** (1995) 367–376.

Table 15 Composition of glasses used in glass ionomer cements

Components	Proportions (wt%)
SiO_2	24.9
Al_2O_3	14.2
CaF_2	12.8
Na_3AlF_6	19.2
AlF_3	4.6
$AlPO_4$	24.3

6.3.2 *Bioceramics and fluorinated biomaterials*

Fluoride is used in ceramics, composites, or deposits for orthopedic applications. The presence of fluorides facilitates the shaping of these biomaterials, but also gives them a specific biological activity, especially at the bone-implant interface. Fluorides are involved in the composition of bio-glasses and apatites, forming by substitution fluoro-hydroxyapatites, but also the fluorapatite ($Ca_{10}(PO_4)_6F_2$). Because of their greater thermal stability, these compounds make it possible to produce the bioceramics at higher temperatures. The most important biological effect is to promote the adhesion of osteoblasts and their proliferation often leads to the formation of a bone-like deposit on the surface. Their low solubility and biological activity explain a lower resorption rate and greater in-vivo stability, with better implant performance.

In the physiological pH range, fluorapatite is less soluble than hydroxyapatite. In addition, during the synthesis by sol-gel route, it is generally obtained as stoichiometric and its formation is favored in any medium containing Ca^{2+} ions and phosphate groups, thus accelerating the production of cements based on calcium phosphate. Reviews on the various synthetic routes, by C. Rey et al.,[308] and the biological aspects of fluorinated bioceramics have recently appeared,[309,310] focusing on the most important point, which is an osteoblastic proliferation in contact with these materials.

Many glass-ceramic compositions have been tested. The crystallized phase is generally composed of fluorapatite mixed with various silicates or aluminosilicates. For example, in Japan glass-ceramics based on silica and phosphates known under the name "AW GC" (GC for glass-ceramics) are formed of crystalline fluorohydroxyapatite and β-wollastonite ($CaO{\cdot}SiO_2$) in a vitreous matrix. They are used for reconstructions of vertebrae and the iliac crest. These ceramics have a regular structure with small particles and are resistant at the interface with bone tissue.[311] However, it is more difficult

[308] C. Rey, C. Combes, C. Drouet, H. Sfihi, Fluoride-based bioceramics, pp. 279–331, in"Fluorine and Health", A.Tressaud, G. Haufe Eds., Elsevier, Amsterdam (2008).

[309] J.J. Clarkson and J. McLoughlin, Role of fluoride in oral health promotion, Int. Dent. J., **50** (2000) 119–128.

[310] A.D. Wilson and J.W. Nicholson, Acid-base cements, The University Press, Cambridge, 1993.

[311] C. Robinson, S. Connell, J. Kirkham, S.J. Brookes, R.C. Shore, A.M. Smith. The effect of fluoride on the developing tooth, Caries Research, **38** (2004) 268–276.

to use them in contact with metal implants because of their reactivity with this type of surfaces. In Germany, mixtures of fluorohydroxyapatite and phlogopite [(Na,K) $Mg_3(AlSi_3O_{10})F_2$], a type of mica,[312,313] have been clinically tested. Other silicate components have been considered: mica, diopside, carnasite. Glass-ceramics used in powdered form are also used to prepare bio-active cements for bone filling. Deposition by spray or plasma of glass-ceramics on alumina substrates reinforces the bioreactivity of the assembly and improves the ossification of the prosthesis. Various deposits have been made for example on Ti_6Al_4V alloy substrate: Bioverit, or Bioverit composites with titanium particles, or even vitreous fluorapatite-mullite film.[314,315]

6.3.3 Fluorinated silicates in cosmetics

We have seen that nature is an inexhaustible source of lessons in the search for new materials or new features. Lamellar aluminosilicates of the mica type are a striking example. They are used in the composition of bioceramics, glass ceramics for electronics. Depending on the purity level—industrial quality or cosmetic quality—these iridescent and shimmering lamellae can be used either in the manufacture of rubber, tires, paints, mastics, wallpapers, or as components of creams, lipsticks, nail polish, blushes, eye shadow, make-up. Thousands of years ago, beautiful Egyptian women were already using these natural glitters to brighten their complexion. Today, cosmetics companies prefer to use synthetic products such as fluoro-phlogopite or fluormica $K_2Mg_6(Si_3Al)_2O_{20}F_4$ because natural crystals are too sensitive to water absorption. Thus, powders of mica or talc mixed with perfluoroalkylphosphoric acid $(C_8F_{17}CH_2CH_2O)_2P(O)OH$, providing hydrophobic and oleophobic functions are prepared in a formulation containing a fluorinated resin, oil, and a polymer in order to give a cosmetic product known for its softness, and bringing transparency and irradiance.

6.3.4 Fluorinated polymers in cardiovascular surgery

Another field of application of fluorinated biomaterials is related to cardiovascular pathologies. Fluoropolymer prostheses: PTFE and PTFE-expanded (PTFEe) can consolidate arterial lesions.[316] The perfect biocompatibility of these materials with blood plasma and cells as well as their excellent mechanical characteristics are perfectly

[312] L. Li, The biochemistry and physiology of metallic fluoride: action, mechanism, and implications, Crit. Rev. Oral Biol. Med., **14** (2003) 100–114.

[313] J.F. Shackelford, Bioceramics-current status and future trends. Materials Science Forum, **293** (1999) 99–106.

[314] X. Ranz, Développement et caractérisation de dépôts d'apatite obtenus par projection plasma sur prothèses orthopédiques. PhD Thesis, INPT, Toulouse, France, 1996.

[315] A. Ignatius, M. Peraus, S. Schorlemmer, P. Augat, W. Burger, S. Leyen, L. Claes. Osseointegration of alumina with a bioactive coating under load-bearing and unloaded conditions. *Biomaterials*, **26** (2005) 2325–2332.

[316] C. Baquey, M.C. Durrieu, R. G. Guidoin, Fluorinated biomaterials for cardiovascular surgery, in "Fluorine and Health", A. Tressaud, G. Haufe Eds., pp. 379–406, Elsevier, Amsterdam (2008).

adapted to this type of intervention. The use of polymeric biomaterials has led to considerable progress, through the use of arterial substitutes based on woven, knitted, or non-woven synthetic textiles. These new techniques have made obsolete the replacement of arterial segments damaged by an autologous vein transplant whose availability is necessarily limited. Although ethylene polyethylene terephthalate (PET) is dominant in this type of material, it is highly competitive with ePTFE. Known for its chemical inertness and high thermal stability, the latter has many advantages (biostability, sterilization by heat) compared to its main competitor. Among the essential properties that these materials must exhibit are an excellent suturability, a hemo- and biocompatibility with respect to the surrounding tissues, a mechanical strength withstanding for years the stresses exerted by the pulsating blood pressure.

The important data in terms of relationships between the blood and the biomaterial interface, or more precisely between macromolecules of the plasma and blood cells with the wall of the prosthesis, are the morphology and the superficial physicochemical characteristics of the prosthetic wall on the one hand, and the conditions of flow of blood into the prosthesis on the other hand. For example, blood platelets tend to adhere to foreign surfaces, which can cause them to aggregate and generate thrombosis. It is preferable that the surface in contact with the blood is as smooth as possible. It is also important that the dynamic mechanical properties of the biomaterial be closest to those of the treated vessel so that the blood flow does not suffer disturbances, creating areas of turbulence. One of the interests of the textile morphology of microporous fluoropolymers (Teflon, Gore-Tex, Impra, Vitaflon) is to allow infiltration of the material by blood capillaries which constitute a source of cells for endothelialization supply of oxygen and in nutrients of the newly formed arterial layer. The walls of these flexible tubes appear to consist of layers of short polymeric ribbons (up to a few tens of microns) which intersect to form denser areas of matter and connect many fine filaments or fibrils. According to the manufacturers, more than 2 million microporous Teflon prostheses have been put in place worldwide since their commercial launch. In addition, the PVDF polymer, for example Centidene sold by Centenial Surgical Suture Ltd., allows high-strength sutures under tension.

7 Plant protection products and insecticides

It should be emphasized that fluorinated molecules are also present in the field of biocompatible materials, and in the phyto-sanitary field. Thus, in agrochemicals they represent from 3% in 1970 to about 50% of the market today. Agrochemicals such as herbicides, insecticides, and fungicides have the same "fluorine effect" as other biomolecules. These exceptional properties are mostly related to increased lipophilicity and steric effects, because given its small size, fluorine can easily bind to an active site. The particular electronic effects of such bonding are due to the extreme electronegativity of fluorine, as well as its enhanced thermal stability and oxidation resistance. The most commonly used fluorine groups in agrochemicals are fluorinated aromatics (37%) and trifluoromethylated aromatics (32%). Some fluorinated esters are potent mosquito products.[317] Their outstanding performance is reflected in a steady increase in the number of fluorinated compounds for agrochemicals: 23 out of a total of 543 (4%) (Pesticide Manual, 1977 Edition), and 126 out of 858 (14%) (Pesticide, 2006 Edition Manual). In terms of utilization of fluorinated molecules, 45% are herbicides, 33% insecticides, and 18% fungicides.

One of the most striking examples is Fipronil, launched by Rhone-Poulenc in 1993 and currently marketed by BASF-Agro, which remains one of the most effective insecticides. With its broad spectrum, this molecule, *N*-phenylpyrazole, can treat crops and animals. It acts on the gamma-aminobutyric acid receptor (abbreviated as GABA) to block the chloride chain. γ-Aminobutyric acid is the main neurotransmitter inhibitor of the encephalic nervous system. This product is used both to destroy harmful insects such as termites (Terminator) and fleas and ticks in dogs and cats (Frontline). According to the Phillips McDougall-2002 report, it was sold for $358 million in 2001.[318,319] In France, due to the abnormally high mortality observed in hives, treatments of sunflower seeds and corn with products such as Gaucho and Regent TS (which is based on fipronil) have been implicated, in particular, by beekeeper unions. Although no direct link could be established between excess mortality and this type of treatment, the use of these products was suspended on sunflower and corn in 2005. However, by Directive 2007/52/EC of August 16, 2007, the European Community finally listed fipronil as an insecticide for seed treatment, asking member states to pay special attention to the protection of swarms and to carry out further studies on the assessment of risk for bees, especially in the brood.[320] However, we must remain very vigilant about the harmfulness of certain products. For example, the question arose as to whether maintain the use of sulfuryl fluoride (SO_2F_2), considered as a substitute for methyl bromide (CH_3Br), for post-harvest fumigations of crops. Residues of this product metabolized to inorganic fluorides were found on nuts and grapes.[321]

[317] Mori T., M. Sugano, Y. Shono, N. Matsuo, Actes du Congrès ICFC, May 2009, Kyoto, Japan.

[318] F. Colliot, K.A. Kukorowski, D.W. Hawkins and D.A. Roberts, *BCPC Conf. Pests Dis.*, **1** (1992), 29.

[319] G. Theodoridis, Fluorine-containing agrochemicals, Fluorine and the Environment, Vol. 2, pp. 121–175, A. Tressaud, Ed. Elsevier (2006).

[320] J.R. Bloomquist, Ion channels as targets for insecticides, *Annu. Rev. Entomol.*, 1996, 41, 163–190.

[321] Hileman B., Fluoride concerns surface once again, Chem. Engin. News, August 2003, 22–23.

Fluorinated compounds in our environment: Fluorine, friend or foe for humanity?

3

Chapter Outline

Fluorine. https://doi.org/10.1016/B978-0-12-812990-6.00003-9

1 Genesis and cosmo-chemistry of fluorine

1.1 The discovery of fluorinated molecules in interstellar space

Fluorine has a unique place in the understanding the history of the universe as quoted by H. Jönsson.[322] Interstellar fluorine was reported as early as 1980 by the Copernicus satellite in the spectrum of Delta Scorpii (B1). However, a decisive discovery occurred in August 1997 when hydrogen fluoride (HF) gas was observed in a giant cloud of gas in the interstellar space near the center of the Milky Way.[323] The discovery of molecules containing fluorine was considered as one of the major advances obtained from the first years of the European mission ISO (Infrared Space Observatory). Launched in November 1995, the observatory was an incomparable instrument for observing the entire universe in the infrared domain. Although a hundred or so different types of molecules have been discovered in interstellar space over the last 30 years, this is the first time that a new interstellar molecule (HF) has been demonstrated by IR spectrometry. The HF molecules absorb at the characteristic wavelength of around 2.33 μm. The gas cloud was 20,000 light-years from Earth in the southern part of the Sagittarius constellation. Known by astronomers as Sagittarius B2, this cloud consists essentially of hydrogen. The environment conditions are extreme, with temperatures of −223°C (50 K), pressures 10–14 lower than on Earth, and the concentration of HF a billion times lower than that of hydrogen. These HF molecules are probably obtained by chemical reactions between F_2 and H_2 molecules, which are the main components of interstellar gas. It should be noted that at these temperatures, the interstellar fluorine is probably solid and nongaseous, bound to minerals in cosmic dust grains. Another diatomic species, CF^+, was also found in interstellar space, but always less so than HF.[324]

1.2 The enigma of the genesis of fluorine

The origin of fluorine in the universe is uncertain, but there are three production mechanisms that can be proposed: asymptotic giant branch (AGB) stars, ν nucleosynthesis in Type II supernovae, and/or the winds of Wolf–Rayet stars.[325]

[322] H. Jönsson, Fluorine in space, 9 July (2015), https://www.chemistryworld.com/opinion/fluorine-in-space/8740.article.

[323] Neufeld D.A., Zmuidzinas J., Schilke P., Phillips T.G., (1997). Discovery of Interstellar Hydrogen Fluoride. Astrophys. J., Lett., **488**, L141. "There is a whole field, called astrochemistry, where you are trying to understand where these molecules come from and what the presence of particular molecules is. Inflammation of fluoride molecules in the field of hydrogen fluoride and hydrogen fluoride In the context of hydrogen fluoride Unlike most atoms, fluorine atoms are extremely reactive and attack the relatively inert hydrogen molecules that are the main constituent of the interstellar gas. David Neufeld, "Johns Hopkins Gazette", September 1997.

[324] D. A. Neufeld, P. Schilke, K. M. Menten, M. G. Wolfire, J. H. Black, F. Schuller, H. Müller, S. Thorwirth, R. Güsten, S. Philipp, First astronomical detection of the CF^+ ion, astrochemistry: recent Successes and Current Challenges Proceedings IAU Symposium No. 231, 2005; Neufeld, D. A., Wolfire, M. G., & Schilke, P. (2005), ApJ, **628**, 260.

[325] H. Jönsson, N. Ryde, G. M. Harper, M. J. Richter, K. H. Hinkle, Fluorine in the solar neighborhood: is it all produced in asymptotic giant branch stars? Astrophys. J., Lett., **789**: L41, (2014). N. Ryde, H. Jönsson, The cosmic origin of fluorine: an astronomer's view on fluorine synthesis, in "Modern Synthesis Processes and Reactivity of Fluorinated Compounds," Vol. 3 "Progress in Fluorine Science," A.Tressaud series Ed., Chapter 1, 1–6, Elsevier (2016); doi:10.1088/2041-8205/789/2/L41.

Like all other elements, the appearance of fluorine in the Universe has been the product of nuclear reactions in the overheated hearts of stars during the various stages of their evolution.[326] Three hypotheses have been proposed to explain the origin of this element which is so difficult to observe: either due to neutrinos, or during the fusion of nitrogen in the red giant stars (red-giants), or when the massive stars, Wolf-Rayet stars[327] burn their outer layers.[328] Because of its rarity and extreme reactivity, fluorine has a very special position compared to other elements, both at astronomical and chemical levels. It is extremely fragile and difficult to characterize. While stars regularly produce all elements, fluorine atoms, as soon as they are formed, are destroyed by annihilation with hydrogen and helium, the most commonly produced elements. At the heart of the star, the hydrogen nuclei transform the fluorine into oxygen and helium, while the ^{4}He nuclei convert it to neon.

In 1988, Woolsey and Haxton suggested that fluoride could be created and stabilized by supernovae in stellar explosions that eject fluorine into space quickly enough to prevent its annihilation.[329] In 2002, these authors speculated that most of the fluorine in the Universe would come from neutrinos in supernovae. When a massive star such as Antares turns into a supernova, a storm of 10^{58} neutrinos is produced. Before the explosion, a huge stock of ^{20}Ne is created, each containing 10 protons and 10 neutrons. During the explosion, some neutrinos, when crossing the ^{20}Ne layers, create a proton and thus yield ^{19}F, or create a neutron, leading to the unstable radioactive isotope ^{19}Ne which then decays in ^{19}F. However, such a hypothesis remains speculative because such a process has not yet been observed during supernova explosions. In 1992 Jorissen et al. announced the discovery of fluorine not in a supernova but in dozens of red giant stars.[330] To date, fluorine element has been reported only in the Sun, Betelgeuse, the envelope of a carbon star and in a couple of nebulae; however, it has been observed in more than 65 giant stars. Some of these are ordinary giants like K-Arcturus in Bootes and M-Mirach in Andromeda. The fluorine abundance in most K and M type stars is of the order of that found in the sun. On the other hand, the rate of fluorine found in carbon stars is more than 65 times, which means that fluoride is created by these same stars as predicted by Jorissen's thesis. This discovery was described by GJ Matthews who writes in Nature: "Fluoride is probably formed in the flashes of unleashed fusion of a swollen red star".[331] Thus, in the last instants of the life of the star, fluoride-enriched blasts could have been expelled to outer space.

[326] Snow, T.P., York, D.G., The Detection of Interstellar Fluorine in the Line of Sight Toward Delta Scorpii. Astrophys. J., Lett., **247**, (1981) L39.

[327] Named in honor of French astronomers Charles Wolf and Georges Rayet who, in (1867), discovered in the Swan constellation the first 3 stars of this type.

[328] For more details on the origin of fluoride, see the books of astronomer Ken Croswell, author of "Magnificent Universe," "Alchemy of the heavens," and "Planet Quest," as well as his article in "Sky & Telescopes,", Sept. (2003), p30. [http://www.kencroswell.com].

[329] Woosley, S.E., Haxton, W.C., Supernova Neutrinos, Neutral Currents and the Origin of Fluorine. Nature, **334** (1988) 45. Woosley, S.E., Heger, A., and Weaver, T.A., The Evolution and Explosion of Massive Stars. Rev. Modern Physics, **74** (2002) 1015.

[330] Jorissen, A., Smith, V.V., Lambert, D.L., Fluorine in Red Giant Stars: Evidence for Nucleosynthesis, Astronomy and Astrophysics, **261** (1992) 164.

[331] Mathews, G. J., Flashing Giants Forge Fluorine. Nature, **359** (1992) 18.

Two nuclear fission reactions would be responsible for the process of creating fluoride. In a first step, the hydrogen would turn into helium when the star dies, at the same time that the carbon and the oxygen would become ^{14}N. During the ensuing burning of the helium layer, ^{14}N would react on helium to give rise to ^{19}F. The progenitors of the red giant stars of the Milky Way are equivalent to 1–8 masses of the Sun. More massive, these stars become supergiant and explode as supernovae. It is the giants of <4 suns that produce the most fluorine, because in the more massive stars the explosion of the helium layers is too powerful and consumes more fluorine than it produces. The few results obtained for the nebulae tend to confirm these hypotheses: W. Liu thus found a significant fluorine abundance in the nebula NGC 4361 in the constellation Crow (Corvus) whose stars correspond to 1–2 suns.[332]

At the same time, Meynet and Arnould proposed to consider the massive Wolf-Rayet type stars instead. During the last stages of their life—before exploding, it seems, as a supernova—among the elements that rise from the star's core to the outer layers ^{14}N is produced while the hydrogen is burned in helium. As in the case of the red giants, it is during the combustion of the helium core that ^{14}N is transformed into ^{19}F and is ejected into the interstellar space thanks to the strong winds that accompany the final explosion. According to the calculations of Meynet and Arnould,[333] the rate in ^{19}F in the winds of the Wolf-Rayet stars would be 10–70 times greater than in the sun. However, no fluorine has yet been observed in Wolf-Rayet stars.

New evidence for nucleosynthesis of fluorine by neutrino appeared in September 2003: K. Cunha et al. discovered hydrogen fluoride in metal-poor stars.[334] Using the Gemini South 8.1 m telescope at the Cerro Pachón Observatory, Chile, they obtained IR fluoroscopy data in nine stars of the Large Magellanic Cloud (Fig. 60). The observations provided evidence of the involvement of neutrinos in the synthesis of fluorine during supernova supergiant star explosions. Fluorine has even been detected in the Omega Centaurus cluster in which very little iron is present (5%–10% in relation to the sun). Moreover, the decrease of the F/O ratio when the oxygen level decreases (Fig. 61) seems in good agreement with the various modes of fluorine production: (i) helium combustion sequences in the giant stars; (ii) ejection in the winds of the Wolf-Rayet stars; (iii) nucleosynthesis by neutrinos during the collapse of supernovae. A striking illustration of the ejection of matter by Wolf-Rayet stars is provided by nebulae made of ejecta, as M1–67 nebula photo, taken by the Hubble Space Telescope (Fig. 62).[335] It was thus possible to conclude that the fluorine originated from stars of high masses, and was produced either when these stars explode propelling the neutrinos through the neon atoms, or during nuclear reactions occurring during the

[332] Liu, XW, Temperature Fluctuations and Chemical Homogeneity in the Planetary Nebula NGC 4361. Monthly Notices of the Royal Astronomical Society, **295** (1998) 699.

[333] Meynet, G., Arnould, M., Synthesis of ^{19}F in the He-Burning Zones of Massive Stars. In Nuclei in the Cosmos, edited by F. Kappeler and K. Wisshak (Bristol: Institute of Physics Publishing) (1993) p. 503; Meynet, G., and Arnould, M., Synthesis of ^{19}F in Wolf-Rayet Stars. Astron. Astrophys., **355** (2000) 176.

[334] Cunha K., Smith V.V., Lambert D.L., Hinkle K.H., Fluorine abundances in the large Magellanic cloud and -Centauri: evidence for neutrino nucleosynthesis, Astron. J., **126** (2003) 1305–1311.

[335] Grosdidier Y., Moffat A.F.J., Joncas G., Acker A., HST WFPC2 / Hα Imagery of the nebula M1–67, Astrophys J. **506** (1998) L127–L131.; Grosdidier Y., Moffat A.F.J., Blais-Ouellette S., Joncas G., Acker A., Hubble space telescope imagery and Canada-France Hawaii telescope Fabry-Perot Two-dimensional spectroscopy in Hα of the ejected nebula M1–67: turbulent status. Astrophys. J. **562**, pp.753–769, 2001.

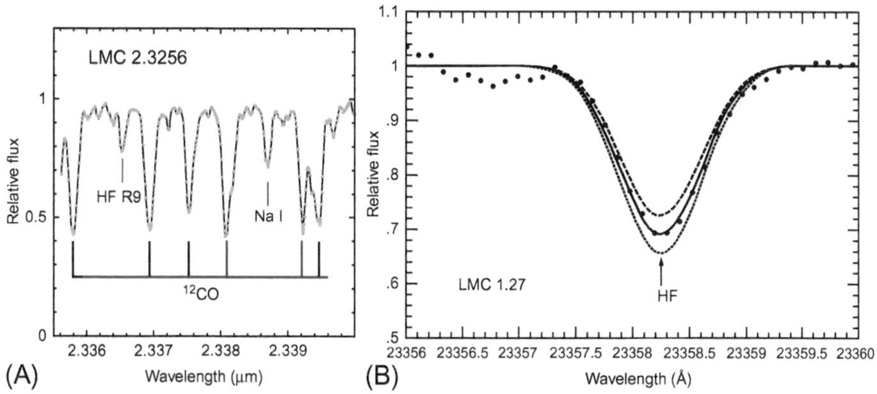

Fig. 60 (A) Near-infrared high-resolution spectrum (recorded by PHOENIX) of a red giant star showing a spectral absorption line of the HF molecule (indicated on the left spectrum) used to evaluate the abundance of fluorine in the stellar atmosphere; (B) observed spectrum (*dashed*) and calculated (*solid lines*) of a giant star LMC (Large Magellan Cloud). Three ^{19}F abundances are presented. An abundance of A(F) = 3.93, with the Sun's hydrogen abundance being 12.0, on a logarithmic base 10-scale was derived [from Meynet et al. (333)].

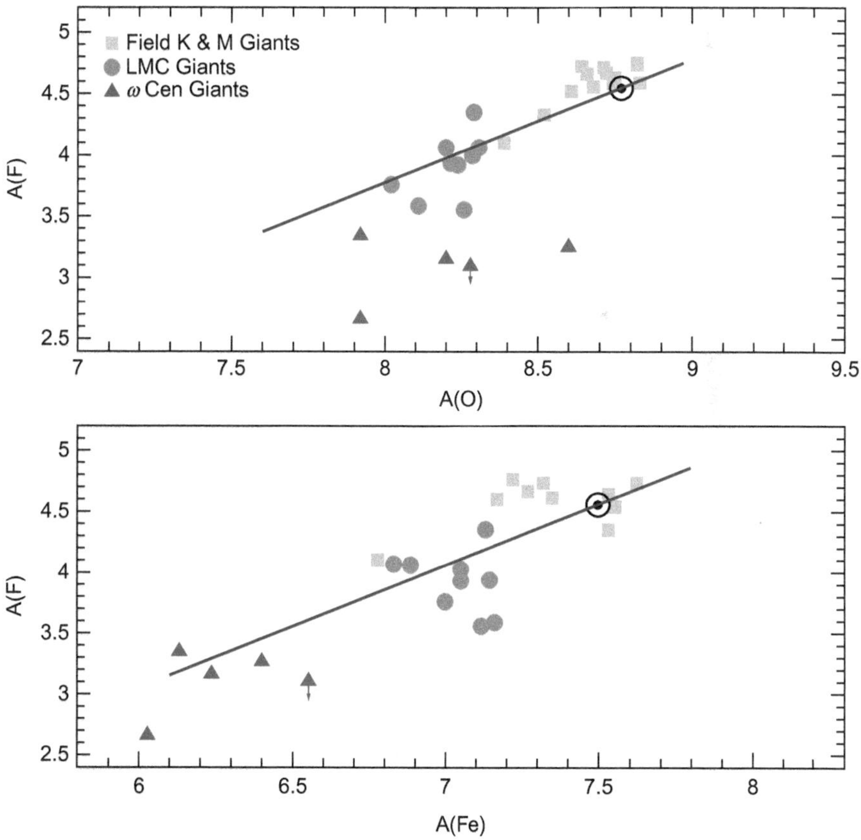

Fig. 61 Abundance logarithmic A (F) of fluorine as a function of oxygen A (O), and iron A (Fe). The colors symbolize the location of the *red* giants: Milky Way (*blue squares*), LMC (*green circles*), ωCentauri (*purple triangles*). The *circled black dots* correspond to the abundance of fluorine in the Sun [from Meynet et al. (333)].

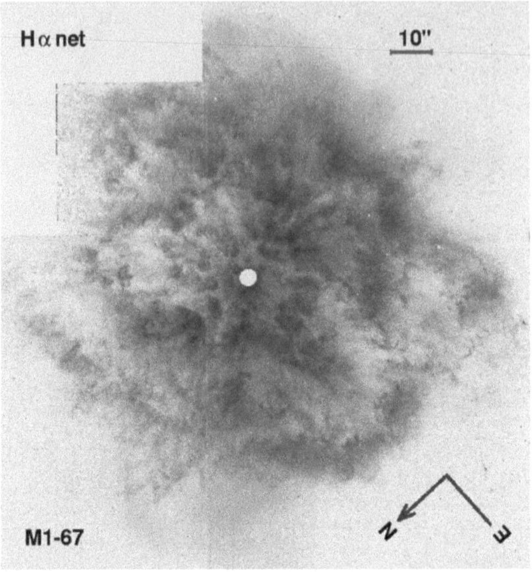

Fig. 62 Deep WFPC2/Hα log-scale image of M1-67 nebula from four combined exposures totaling nearly 3 h. The field stars have been subtracted out and the saturated central star WR 124 has been replaced by a white disk. Image taken by the Hubble Space Telescope.
From Grosdidier Y., Moffat A.F.J., Joncas G., Acker A., HST WFPC2/Hα Imagery of the nebula M1-67, Astrophys J. 506 (1998) L127–L131.

burning of the helium layers in giant stars with propulsion of the fluorine formed towards the external surface, that is, during both processes; on the other hand, it seems that the contribution of the red giants would not be preponderant. In any case, and to use Hubert Reeves' expression, fluorine is also a child of "star dust" (see below "How fluorine was created in the Universe?").

How fluorine was created in the Universe?

In giant stars and Wolf-Rayet stars the nuclear reaction sequences are complex and begin when the ^{13}C isotope combines with ^{4}He with loss of a neutron:

$$^{13}C + {}^{4}He \rightarrow {}^{16}O + \text{neutron}$$

The neutron then reacts with nitrogen or aluminum with the release of a proton:

$$^{14}N + \text{neutron} \rightarrow {}^{14}C + \text{proton}$$

$$^{26}Al + \text{neutron} \rightarrow {}^{26}Mg + \text{proton}$$

These protons will eventually allow the production of fluorine from a sequence that starts when ^{14}N, formed during the combustion of hydrogen reacts with helium to lead to the unstable ^{18}F isotope that decays to ^{18}O and a positron:

$$^{14}N + {}^{4}He \rightarrow {}^{18}F$$

$$^{18}F \rightarrow {}^{18}O + \text{positron}$$

The proton transforms ^{18}O into ^{15}N and finally the creation of fluorine results from an ^{15}N reaction with helium:

$$^{18}O + \text{proton} \rightarrow {}^{15}N + {}^{4}He$$

$$^{15}N + {}^{4}He \rightarrow {}^{19}F$$

It has been also proposed that elements such as fluorine were produced in the early Universe mostly under the form of HF, with an abundance of HF is predicted to be close to 3.75×10^{-17} whereas the initial abundance of neutral fluorine F is 10^{-15}.[336] The origin of the genesis of fluorine in the Universe and particularly in the Milky Way is always subject to discussion.[337]

1.3 First discovery of fluorine on Mars with ChemCam on-board MSL-curiosity

Fluorine has been recently detected for the first time in the Gale Crater on Mars with the ChemCam instrument on board Mars Science Laboratory (MSL)-Curiosity (Fig. 63A).[338] This detection was made possible through the identification of Ca–F

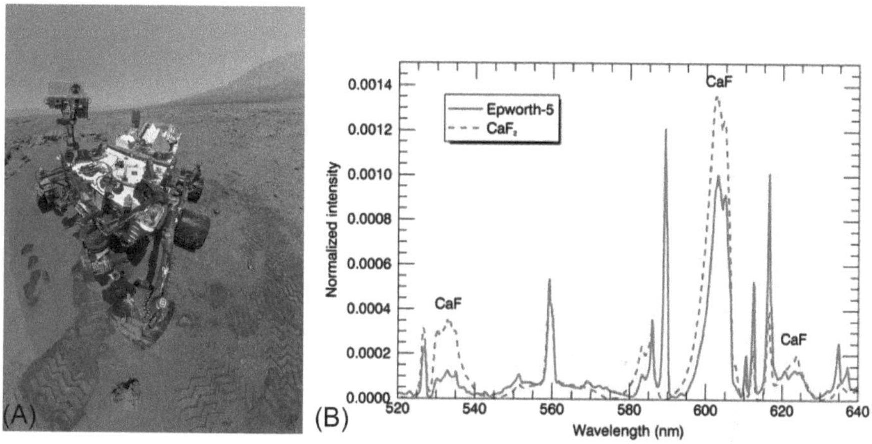

Fig. 63 (A) NASA's Curiosity rover on Mars. Self-Portrait by Curiosity Rover Arm Camera (credit: NASA https://www.nasa.gov/mission_pages/msl/multimedia/pia16239.html);
(B) comparison of spectrum of Epworth-5 soil sample with a pure CaF_2 spectrum obtained under Martian conditions.
(from O. Forni et al., First fluorine detection on Mars with Chem-Cam On-Board MSL, 45th Lunar and Planetary Science Conference (2014)).

[336] D. Puy, V. Dubrovich, A. Lipovka, D. Talbi, P. Vonlanthen. Molecular fluorine chemistry in the early Universe, Astronomy and Astrophysics—A&A, EDP Sciences, **0709** (2007) pp. 4083.

[337] A. Renda, Y. Fenner, B. K. Gibson, A. I. Karakas, J. C. Lattanzio, S. Campbell, A. Chieffi, K. Cunha, V. V. Smith, On the origin of fluorine in the Milky Way, Mon. Not. R. Astron. Soc. 000, 1–6 (2004). Renda, A., Fenner, Y., Gibson, B. K., Karakas, A. I., Lattanzio, J. C., Campbell, S., Chieffi, A., Cunha, K., Smith, V. V., The Evolution of Fluorine in Galactic Systems, Nuclear Physics A, **758** (2005) pp. 324–327.

[338] O. Forni, et al., Fluorine as seen by ChemCam in Gale Crater-Mars, Proceedings Goldschmidt 2014/O. Forni et al., First fluorine detection on Mars with Chem-Cam On-Board MSL, 45th Lunar and Planetary Science Conference (2014)/O. Forni et al., First detection of fluorine on Mars: Implications for Gale Crater's geochemistry, Geophys. Res. Letters, **42** (2015) (4) 1020–1028.

molecular-band emissions in the Laser-Induced Breakdown Spectroscopy (LIBS) that appear when recombination occurs in the cooling plasma. These emission lines enable a much more sensitive detection of fluorine and halogens with respect to the usual atomic emissions. Two principal F-based types of minerals have been identified.

1. Samples with high aluminum and low calcium contents, in which the F-bearing phase is an aluminosilicate, might be topaz or muscovite. It is found in conglomerates and may indicate petrologically evolved sources. This was the first time that such a petrological environment is found on Mars.
2. Samples with high calcium contents, in which the main F-bearing minerals are likely to be fluorapatite and/or fluorite. When compared to a pure fluorite spectrum (Fig. 63B), preliminary estimate of the fluorine content from Epworth-5 soil sample is about 10wt%. Fluorapatites are found in some sandstone and may be debris or disintegrated materials, while fluorites are also found in conglomerates, possibly indicating low-temperature alteration processes.

Organic molecules have been recently discovered on Sheepbed mudstones, Gale Crater.[339] Based on the thermochemical modeling proposed by Bridges et al.[340] the presence of fluorine at Gale Crater allows to simulates which mineral phases fluorine is likely to have precipitated as in an open system to determine whether it is related to an evolved magma composition in the Martian crust, secondary mineral alteration in Gale Crater or biological activity.[341] This could aid identifying the sources and pathways of important species responsible for Martian diagenetic and hydrous alteration, which in turn could provide further scope on determining the habitability and paleo-environment of ancient Mars.

(Mars Science Laboratory (MSL) is a robotic space probe launched by NASA on November 26, 2011, which successfully landed in Gale Crater on August 6, 2012. ChemCam is an active remote sensing instrument using laser pulses to remove dust and to profile through weathering coatings of rocks up to 7m away. Laser-induced breakdown spectroscopy (LIBS) obtains emission spectra of materials ablated from the samples in electronically excited states. When the plasma cools down, elements can recombine and molecular emission lines are observed. Recent experiments have shown that, when occurring, these molecular emissions can be much brighter than the associated atomic lines especially when halogens and REE are concerned.)

1.4 *Presence of fluorine in Martian and lunar meteorites found on Earth*

The effectiveness of resonant nuclear reaction methods has made it possible to considerably advance the characterization of light elements, including fluorine, in Martian or lunar meteorites. It is the reaction ^{19}F (p, αγ) ^{16}O that is used (see first

[339] Freissinet, C. et al., Organic molecules in the Sheepbed Mudstone, Gale Crater, Mars. J. Geophys. Res. Planets, **120** (3) (2015) pp. 495–514.

[340] Bridges, J. C.; Schwenzer, S. P.; Leveille, R.; Westell, F.; Wiens, R. C.; Mangold, N.; Bristow, T.; Edwards, P., Berger, G., Diagenesis and clay mineral formation at Gale Crater, Mars, J. Geophys. Res. Planets, **120**(1) (2015) pp. 1–19.

[341] Bedford, C.C., Schwenzer, S.P., Bridges, J.C. (2016). An Investigation into Fluorine within Gale crater and its Implications for Martian Geology. UKPF 13th Annual Early Career Scientists' Meetings, January 2016.

Section—Chapter 2.2). Due to the dispersion of the results related to the very small quantities[342] and the possible contamination by terrestrial pollution, analyzes on standard sample are necessary. By way of example, in the USGS geochemical references, the GSP-1 granodiorite contains 0.32% by weight of fluorine.[343] Measured in 21 meteorites, the average fluorine level is between 20 and 90 ppm.[344] In addition to an often-nonhomogeneous distribution, one of the important points is the inverse relationship between the fluorine concentration in the magnetic phase and the metal content in the meteorite. Allen and Clark thus suggested that in ordinary chondrites, fluorine was associated with the aluminum-based phase, usually plagioclase, or even with CaF_2 or apatite.[345] It should be noted that amphiboles containing low levels of fluorine and chlorine have been detected in Martian meteorites. Fluoride was also found in the three types of lunar glasses presumably of volcanic origin that were brought back to Earth by the various Apollo expeditions: Apollo-15 green-emerald, Apollo-17 orange and Very Low Titanium (VLT) samples.[346] These glasses help improving our knowledge of the nature of the basaltic reservoirs of the Moon. Chemical and petrological data suggest that these partially melted samples came from a reservoir about 400 km deep from the Moon's surface. Moreover, the in situ chemical analysis carried out during the Surveyor-7 mission using α particle scattering, experiments led to high fluorine contents—up to 0.27% by mass—and thus allows to reinforce the volcanic origin of lunar basaltic samples. Recall that the measurement of a very low level of fluorine pollution on the famous Martian meteorite ALH-84001 allowed to evaluate a very short residence time of this object on the icy surface of Antarctica, thus providing an argument in favor of the Martian origin microorganisms found on the surface.

1.5 Determination of the residence time of meteorites on Antarctica surface through ion beam analyses

The abundance of halogens in chondritic meteorites is of fundamental importance for interpreting geochemical behavior within the mantles of asteroids and the terrestrial planets.[347,348] Given the very low temperatures and the absence of industrial pollution, Antarctica is the place on Earth where the most preserved meteorites can be found. A remarkable concentration of specimens has stayed there for many thousands of years (up to 7×10^5 years) with very low interactions with anthropogenic pollution

[342] Reed, G.W., Fluorine, in "Handbook of Elemental Abundance in Meteorites," B. Mason, Ed., Gordon & Breach pp. 103–108 (1971).

[343] Flanagan F.J., Values for international geochemical references, Geochim. Cosmochim. Acta **37** (1973) 1189–1200.

[344] Allen R.O., Clarck P.J., Fluorite in meteorites, Geochim. Cosmochim. Acta **41** (1977) 581–585.

[345] Allen R.O., B. Mason, Minor and trace elements in some meteoritic minerals, Geochim. Cosmochim. Acta **37** (1973) 1435–1456.

[346] Delano J.W., Livi K., Geoch. Cosmochim. Acta **45** (1981) 2137; Delano J. W., Lindsley D.H., Lunar Planet. Sci. **14** (1983) 156.

[347] R. H. Jones, A. J. Brearley, J. A. Lewis, Bulk fluorine and chlorine contents of chondritic meteorites. Proceedings, 79th Annual Meeting of the Meteoritical Society (2016).

[348] Brearley A. J., Jones R. H., in The Role of Halogens in Terrestrial and Extraterrestrial Geochemical Processes, ed. D. Harlov, Springer (2016).

and weathering because of the very low ambient temperatures. These conditions can be compared with those of western USA, where chondrites can no longer be recognized after about 3600 years.[349] As long as the specimens are located at the ice surface, contaminants such as halogens from the sea or volcano emissions interact with the meteorite crust and diffuse into the interior. The major sources for F, Cl, and Br pollution are spray aerosols arising from the salted sea and for Cl and F from volcanic emissions (from Mt. Erebus foe instance). The rate of halogen enrichment therefore gives valuable information on the residence time of a meteorite at the surface of Antarctica's ice. Fluorine is the best suited element among halogens because the formed fluoride compounds are weakly soluble and thus weakly leached by water. It has been shown indeed that most Cl, Br, and I from an Antarctic meteorite can be leached by water and interpretations of isotopic data such as ^{36}Cl could be strongly affected.[350] On the contrary, neither the original meteorite nor the weathering products release any fluoride in a soluble form. So, fluorine enrichment has been measured using the wide range of nuclear methods, as illustrated by Döbeli's works.[43] This allows to evaluate the time spent on the ice of Antarctica and the duration of its exposure to the atmospheric pollution. The proposed hypotheses can be accounted for if some precautions are taken, such as:

- use of reference meteorites on Earth whose age is perfectly known,
- good knowledge of fluoride ion diffusion laws in the matrix,
- morphology, grain size and porosity of the meteorite,
- position of the objects with respect to the various sources of gaseous pollution: distance from the sea (halogenated aerosols) Mt. Erebus (fluorinated emissions),
- modification of fluorinated fluxes over the centuries.

Fluorine is mainly sorbed on the external part (about 50 μm) of the meteorite and diffuses very slowly into the interior. In early procedures, the surface layers were stepwise removed, fluorine was isolated by pyrohydrolysis and further analyzed by ion-sensitive electrode. Because of the very small amount of fluorine present at the surface of the meteorites, NRA methods appeared to be, in this case also, the most relevant choice for analysis (see Section 2.2). The F enrichment is measured by means of PIGE $^{19}F(p,\alpha,\gamma)^{16}O$ and $^{19}F(p,p'\gamma)^{19}F$ and $^{19}F(p,p'\gamma)^{19}F$ nuclear reactions. A higher spatial resolution and a non-destruction of the samples can be quoted among the interests of these methods. This type of analyses, initially proposed by Allen and Clark,[343] are now currently used for the determination of the degree of F-enrichment in several chondrites found on Allan Hills in Antarctica.[351] The degree of F-enrichment is

[349] Boeckl R., Terrestrial age of nineteen stony meteorites derived from their carbon content, Nature **236** (1972) 25–26.

[350] Dreibus G., Wänke H., Halogens in Antarctic meteorites, Meteoritics **18** (1983) 291–292; V Gooding J. L., Mineralogical aspects of terrestrial weathering effects on chondrites from Allan Hills, Proc. 12th Lun. Planet. Sci. Conf., (1981) 1771–1787.

[351] Langenauer M., Krähenbühl U., Depth profiles and surface enrichment of the halogens in four Antarctic H5 chondrites and in two non-Antarctic chondrites, Meteoritics **28** (1993) 98–104; Langenauer M., Krähenbühl U., Halogen contamination in Antarctic H5 and H6 chondrites and relation to sites of recovery, Earth Planet. Sci. Lett. **120** (1993) 431–442.

determined from the net F enrichment, that is the total concentration minus the interior concentration, and also from the total amount of fluoride that has diffused trough the surface layer to the interior. On the other hand, the enrichment can be approached from Fick's second diffusion law, which is used to describe one-dimensional diffusion processes, that is, with a concentration gradient along one axis only. In this case, diffusion of a trace element through a porous solid is dependent on the molecular diffusion in the aqueous phase, sorption phenomena on the solid, and on the constrictivity, depending on the ratio of the diameter of the diffusing particle to the pore diameter of the solid. The terrestrial age of the meteorites was also determined by ^{14}C or ^{36}Cl measurements. The amount of halogen, and more particularly fluorine, that was sorbed in chondrites on different antarctical sites was determined, allowing an evaluation of their exposure time onto Antarctic ice. Fluorine profiles obtained from NRA γ-spectra of two meteorites, ALHA-79025 and ALH-84001, are compared in Fig. 64B, showing the weak F-concentration/residence time of ALHA-84001: in ALHA-79025, 380 ± 20 mg/g are found on the exterior crust and about 9 mg/g in the interior, whereas in ALH-84001 the concentration does not exceed 10 mg/g.[352] The maximum exposure time on the ice can be determined if a meteorite whose terrestrial age is known with high accuracy is taken as a reference. ALH-84082 meteorite was selected as a reference because ^{14}C analyses gave precisely its age on the Earth: 23×10^3 years. Using such a reference, no maximum exposure age larger than the measured terrestrial age resulted for all the other meteorites. The maximum exposure times of fluorine enrichment in several meteorites, normalized to that of ALH-84082, are grouped in Table 16.

Fig. 64 (A) ALH-84001 meteorite found in Allan Hills site, Antarctica; (B) comparison of NRA γ-analysis of two meteorites from Antarctic sites: ALHA-79025 and ALH-84001 chondrites, showing the weak F-concentration/residence time of ALHA-84001.

[352] Noll K., M. Döbeli M., Tobler L., Grambole D., Krähenbühl U., Fluorine Profile in achondrites and chondrites from Antarctica by Nuclear Reaction Analysis, Meteorit. Planet. Sci. **32** (1997) A101; Noll K., M. Döbeli M., Krähenbühl U., Fluorine Profile in Antarctic Meteorites by Nuclear Reaction Analysis (NRA). Fresenius J. Anal. Chem **361** (1998) 713–715.

Table 16 Fluorine enrichment of some Antarctic chondrites[a] normalized to ALH 84082 (being taken as one), with deduced estimated exposure times on ice surface

	ALH 84082	**ALHA 76008**	**ALHA 77294**	**ALHA 77182**	**ALH 84069**	**ALHA 79025**	**ALHA 77271**
Terrestrial age (10^3 years)	23.5	100	10	250	?	>38	70
F enrichment normalized to that of ALH 84082	1	0.05	0.13	0.42	0.55	1.08	1.32
Upper limit of exposure time on ice surface (10^3 years)	23.5	1.2	3.1	9.9	12.9	25.4	31

[a]Chondrites are called after their origin: ALH, ALHA, for Allan Hill Antarctic.

However, the limits of the method should be pointed out. For example, the fluorine content of the surrounding ice should be perfectly known. Nevertheless, different parts of a meteorite can show different F enrichments, and different types of meteoritic structures take up F differently because of different chemical composition, grain size, and porosity. Other data should also be included, such as the distance relative to the sea, the elevation, the contamination process of other meteorites coming from different places of Antarctica. In addition, does the fluorine flux to Antarctica vary with time, as deduced from analyses over a long period (1983–1991)? Could F-rich water, as from a lake beneath the glaciers, have come into contact with the meteorite while it was buried into the ice?

The most peculiar event concerning Antarctic meteorites is the discovery of ALH-84001 in 1984, again in Allan Hills (Fig. 64A). Nine years later, scientists determined, on the basis of chemical analysis, that this 4.5-billion-year-old rock had been blasted away from Mars and has floated through space to fall finally on Earth about 13,000 years ago. A research team from NASA's Johnson Space Center and Stanford University examined the meteorite for 2 years, and in 1996, McKay et al. claimed the discovery of traces of so-called "nanofossils,"[353] that is, microscopic structures suggestive of fossilized bacteria. In particular it was found: (i) abundant polycyclic aromatic hydrocarbons (PAHs) on the fracture surfaces, (ii) these hydrocarbons being the same as breakdown products of dead micro-organisms on Earth, (iii) mineral phases consistent with by-products of bacterial activity, (iv) tiny carbonate globules which may be microfossils of the primitive bacteria, all within a few hundred-thousandths of an inch of each other. Taken together, all these findings were thought to be strong evidence pointing to primitive bacterial life on Mars.

One of the major questions raised by the discovery of these possible biogenic features on the Martian meteorite ALH-84001 was to know whether these PAH's, bacteria-shaped objects arose from terrestrial contamination and alteration during its 13,000 years on Earth. The knowledge of the residence time of the meteorite on Antarctic ice surface was thus a decisive parameter. The determination of fluorine amount present in the fusion crust of a meteorite appears to be a convenient way to measure the level of terrestrial contamination and to evaluate the time spent by the meteorite at the ice surface. The fusion crust, which is formed on a meteorite when it passes through the Earth's atmosphere, is a thin layer of glass that contains essentially no fluorine. On the contrary, when in contact with atmospheric pollutants on Antarctic ice surface, the fusion crust absorbs steadily fluorine from water. The time spent by ALH-84001 on the ice surface can be calibrated using known exposure times of other meteorites. The diogenite meteorite ALHA-77256 is a good calibration for ALH-84001, because they have both similar mineralogy, chemical composition and total time passed on Earth: 11,000 years versus 13,000 years respectively, and were both found in Allan Hills. The surface of ALHA-77256 is enriched by 114 ppm F, while the surface of ALH-84001 has almost no detectable fluorine, within the signal background,

[353] McKay D. S., Gibson E. K., Thomas-Keprta K. L., Vali H., Romanek C. S., Clemett S.J., Chillier D. F., Maechling C. R., Zare R. N., Search for past life on Mars: Possible relic biogenic activity in Martian meteorite ALH-84001. Science, **273** (1996) 924–930.

which indicates a very short exposure on the ice surface: if those 114 ppm represent 11,000 years exposed on the ice, then ALH 84001 was exposed for <500 years only. This short time deduced from fluorine enrichment is therefore a severe limit to the rate of terrestrial contamination of ALH-84001 on Earth.[354]

Years later, the debate on ALH 84001, or on a newly found meteorite Yamato 000593[355] is still open, some teams supporting the initial findings, while other researchers saying that these "nanofossils" are actually of geological formations or even from Earth contamination rather than biological artifacts. Following A.H. Treiman, a temporary conclusion could be "From my perspective, the strongest conclusion is that ALH-84001 contains polycyclic aromatic hydrocarbons (PAHs) that formed on Mars. These PAH molecules may be related to Martian micro-organisms, as McKay and co-workers suggest. The PAHs might also have formed without assistance from living organisms, in what might be called a prebiotic organic chemistry. Proof of a prebiotic organic chemistry system in Mars would be nearly as exciting as proof of life itself."[356]

2 Fluorine and our atmosphere

2.1 The fragile balance of our planet

Although the atmospheric environment of our planet seems immutable to us, this equilibrium is the result of an immense set of biological, chemical and physical processes. These mechanisms occur at both the troposphere (up to 15,000 m from the ground) and the stratosphere (15,000–50,000 m). Their number and their diversification make utopian any quantified representation of these effects. For example, how can assumptions such as the "butterfly effect" be incorporated into these processes?[357] That is how "small causes can have larger effects." Nevertheless, in a balance that may prove to be very fragile, the man who already seems to control many aspects of his destiny must preserve intact this formidable heritage.

[354]Krähenbühl U., Noll K., Döbeli M., Grambole D., Herrmann F., Tobler L., Exposure of ALH-84001 and other achondrites on the Antarctic ice, Meteorit. Planet. Sci. **33** (1998) 665–670. Noll K., Döbeli M., Tobler L., Krähenbühl U., Fluorine profiles in achondrites and chondrites from Antarctica, Meteorit. Planet. Sci. **32** (1997) A101.

[355]Lauren M. White, Everett K. Gibson, Kathie L. Thomas-Keprta, Simon J. Clemett, David S. McKay. Putative Indigenous Carbon-Bearing Alteration Features in Martian Meteorite Yamato 000593. Astrobiology, **14** (2) (2014) 170.

[356]Treiman A. H., Fossil Life in ALH-84001? Lunar and Planetary Institute (1998) Recent Scientific Papers on ALH-84001 Explained, with Insightful and Totally Objective Commentaries http://www.lpi.usra.edu/lpi/meteorites/alhnpap.html).

[357]This expression was coined by E. Lorenz, a meteorologist at MIT who proposes that every action, even insignificant, may eventually have colossal consequences: "how can a tiny disruptive phenomenon trigger an avalanche"? or in a more poetic way how can "the flapping of a butterfly's wings in Brazil triggers a tornado in Texas?" Beyond the spectacular metaphor, this theory has its source in the work of H. Poincaré on chaotic phenomena and takes into account the fact of the sensitive dependence on initial conditions in which a small change in one state of a deterministic nonlinear system can result in large differences in a later.

2.1.1 Stratospheric phenomena

In the upper layers of the atmosphere, the stratospheric ozone shield protects the Earth from UV rays.[358] A consequent decrease of this protective layer would have multiple and very serious effects: slowing down the growth of plants, decrease of the immunity to infectious diseases, increase of skin cancers and ocular affections, reduction of the immune defenses, etc. In the early 1980s, significant thinning of the ozone layer was effectively observed over Antarctica. Such a phenomenon had been predicted a few years earlier by Molina and Rowland who suggested how the destruction of stratospheric ozone could take place.[359]

Human activity accounts for 75%–85% of the chlorine presence in the stratosphere. An additional 15%–20% is due to chloromethane from natural sources and biomass combustion, and volcanic emissions provide the remaining few percent. Sources of stratospheric chlorine are quite different from those of tropospheric chlorine from marine aerosols, volcanic emissions and volatile organic compounds. It should be noted that most tropospheric chlorine does not convert to stratospheric chlorine, as it is rapidly decomposed by the oxidizing species present in the lower atmosphere and leads to soluble molecules such as HCl that are rapidly driven by rainwater. By the 1970s, it had even been suggested that the 68 t of HCl released into the atmosphere at each space shuttle launch could also have a detrimental effect on the quality of the stratosphere. In fact, 10 launches correspond to 0.06% of all CFCs emitted each year (1.2 Mt in the 1980s). The analyzes showed that almost all the chlorine was eliminated in the troposphere and did not reach the stratosphere.[360] This is not the case for CFCs, whose stability in the troposphere ranges from 50 years to over 200 years and can thus extend their action in the higher layers. Nevertheless, it should be pointed out that fluorine is not concerned, or very little, in the stratospheric phenomena and the fluorinated end-of-process products are essentially HF and COF_2.[361]

2.1.2 Tropospheric phenomena

Concerning the reactions occurring closer to us in the troposphere, the equilibrium which protects our planet has been assimilated by S. Arrhenius[362] to that of a vast greenhouse. It is this greenhouse effect that has created and maintained the conditions that led to the emergence of life on Earth. Thanks to solar energy, the combination of the four essential elements of the life cycle: carbon, hydrogen, oxygen and nitrogen has led to the development of all organic and biological molecules, with contribution of various mineral constituents. Our atmosphere allows solar radiation

[358] Information on the depletion of the ozone layer and the increase of the greenhouse effect can be found on many sites, but in particular on that of the World Meteorological Organization (Global Ozone Research and Monitoring Project: http://www.wmo.ch/web/arep/gaw/o3reports.html; http://www.cmdl.noaa.gov/index.php).

[359] F.S.Rowland, Chlorofluorcarbons and the depletion of stratospheric ozone, Am. Sci., **77** (1989) 36.

[360] Prather M.J. et al., The Space shuttle's impact on the Stratosphere, J. Geophys.Res., **965** (1991) 18583.

[361] P. Ricaud and F. Lefèvre, Fluorine in the Atmosphere 1–34, Fluorine and the Environment, Vol.1, A. Tressaud Ed., Elsevier (2006).

[362] Arrhenius, S., L'évolution des mondes (1906).

to pass through; on the other hand, the re-emitted IR radiations are partially absorbed by water, carbon dioxide, gaseous elements present in the atmosphere and more particularly in the troposphere. It is this energy which, retransmitted towards the surface of the Earth, allowed this surface to reach an average temperature of 15°C.

Remember that the range of the various pollutants depends on their nature: the suspended particles are deposited within a radius of a few kilometers from the source, while the aerosols and liquid micro-droplets that are for example responsible for acidification can travel for hundreds of km. The end-of-life products are HF or CaF_2 for the F species, and CF_3COOH for the trifluoromethylated groups. The atmospheric pollutants, which are responsible for the greenhouse effect and to some extent to the depletion of the ozone layer are grouped in Table 17. The fluorinated species essentially concern the first two types of pollution and incidentally acidification by HF in some volcanic areas. The impact of halocarbons: CFCs and PFCs on the greenhouse effect and the ozone layer depends on their average residence time in the atmosphere, which varies between 1 and 10,000 years depending on the species. The global system of observation of atmospheric phenomena is coordinated by the World Meteorological Organization with the use of all the arsenal of modern technologies.

2.2 The total fluoride cycle: Evaluation of average fluorine levels in our environment

Because of its extreme reactivity, fluorine originating from the interior of the earth's crust and emitted through volcanic projections is immediately converted into fluorides. Most of the fluorine is present in the earth's crust and minerals and between 0.06 and 0.08 wt% by mass. In the oceans, the ratio is very low: 7×10^{-5} and corresponding to a level of 0.0014 wt% by mass (1.4 ppm), compared with 18.98 wt% chlorine. Although atmospheric fluorine may be derived from marine aerosols, it is questionable whether this is indeed to be considered because of the very low levels of this element.[363]

There are several major sources of fluorine in the atmosphere, the most important of which are fluoride, but several types of fluoride, particularly those that deal with high tonnages. Containing fluorides, such as blast furnaces producing aluminum, iron or steel, coal-fired power plants, phosphates and ceramics industries (ceramics, bricks, tiles, etc.). In these sites, the fluorinated residues are either in solid form or in the form of HF hydrogen fluoride vapors, which must be neutralized. Once adsorbed onto very small particles, fluorine can remain in the atmosphere for very long periods. The impact on geological dependence is a function of the nature of the pollutants, the aerosols and liquid micro-droplets that are responsible, for example, for acidification can travel over several hundred kilometers. Gases, which affect the greenhouse effect and/or the thinning of the ozone layer, act around the globe. Their impact depends on their average residence in the atmosphere, which varies between 1 and 10^4 depending on the species. This chapter presents the different components of fluorine and fluorinated gas emissions, as well as the contribution of perfluorocarbons.

[363] Highly valuable information on the impact of fluoride and fluorides on our biosphere can be found on the Environment/Environment Canada website: http://www.ec.gc.ca/Environnement/ Environment Canada: http://www.ec.gc.ca/.

Table 17 Various atmospheric pollutants

Types of pollution	Major impacts	Involved gases and pollutants	Formulas, acronyms
Greenhouse effect	Water resources Ecosystems Coastal erosion Agricultural production	Carbon dioxide Methane Azote protoxide Stratospheric, tropospheric ozone Halocarbons Sulfur hexafluoride	CO_2 CH_4 N_2O O_3 CFC, HCFC, PFC, HFC SF_6
Destruction stratospheric ozone layer	Health Decrease in agricultural yield	Stratospheric ozone CFC, HCFC, and related gases	O_3 CFC, HCFC, PFC, HFC
Acidification	Health Acidification waters and soils Forest decline Degradation of buildings	Sulfur dioxide Nitric oxides Ammoniac	SO_2 NO_x NH_3
Airborne particles	Health Danger by visibility lowering	Airborne particles $\Phi < 10\,\mu m$ $\Phi < 2\,\mu m$ Persistent organic pollutants	 PM10 PM2 POP, PAH
Photochemical pollution	Health Decrease in agricultural yield Forest decline Danger by visibility lowering	Organic volatile products (precursors) Nitric oxides (precursors) Tropospheric ozone	COVNM NO_x O_3
Dust deposition	Degradation of the living environment Accumulation in ecosystems	Sedimentable dust $\Phi > 10\,\mu m$ Heavy metals	 Hg, Pb, Cd, Cr...
Bioaccumulation	Health Accumulation in ecosystems and living beings	Persistent organic pollutants Heavy metals	POP, HAP Hg, Pb, Cd, Cr, ...

Analyzed impacts result either from gases emitted (ex: SO_2, CO_2, and NO_x) from a natural or anthropogenic source, or from gas arising from physical–chemical reaction taking place in the atmosphere (O_3).

2.2.1 The global cycle of fluorine

A schematic view of the fluorine cycle in our environment is given in Fig. 65.[364] According to first estimates,[365] fluoride emission levels to the atmosphere were undervalued and the different sources (dust, oceans, volcanoes, and industries) could not quantitatively account for the amounts of fluoride observed in precipitations. The various fluorine emissions must in fact be balanced with the quantities of fluorine measured in dry deposition and precipitation. In a stratospheric ozone assessment for the World Meteorological Organization (WMO-Global Ozone Research), the different atmospheric fluxes of fluorine were recorded and compared with values measured in precipitations (Table 18).[366] From this table, it can be noted that precipitations of about ~5.6 Tg $year^{-1}$ are required to balance the exchanges. This value, which corresponds to 11.7 ppb, is in relatively good agreement with the precipitation results measured in the USA, both in coastal areas (Delaware coasts) and inland (central Virginia), which provide an average fluorine of 8.1 ppb (within a scale of 1–24 ppb).[367] It can also be noted that the main contribution, coming from the 600 active volcanoes, is very localized around these sites and decreases extremely rapidly when one moves away from them a few kilometers (Table 19). But the eruptions cause much greater damage to our biosphere through the thermal shock of the explosion, ash deposits, mudslides, CO_2 asphyxiation, SO_2 and Cl_2 poisoning, fluorine emissions having a much-limited impact (Table 20)[368,369,370,371,372,373]

[364] S.·Tavener and J. Clark—Fluorine: Friend or Foe? A Green Chemist's Perspective, pp. 177–202. "Advances in Fluorine Science" series, Vol.2, A. Tressaud Ed., Elsevier Amsterdam (2006).

[365] Bewers, J. M., North Atlantic Fluoride Profiles, Deep Sea Res. **18** (1971) 237–241; Bewers, J. M., Miller G. R., Kester D. R., Warner T. B., Rapid changes in the fluoride to chlorinity ratio south of Greenland, Nature **242** (1973) 142–143.

[366] Friend J.P., Natural chlorine and fluorine in the atmosphere, water and precipitation, in "Scientific Assessment of stratospheric ozone: World Meteorological Organization Global Ozone Research", Report N° 20-Vol II (1989) 432–448.

[367] Barnard W.R., Nordstrom D.K., Fluoride in precipitation II. Implications for the geochemical cycling of fluoride, Atmosph. Environ., **16** (1982) 105–11. NB—Average fluorine levels in soils, water and air, as well as precipitation results have been measured in remote sites of fluoride-rich deposits or any fluorinated pollution.

[368] Tappen C.M., Webster J.D., Analysis of silicate melts inclusions from Augustine Volcano, Alaska, Eos. Trans. AGU, **84** (46) Fall Meet. Suppl., Abstract V42B-0343 (2003).

[369] Ablay, G., Ernst G. G. J., Marti J., Sparks R. S. J., The −2 ka subplinian eruption of Montaña Blanca, Tenerife, Bull. Volcanol. **57** (1995) 337–355.

[370] Shinohara H., Iguchi M., Hedenquist J.W., Koyaguchi T. Preface of a special issue devoted to Satsuma-Iwojima volcano, Earth Planets Space, **54** (2002), 173–174. Mori T., Sato M., Shimoike Y., Notsu K., High SiF_4/HF ratio detected in Satsuma-Iwojima volcano's plume by remote FT-IR observation, Earth Planets Space, **54** (2002) 249–256.

[371] Yoshida M., Uchikawa K. Hirabayashi J., Fluorine and chlorine contents in the lavas and ejecta of the 1986 eruption of Izu-Oshima volcano. J. Volcanol. Geotherm. Res., **63** (1994) 231–241.

[372] Uchikawa K., Yoshida M., Hirabayashi J., Fluorine and Chlorine Contents of Volcanic Rocks of the 1991–92 Eruption of Unzen Volcano, The volcanological society of Japan, Abstract B-34, p. 104 (1992).

[373] Matthews S. J., Gardeweg M. C., Sparks R.S.J., The 1984 to 1996 cyclic activity of Lascar Volcano, northern Chile: cycles of dome growth, dome subsidence, degassing and explosive eruptions Bull. Volcan. **59** (1997) 72–82.

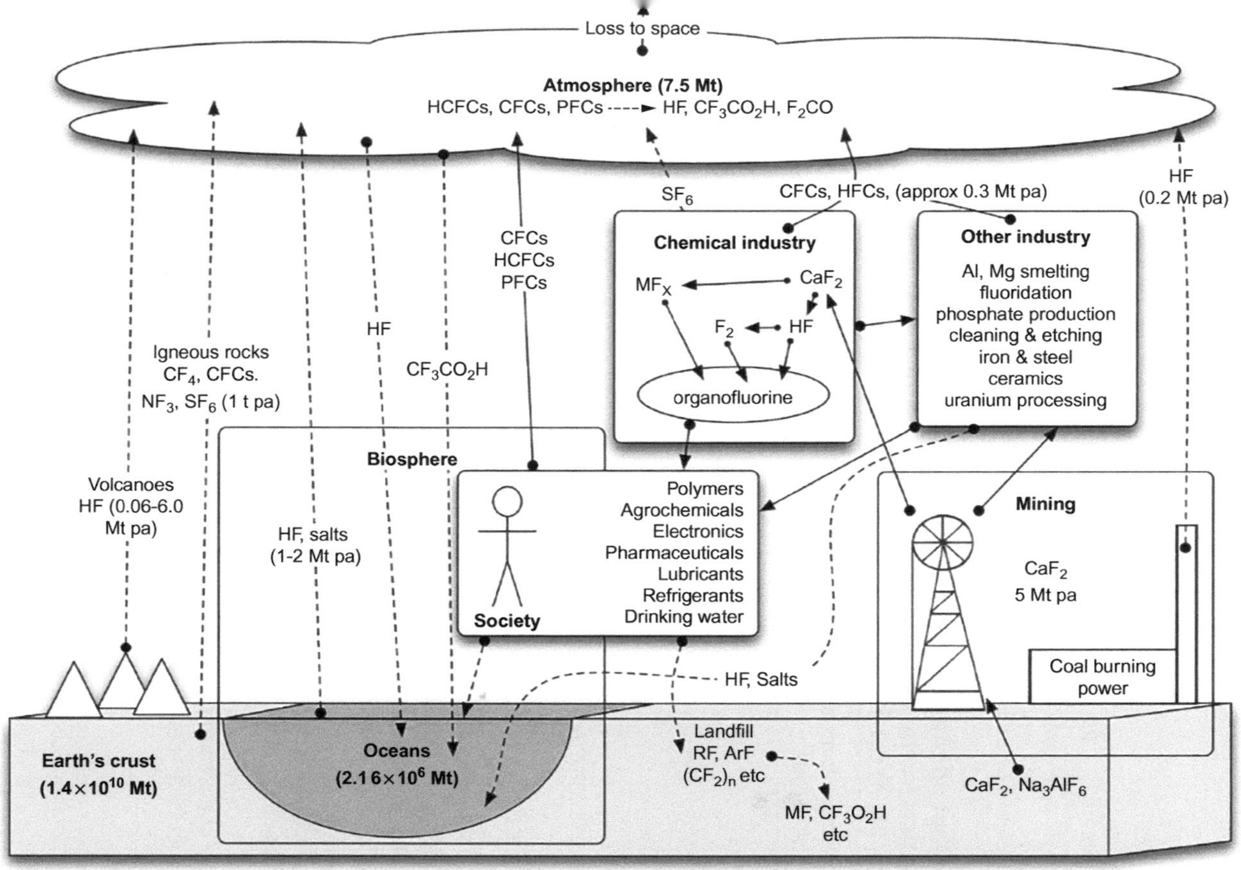

Fig. 65 A schematic view of the "Fluorosphere" representing the exchanges and cycles of fluorine in the atmosphere, according to S. Tavener and J. Clark.
From S. Tavener and J. Clark - Fluorine: Friend or Foe? A Green Chemist's Perspective, pp. 177–202 "Advances in Fluorine Science" series, Vol. 2, A. Tressaud Ed., Elsevier Amsterdam (2006).

Table 18 Maximum fluxes of fluorides and solid depositions

Flux ($Tg\,year^{-1}$)	Seawater	Soils	Volcanoes	Anthropogenic origine	Total
Flux of fluoride emissions toward atmosphere	1	0.5	6	0.5	8
Dry solid deposits	0.8	0.3	1.3		2.4
Precipitated fluorine content required for equilibrium between emissions and deposits					5.6

Table 19 Dependence of HF, HCl, SO_2 (in ppbv) concentrations on the distance to the initial crater (Eruption of 1995, Soufriere, Montserrat, West Indies)

Reported distance from the Soufrière crater (m)	HF (ppbv)	HCl (ppbv)	SO_2 (ppbv)
0	192.31	29,207.80	729.71
20	22.01	532.95	168.23
1400	3.06	197.55	11.18
3400	1.22	52.02	6.09
4900	0.81	13.98	5.89
6700	0.02	0.66	0.54

2.2.2 Assessing the rates of fluorine present in our environment

2.2.2.1 Evaluation of the fluorine content in air

Very few global measurements are available; fluoride levels in the atmosphere have been estimated at about $5\,ng\,m^{-3}$ by the World Health Organization[374] (2002 WHO report). Without any source of near pollution, this rate can decrease down to $0.5\,ng\,m^{-3}$, but can reach much higher values in other cases (e.g., 30–40$\,ng\,m^{-3}$ in the Netherlands) without any close source of contamination, 20–2000$\,ng\,m^{-3}$ in certain areas of the United States and Canada[375] to peak at 16,000–46,000$\,ng\,m^{-3}$ in some provinces of China.[376]

The main contribution in the atmosphere is of volcanic origin, and has been estimated at about 7000–8600$\,kt\,year^{-1}$.[377] It should be noted that the very small amounts

[374] Slooff W et al., eds. Basis document fluoriden. Bilthoven, Netherlands, National Institute of Public Health and Environmental Protection, 1988 (Report no. 758474005).

[375] Thompson R.J., et al., Fluoride concentrations in the ambient air, J. Air Pollution Control Assoc. **21** (1971) 484–487.

[376] Cao S.R., Li Y.F., The evaluation of indoor air quality in areas of endemic fluorosis caused by coal combustion. In: Proceedings XIXth Conf. Intern. Soc. Fluoride Res., Kyoto, Japan, (1992), Kyoto.

[377] Halmer M.M., Schminke H.U., Graf H.F., The annual volcanic gas input in the atmosphere, J. Volcan. Geotherm. Res. **115** (2002) 511–528.

Table 20 History of volcanic eruptions with significant fluoride emissions: types of eruptions, environmental degradation and loss of life

Volcano	Eruptions	Types of eruptions and environmental consequences
Augustine, Alaska	Six eruptions over 200 years	Analysis of silicate glassy inclusions of felsic crystals from the tephra layer dated to about 1400 years ago showed fluorine levels up to 510 ppm associated with barium (Tappen C.M., Webster J.D., Analysis of silicate melts inclusions from Augustine Volcano, Alaska, Eos. Trans. AGU, **84** (46) Fall Meet. Suppl., Abstract V42B-0343 (2003))
Montaña Blanca, Teide el Pico Viejo, Tenerife, Canaries	About 2000 year ago	The eruptions of Canary volcanoes were reconstructed in detail: extrusion from a crack of phonolite-type lava on about 0.022 km^3. The eruption then entered an explosive sub-Plinian phase during which, for 7–11 h, 0.25 km^3 of pumice stone of phonolitic type were deposited from a sub-Plinian column of 15 km high. The glass inclusions in the felsic crystals showed that the pumice came from the upper part of the magmatic chamber saturated with water. Beside 3.0–4.5 wt% H_2O, a significant fluorine content was found: about 0.35 wt%. According to Ablay, a volatile-rich magma is still accumulated in these systems (Ablay, G., Ernst G. G. J., Marti J., Sparks R. S. J., The −2 ka subplinian eruption of Montaña Blanca, Tenerife, Bull. Volcanol. **57** (1995) 337–355)
Laki, Iceland	1783	Lava fountains leaping from a crack reached a height of 1400 m, ejecting 8600 m^3 every second. In 4 days, the lava had advanced over 35 km from several faults. The jets of lava and burning gas lasted 7 months. The main contaminant was fluoride, which was immediately absorbed by the plants. The clouds of aerosols caused a cooling of 1 °C in North America and especially the death of 9000 people in Iceland

Continued

Table 20 **Continued**

Volcano	Eruptions	Types of eruptions and environmental consequences
Hekla, Iceland	→1991	55 eruptions since 1104. Description of the latest eruptions detailed in the text
Tambora, Sumbawa Island, Indonesia	1815	>88,000 people died. Most of the deaths were the result of massive ash rains and acid rain (halogenated) that covered the fields and polluted the sources of the island and nearby islands
Momotombo, Nicaragua	1905	The eruption of 1605–06 destroyed Leon, which was the capital of the country at the time; (0.26 vol% fluorine evaluated in the gas)
Katla, Iceland	1918	The clouds and ash rain lasted between 13 days and 4 months according to sources. The ash clouds are thought to have reached 15 km in altitude
Sakurajima and Satsuma-Iwo Jima, Kyushu, Japan	State of constant eruption since 1955	About 200 explosive eruptions and 10–30 million m^3 of ash per year. 3 million tons of magmatic gas, including HF, emitted per year for the Sakurajima volcano. 135 t of chlorine and 9 t of fluoride are emitted each day from the Satsuma-Iwo Jima volcano (Shinohara H., Iguchi M., Hedenquist J.W., Koyaguchi T. Preface of a special issue devoted to Satsuma-Iwo Jima volcano, Earth Planets Space, **54** (2002), 173–174; Mori T., Sato M., Shimoike Y., Notsu K., High SiF_4/HF ratio detected in Satsuma-Iwo Jima volcano's plume by remote FT-IR observation, Earth Planets Space, **54** (2002) 249–256).
Poàs, Nicaragua	1989, 1990, 1994, 1999	1955: Strombolian eruption; 1994: phreatic eruption and since fumaroles activity. The active crater has a hot acid lake (Laguna Caliente) containing SO_2, HCl, HF, with an average temperature of 35°C and a pH below 1.5. During the last explosions, the acid gases transported by the winds had a direct impact on the health of the populations and on the cattle
Mt. Saint Helens, Washington, United States	Mai 1980	The upper part and the heart of the volcano were blown and the whole area devastated. 57 people were killed and thousands of forest animals perished

Table 20 Continued

Volcano	Eruptions	Types of eruptions and environmental consequences
Izu Oshima, Japan	Nov. 1986	The eruption of the Izu-Oshima volcano from a fault produced a cone of slag and three lava flows of clastic origin with a lava fountain of 1.6 km high. Finally, 7.9 tons of products were ejected. (Yoshida M., Uchikawa K. Hirabayashi J., Fluorine and chlorine contents in the lavas and ejecta of the 1986 eruption of Izu-Oshima volcano. J. Volcanol. Geotherm. Res., **63** (1994) 231–241)
Unzen, Kyushu, Japan	1990–1994, 1996	1792: An avalanche and a tsunami due to the collapse of the lava dome of Mount Mayuyama caused the death of 14,524 people; 1990–1994: triggered by a phreatic eruption, $>$10,000 pyroclastic flows were counted during this period. 43 people were killed in June 1991, including vulcanologists M. & K. Krafft and H. Glicken. The fluorine and chlorine levels were similar: 230–285 and 220–355 ppm respectively. There are no significant differences between the contents of the pyroclastic flow and that of the lava dome. On the other hand, the pyroclastic flows of Asama volcanoes (Tenmei and Tenjin) contain less fluorine and chlorine. The last eruption, in February 1996, resulted in throwing stones and ash. (Uchikawa K., Yoshida M., Hirabayashi J., Fluorine and Chlorine Contents of Volcanic Rocks of the 1991–92 Eruption of Unzen Volcano, The volcanological society of Japan, Abstract B-34, p104 (1992))
Soufrière Hills, Montserrat, Antilles	August-Sept. 1995	Ash clouds reached an altitude of about 10 km and the amounts deposited at Plymouth on 17–18 September are estimated at 600,000 t. The high rate of fluoride detected in the ash suggests that fluorosis problems could affect livestock grazing around the volcano

Continued

Table 20 **Continued**

Volcano	Eruptions	Types of eruptions and environmental consequences
Ruapehu, New Zealand	1953, 1969, 1975, 1995, 1996	1953: 1.6 million m^3 water brutally released into the atmosphere caused a train disaster where 150 people perished. 1995: steam explosion; a study of the toxicity of fluoride was studied on a pregnant sheep grazing on pasture covered with ash; 1996: narrow plume with low ash emission
Merapi, Java	1994, 1997, 1998, 2001	November 1994: the collapse of the lava dome transforming into pyroclastic flows killed 64 people; January 1997: ash plume; July 1998: a plume of ash containing halogenated gas reached 3 km high and fell back on villages >60 km away. February 2001: pyroclastic flow over 7 km.
Popocatépetl, Mexico	Feb. 1997, Nov. 2000	Feb. 1997: plume 8 km high. Rapid increase of the SiF_4/SO_2 ratio observed before the eruption; this ratio decreased by a factor of 10 after this eruption. Nov. 2000: strongest eruption in 1200 years.
Lascar (Piedras Grandes, Soncor), Chili	1984–2000	The volcanic activity of the Lascar volcano, the most active in the central Andean zone, began 3000 years ago and featured various types of eruptions 26,500, 9200, and 7100 years ago. Since 1984, several demonstrations took place, as in April 1993 and July 2000. In July 2000, an eruptive column 4–5 km high was observed on the summit and a few mm of very fine ash (presence of fluorides) covered the city of Jama. (Matthews S.J., Gardeweg M.C., Sparks R.S.J., The 1984 to 1996 cyclic activity of Lascar Volcano, northern Chile: cycles of dome growth, dome subsidence, degassing and explosive eruptions Bull. Volcan. **59** (1997) 72–82)
Mt. Pago, Papouasie-New Guinea	August 2002	The presence of fluorine and SO_2 in the rain of very fine ashes mixed with silica particles increased the physiological risks associated with acid rain; Dec. 2003: small amounts of fine ash.

Table 20 Continued

Volcano	Eruptions	Types of eruptions and environmental consequences
Eyjafjallajökull, Iceland	April 2010	The eruptive activity of the Icelandic volcano Eyjafjöll, whose grey ash column could be up to 10km, has completely blocked all air traffic in Europe in April 2010. The plume was rich in ash from magma containing mainly feldspars, pyroxenes and/or amphiboles and containing among others fluorine. 850mg of fluorine were analyzed per kg of ash. According to Icelandic scientists, this represents 60% of the amount found in the most polluting ash of the Hekla.
Grímsvötn, Iceland	May 2011	The global rock analyzes indicate a characteristic composition of a basalt, with 50–51% silica (SiO_2). The leached parts of the samples contain 5 to 10mg. kg^{-1} of water-soluble fluorine. Analysis of the particle size distribution indicates that about 10% of the volume of the analyzed samples is below 10μm

of CFCs or perfluorinated compounds: CF_4 and SF_6 recently discovered in some volcanic emissions may be considered negligible compared to anthropogenic emissions. In industrialized areas, fluoride comes from essentially from industrial sources. It was 155kt of fluorinated dust that was sent to the atmosphere in 1968, of which 10% came from the aluminum industry. These various assessments, however, contain a large number of uncertainties and it should be noted that there is in fact no truly reliable quantitative analysis of these exchanges involved in our atmosphere. The atmospheric rate of hydrogen fluoride (CAS No. 7664-39-3) measured in the different states of the USA can however be consulted on the website http://scorecard.goodguide.com/.

2.2.2.2 Evaluation of the fluorine content in seawater

In the oceans, the F/Cl ratio is very low: 7×10^{-5}, corresponding to only 1.3ppm wt fluorine. On the other hand, close to rocks rich in fluoride-containing rocks, this rate can go up to $10\,mg\,L^{-1}$. This average value of fluoride in the oceans, $1.3 \pm 0.1\,mg\,L^{-1}$, varies little with depth. On this basis, given that the total ocean volume is 1.4×10^{21} L, the amount of fluorine present in this reservoir can be estimated at 1.8×10^{12} t. Some local high fluorine values that are sometimes found in the literature must be taken with caution because they may not take into account the presence of industrial discharges. The concentration of fluoride is controlled by the equilibrium between the species in solution: F^- (48%–51%), MgF^+ (47%–49%), CaF^+ (2%), NaF (1%) with percentages provided in parentheses, and the MgF_2 and CaF_2 solids. The difference in speciation of fluoride ions in seawater relative to freshwater is due to the significant presence of magnesium.[378,379] The residence time of fluoride in the oceans is estimated at 2–3 million years.[380]

A fluoride stream is regularly introduced into the oceans by river water, sub-aquatic volcanic emissions and industrial pollution. At the same time 20 kt of fluorine are released into the atmosphere as aerosols. Because the fluoride content in the oceans remains constant, other processes occur in parallel: incorporation into carbonates and phosphates of marine organisms, formation of fluorapatites in sediments.[381]

2.2.2.3 Fluorine content in freshwaters

The amount of fluoride in the world's unsalted waters (rivers, lakes) ranges from 0.01 ppm (detection limit) to 2 ppm (or $mg\,L^{-1}$), but strongly depends on the type of geological strata crossed by the water. When the water is not in contact with fluoridated rocks, the average concentrations are 0.1–0.4 $mg\,L^{-1}$. Among the various fluorinated species contained in water, SF_6 can be measured with great accuracy and rates of the order of a few dozen femto-moles L^{-1} was found when water flowed along fluorine-rich rocks[382].

The concentrations found in rivers can rapidly change over very short distances depending on the rocks crossed, for example, from <0.05 to >8 $mg\,L^{-1}$ in a few kilometers in Norway[383]; this property has made it possible to use the fluorine content in water as an indicator of the presence of fluorinated deposits.[384]

Studies in the desert areas of northwestern China have shown that the highest rates are found in arid plateaus where fluorides are concentrated in freshwater through a mechanism of evaporation and leaching of rocks. This increase occurs when altitude decreases and when Heihe River flows from the mountainous areas to desert basins, irrespective of the type of rock crossed.[385] It has also been noted that low levels of SO_4^{2-} and Ca^{2+} ions favor in alkaline waters the dissolution and migration of F^- ions.

Rates in some African volcanic lakes can reach up to 70 $mg\,L^{-1}$, but record levels have been found in the East African Rift with 300 and 700 $mg\,L^{-1}$ respectively in Ethiopia and Tanzania. The solubility of the fluoride ion is controlled by equilibrium with

[378] Stumm W., Morgan J.J., Aquatic chemistry, J. Wiley & Sons, New York (1981).

[379] Kester, D.R., Marine pollution: Winning or losing? Maritimes **33** (2) (1989) 1–4.

[380] Carpenter R., Factors controlling the marine geochemistry of fluorine, Geochim. Cosmochim Acta **33** (1969) 1153–1167.

[381] Cadle, R. D., A comparison of volcanic with other fluxes of atmospheric trace gas constituents, Reviews of Geophysics and Space Physics, **18** (1980) 746–752.

[382] Harnisch J. and Eisenhauer A., Natural CF_4 and SF_6 on Earth, Geophys. Res. Lett., **25** (1998) 2401–2404.

[383] Banks D., Frengstad B., Midtgård A.K., Krog J.R., Strand T., The chemistry of Norvegian groudwaters, Sci. Total Environ. **222** (1998) 71–91.

[384] Lalonde J., Fluorine: an indicator of mineral deposits, Can. Min. Metall. Bull. (1976) 110–122.

[385] Wang G.X., Cheng G.D., Fluoride distribution in water and the governing factors of environment in arid north-west China, J. Arid Environ. **49** (2001) 601–614.

CaF_2 (see Section 3.4.1.), But the concentration by evaporation leads to a precipitation of calcium carbonate (calcite) which increases the solubility of calcium carbonate.[386] In regions where the fluoride level in drinking water exceeds 1.5 mg L^{-1}, there is a risk of fluorosis for both humans and animals. UNICEF has compiled a list of some 30 countries that fall into this category, countries where high fluoride levels are often associated with particularly difficult living conditions. The effect of an excess of fluoride on the health of the man will be treated thereafter.

World Health Organization guidelines on water quality can be consulted in regular reports,[387] and effects on plants and animals have been extensively addressed by L. Weinstein and A. Davison.[388]

2.2.2.4 Fluorine in rocks and various types of soils

Fluorine is present to <0.1 wt% of the upper part of the Earth's crust, the lithosphere. It is involved in rocks, coal or clays, but also in the soil. In samples collected down to 8 cm, the concentration of fluoride ions is about 20–500 ppm, which leads to an average rate of 190 ppm. Concentration tends to increase with depth. The main fluorine reservoirs are fluorite, fluorapatite, bastnaesite, and fluoro-aluminates from the cryolite mine. These minerals will be discussed later in this chapter. More detailed studies based on the nature of the minerals in the soil have shown that concentrations ranged from <100 ppm to several 100 ppm (or mg kg^{-1}) depending on the type of mineral. As shown in Table 21, fluorine is associated with clay soils rather than sandy soils.[389,390,391,392,393] It is obvious that for soils rich in fluorinated minerals such as fluorspar, fluorophosphates, or cryolite, these values are largely exceeded. From these lands, fine fluoride particles can be transported in the air as dust. On Earth, fluoride ions are transported either by dry deposition, by more or less fine dust, or by wet, that

[386] Chernet T., Travi Y., Valles E., Mechanism of degradation of the quality of natural water in the lakes of the Ethiopian Rift Valley, Water Res. **35** (2001) 2819–2832.

[387] Guidelines for drinking-water quality, third edition, incorporating first addendum (2006), World Health Organization: (WHO), Geneva.

[388] Weinstein L.H., Fluoride and plant life, J. Occupational Medicine **19** (1977) 49–78; Weinstein L.H., Davison A.W., Fluorides in the Environment, Effects on Plants and Animals, CABI Publ., Cambridge, USA (2004).

[389] Fleischer M., Robinson W.D., Some problems of the geochemistry of fluorine, R. Soc. Canada Special Papers (1963) 58–75.

[390] Zhang, X.P.P., W. Deng, and X.M.M. Yang, The background concentrations of 13 soil trace elements and their relationships to parent materials and vegetation in Xizang (Tibet), China. Journal of Asian Earth Sciences, **21** (2002) 167–174.

[391] Cronin, S.J., Manoharan, V., Hedley, M.J., and P, Loganathan, Fluoride: A review of its fate, bioavailability, and risks of fluorosis in grazed-pasture systems in New Zealand. New Zealand J. Agr. Res., **43** (2000) 295–321.

[392] Loganathan P.; Hedley M.J.; Grace N.D.; Lee J.; Cronin S.J.; Bolan N.S.; Zanders J.M., Fertiliser contaminants in New Zealand grazed pasture with special reference to cadmium and fluorine, a review, Australian J. Soil Res. **41** (2003) 501–532.

[393] Gemmell G.D., Fluorine in New Zealand soils, New Zealand J. Sci. Technol. **27** (1946) 302–306. Nommik H., Fluorine in Swedish agricultural products, soil and drinking water, Acta Polytechnica, **127** (1953) 1–121.

Table 21 Fluorine rate in various types of minerals and soils

Type of soils	Sandstone	Andesite	Limestone	Dolomite	Basalt	Schists	Various types of rocks	Sedimentary soils (hosting phosphates or fluorspar)	Clay soils		Sandy soils	
Fluorine amount (ppm)	180	210	220	260	360	800	390–610	700	540	248–660	68	43–200
Country	United States						Tibet	N^{w} Zealand	N^{w} Zealand	Sweden	N^{w} Zealand	Sweden

Table 22 Deposits of fluorine via humid routes in various cities

Country	City	F^- conc. ($\mu eq L^{-1}$)
China	Beijing	19–21
	Liuzhou, Guangxi	65
	Shaoguan, Guangdong	11
Japan	Tokyo	1.8±1
Great Britain	Plynlimon, Wales	0.3
United States	Lewis, DE	0.3
	Charlottesville, VA	0.6
World (average)		5.3

is by aerosols. Table 22 focuses on the significant differences in fluoride levels deposited by this last way in various cities worldwide,[394,395,396] whereas Table 23 gives the average fluorine contents in air and water, thus indicating considerable differences in mean daily intakes.

2.3 The various fluorinated gases: Halocarbons, perfluorocarbons, SF_6

2.3.1 Environmental impact

Halocarbons are carbon compounds that contain halogens: fluorine, chlorine and/or bromine and possibly hydrogen. CFCs, HCFC, HFCs or halons are among them. To these halocarbons must be added other types of fluorinated gases such as perfluorocarbons and SF_6 which contribute to the increase of the greenhouse effect as GHGs. Table 24 lists the sectors that required the use of these gases these last decades: HFCs, HCFCs, PFCs, SF_6 in the United Kingdom.

Numerous articles have detailed the impact of these gases, in particular in light of the changes in industrial production that followed the Montreal Protocol decisions, corresponding globally to a suspension of CFCs, HCFCs and an increase in HFCs. For European Union member countries, the EEA GHG reports show that in the period 1990–2006, fluorinated gas emissions in 2003 amounted 3% of total greenhouse gas emissions, GHGs, that is, 64 Tg (in CO_2 equivalents) (Fig. 66; Table 25). It can be noted that while emissions from production operations have decreased significantly, those due to HFC consumption have increased due to the ban of CFCs and HCFCs.[397]

[394] Feng Y. W., Ogura N., Feng Z. W., Zhang F. Z., Shimizu H., The Concentrations and Sources of Fluoride in Atmospheric Depositions in Beijing, China Water, Air, and Soil Pollution, **145** (2003) 95–107.

[395] Wang W., Gao S., Wang J., Sakamoto K., Wang Q., Mizoguchi T., Maruyama T., Atmospheric pollution caused by coal combustion and its countermeasure. Fluoride pollution and its control, Procdgs. Intern. Symp. on Acidic Deposition and its Impacts, pp. 265–268, Dec. 1996, Tsukuba, Japan.

[396] Neal C., Fluoride variations in Welsh stream and soil water, Sci. Total Environ. **80** (1989) 213–223.

[397] Annual European Community greenhouse gas inventory 1990–2006; and inventory report 2008, EEA Technical report No 4/2005; site web: http://www.eea.europa.eu/publications/technical_report_2008_6.

Table 23 **Average amounts of fluorine in air, water, and daily uptakes**

Average amounts of fluorine	**Air**	**Air in very polluted Chinese zones**	**Sea water**	**Freshwaters**	**Volcanic lakes of African Rift**
	($<5.10^{-6}$ mg/m^3)	**(0.05 mg/m^3)**	**(1.3 mg/L)**	**(0.01–2 mg/L)**	**(20–700 mg/L)**
Average daily intake of fluorine	Industrialized countries (0.2–2.0 mg per person)	Near industrial sources of fluoride pollution, Netherlands, Sweden (0.06 mg per child)	Areas with high volcanic activity (30 mg/person)	Polluted areas Jiangxi, China (mg/person)	

Table 24 **HFC, HCFC, PFC, SF_6 emissions in UK from various industry fields**

Emissions from	Industry sectors	Percentage (%)
HFC, HCFC	Refrigeration and air conditioning	41
	Manufacture of halocarbons	36
	Aerosols (industry, medical inhalers)	21
	Foams, fire-fighting, solvents	2
PFC	Electronic and miscellaneous Industries	63
	Aluminum	32
	Refrigeration, fireproofing	5
SF_6	Manufacture of magnesium	43
	Electrical insulation	34
	Electronic industry, sports, shoes	23

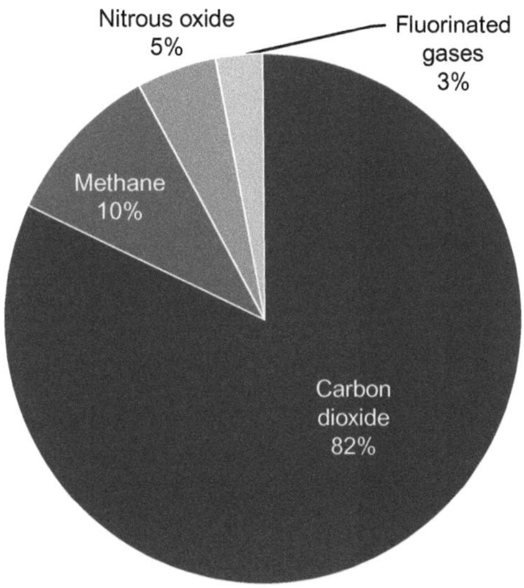

Fig. 66 Inventory of the US Greenhouse Gas Emissions and Sinks. U.S. Environmental Protection Agency (2017) (credit: https://www.epa.gov/ghgemissions/overview-greenhouse-gases).

Table 25 **Emissions of fluorinated gases in the EU (1990–2003)**

Emissions of fluorinated gases (equivalents CO_2)	Base level (Tg)	1990 (Tg)	2003 (Tg)
HFC	41	27	50
PFC	12	16	6
SF_6	15	10	9
Total GHG	4252	4238	4180

Table 26 Characteristic data of some CFCs and HCFCs

Type of gas	CFC-12	CFC-113	CFC-11	Halon-1301	Halon-1211	HCFCs
Formula	CCl_2F_2	CCl_2FCClF_2	CCl_3F	$CBrF_3$	$CBrClF_2$	CH_3CClF_2
Lifetime (years)	100	85	45	65	16	1–26
Global emissions in 2000 (Gg yr^{-1})	130–160	10–25	70–100	~3	~10	340–370
Potential for ozone depletion	1	1	1	12	6	0.02–0.12

2.3.2 HFCs and fluorinated ethers as substitutes for CFCs and HCFCs

CFCs and HCFCs have a significant direct contribution to increasing the energy transmitted back to the Earth's surface. Their replacement by HFCs, whose GWPs are much lower, is a positive step forward. Table 26 groups the lifetimes of the various species, that is, the time required for 60% of the gas to react as well as their potential for ozone depletion (with CFC-11 taken as reference).

Before CFCs were banned and replaced with HCFCs, HFCs were (relatively) little used, mainly in semiconductor manufacturing processes and as CFC additives for refrigerants. Their use now includes domestic and industrial refrigeration, air conditioning, aerosols, medical inhalers, and fire extinguishers. After increasing between 1990 and 1998, their emissions have fallen sharply since 1999. Globally, they have decreased by 24% since 1990, although the HFC-23 concentration more than tripled between 1978 and 1995. Although contributing little to radiative forcing, HFCs are still relatively undeveloped. Possible new phases include hydrofluoroethers, such as C_3F_7–O–CH_3, C_4F_9–O–CH_3, C_4F_9–O–C_2H_3 (Novec 3M range). Their GWP is between 50 and 400 and their lifespan is <5 years. They are only a few days for fluoroketones. In 2013, General Motors planned to market a hydrofluoroolefin (HFO-1234yf) produced by Honeywell and DuPont as a new refrigerant for cars. This molecule has a much lower impact on global warming. Its GWP is 4 compared to CO_2, while that of hydrofluorocarbon currently in service (HFC-134a) is 1340 times higher. Due to the banishment of the most-used CFCs, the emissions reduction that was observed after the peak of 1995–1998 is significant. There are certain areas in which these molecules are still irreplaceable, for example, expansion agents, thermosetting or thermoplastic foams, and fireproof products.[398]

Expansion agents: For many decades, CFC-11 has been widely acclaimed as an expanding agent for rigid polyurethane foam used in building construction because

[398] Jho C., Interactions of fire-fighting foam with hydrocarbon fuel, 5th Reebok Foam Seminar, **2013**, Bolton (G.-B.).

Table 27 Physical–chemical characteristics of products used or potentially usable as expansion agents

Product	CFC-11	HCFC-141b	HFC-134a	HFC-245fa	c-pentane	n-pentane
Formula	$CFCl_3$	CH_3CFCl_2	CF_3CFH_2	$CF_3CH_2CF_2H$	C_5H_{10}	C_5H_{12}
Molar mass ($g\,mol^{-1}$)	137	117	102	134	70	72
T_{boil} (°C)	23.8	32.1	−26.4	15.3	49.0	36.0
λ gas (mW/m. K) 25°C	8.7	9.7	13.5	12.2	12.0	15.0
Inflamm.	No	No	No	No	Yes	Yes
ODP	1	0.11			0	0
GWP_{100}	4000	630	1300	820	11	?
POCP	<0.1	0.1	?	?	?	64.4

λ: thermal conductivity/ODP: (Ozone Depletion Potential); GWP_{100}: (Global Warming Potential), calculated over 100 years; POCP: (Photochemical Ozone Creation Potential) (tropospheric); ?: values not available, but probably close to those of compounds of the same type.

of its high stability. A temporary alternative has been found in recent years with emphasis on HCFC 141b, which is much more unstable than CFC-11. However, the choice of HFCs appears much more promising in terms of sustainability. HFC-134a is already available to replace HCFC 142b/22 and HCFC 141b, and other molecules; HFC-245fa and HFC-365mfc are under study.[399] However, their use should require modifications in the methods of use.[400] The characteristics of these various classes of products used as expansion agents are summarized in Table 27. Note that other HFCs may also be considered, such as HFC-365.[401] Other non-halogenated substitution molecules, such as c-pentane or n-pentane, have some advantages in terms of ODP or GWP100, but they are flammable and contribute to the creation of tropospheric ozone. In fact, hydrocarbons have very short lifetimes in the atmosphere, and their decomposition products, including ozone, are found in the troposphere. Unlike the depletion of the stratospheric ozone layer (between 12 and 25 km from the surface of the Earth), excess ozone in the troposphere (0–10 km) is to be avoided, because it is a lung irritant often related to pollution of industrialized areas.

Fireproof products: Because of their high stability and their insulating nature, haloalkanes or halons have been used to fight fires very effectively. The most-used ones were bromochlorodifluoromethane (Halon 1211: CF2BrCl) and bromotrifluoromethane (Halon 1301: CF_3Br). HFCs such as HFC-23 (CHF_3) and 227ea (CF_3CHFCF_3) are now

[399] Wu J., Albouy A., Mouton D., J. Cell. Plast. **35** (1999) 421.

[400] Albouy A., Tersac G., Substitution of CFCs in rigid polyurethane foams,L'Actualité Chimique, 4, (2000), pp. 20–26 haL-01254727v1.

[401] For instance, Solkane® is proposed by Solvay Co—http://www.solvay-fluor.com/news/.

proposed as substitutes because of their lower toxicity and ODP. But they require a 50% higher concentration for an extinction power equivalent to halons.

Refrigerants: Although HFCs today appear necessary, alternative families of fluids are being tested. Thus, HFC ethers such as partially fluorinated dimethyl ethers (HFE): CF_3OCH_3 (E-143a) and CF_3OCHF_2 (E-125) have lower GWP than HFC-134a or HFC-125.[402] These phases can be easily oxidized by the $OH^{\bullet}$ radicals of the troposphere.[403] However, some of their properties (low refrigeration capacities, economically unviable synthetic routes) counterbalance the minimal advantage of a lower GWP. The Australian government is developing research programs on HFC-152a, which has a lower GWP than most other refrigerants.

2.3.3 Perfluorocarbons

In addition to their application in biomedical technologies and in cosmetics, PFCs are largely used in the stripping and CVD stages in the electronics and semiconductor industry, leak testing, radar cooling circuits, domestic refrigerant blends, fire products, and sport shoes. Note that PFCs are also byproducts formed during aluminum steel operations (see Chapter 8.1.2 in Section 1).

2.3.4 Sulfur hexafluoride

SF_6 applications include electrical insulation of high and medium voltage equipment, protection of molten magnesium in foundries, stripping of Si and N-based surfaces, and use as an atmospheric tracer. A famous brand filled SF_6 cushions in some models of sports shoes to achieve a damping effect under the heel. Studies are underway to limit SF_6 emissions through recovery and recycling in equipment containing molten magnesium, or by substituting lower GWP gases. For example, in Australia, the Cooperative Research Center for Cast Metals Manufacturing (CAST) proposes replacing it with HFC-134a to prevent the oxidation of molten magnesium.

Another molecule has been identified that has a lifespan of around 1000 years: SF_5CF_3.[404] The greenhouse potential of this molecule is equivalent to nearly 18,000 t CO_2. It should be noted that its rate is tiny, 0.1 ppt (10^{-12}), but its increase is problematic because it is about 6% per year.[405] It occurs by recombination of

[402] Sekiya A., Misaki S., The potential of hydrofluoroethers to replace CFCs, HCFCs and PFCs, J. Fluor; Chem. **101** (2000) 215–221; Sekiya A., M. Yamabe, K. Tokuhashi, Y. Hibino, R. Imasu and H. Okamoto, Evaluation and selection of CFC alternatives, pp. 33–87, Fluorine and the Environment, vol. **1**, A. Tressaud Ed., Elsevier(2006).

[403] Tokuhashi K., Chen L.,Kutsuna S., Uchimaru T., Sugie M., Sekiya A., "Environmental assessment of CFC alternative," J. Fluor. Chem. **125** (2004) 1801–1807.

[404] Sturges W.T. et al., Science, **289** (2000), 611.

[405] J. Bartyzel, J. Rosiek, K. Rozanski, W.T. Sturges, Atmospheric concentrations of SF_5CF_3 over central Europe during the period 2001–2008, Fifth International Symposium on Non-CO_2 Greenhouse Gases (NCGG-5), Wageningen, The Netherlands, June 30—July 3, 2009.

$SF_5^{\bullet}$ and $CF_3^{\bullet}$ radicals.[406] The first is produced when using SF_6 in microelectronic etching, and the second is a byproduct of fluoropolymers.[407] Electrochemical processes leading to "triflic acid" (trifluoromethane sulfonate) could also be another source.

2.4 Control and limitation of emissions of fluorinated gases in relation to the risks of climate change

2.4.1 Control of emissions of fluorinated gases

In the last 20 years of the 20th century, the observation of two major phenomena (the thinning of the ozone layer in certain sensitive parts of the stratosphere and the significant global warming, one origin of which could be an increase in the greenhouse effect) gave rise to a collective awareness at the global level that led to the Montreal Protocols and then to Kyoto (see Chapter 8.2.2 in Section 1). It may be considered that, if there had been no rapid recognition of the depletion of the ozone layer following the scientific discovery, the problem could have rapidly become unsolvable.[408] This global awareness has led researchers and companies to surpass themselves in their quest for technological breakthroughs that could allow them to maintain the same safety and efficiency standards as with CFCs, or to find products even more efficient and less harmful, for example, brominated extinguishing products.[409] The 1987 Montreal Protocol and its many recommendations were put in place to control ozone depletion. With these measures taken, it was hoped that decline would occur in the next decade(s), but the eventual reconstruction of the initial potential of the ozone layer would take a much longer time. More than 30 years after the first meeting in Montreal, the elimination of CFCs in developed countries has proven slower than originally planned. CFC and HCFC stocks are still very large, especially in many emerging countries, some of which are allowed to continue manufacturing CFCs. Some producers from developed countries have transferred there, quite illegally, reclassifying their CFC stock surpluses as "recycled" for the occasion. To avoid such anomalies, the European Union decided to ban, from January 2000 forward, all uses of recycled CFCs, but this was not the case in the United States. For example, with respect to the reduction of ozone depleting substances (ODS), if the pre-1987 rate had been maintained, that is, an increase of 3% per annum, the stratospheric ODS would have reached 17 ppb in 2050. The 1987 Montreal Protocol proposals lowered this rate to 9 ppb and the various Montreal-1997 amendments to 2 ppb. The 1997 Kyoto Protocol

[406] Tuckett R.P., Trifluoromethyl sulphur pentafluoride, SF_5CF_3: atmospheric chemistry and its environmental importance via the greenhouse effect, Fluorine and the Environment, vol. **1**, A. Tressaud Ed., Elsevier (2006).

[407] Hites et al., Clarifying the SF_5CF_3, record, Science **290** (2000) 935–936.

[408] Prather M., Midgley P., Rowland F.S., Stolarski R., Nature **381** (1996) 551.

[409] R.E. Banks, B.E. Smart, J.C. Tatlow, Organofluorine Chemistry. Principles and Commercial Applications, Plenum Press, New York, (1994); R. E. Banks, pp. 12, in Organo-fluorine compounds, Vol. **E 10a**, 4th ed. ((eds.): B. Baasner, H. Hagemann, J. C. Tatlow), Georg Thieme, Stuttgart, (1998).

was relative to the increase of the greenhouse effect and the climate change associated with it. The industrialized countries that ratified the Protocol committed themselves to reducing their greenhouse gas (GHG) emissions to the 1990 level between 2008 and 2012, but developing countries were not affected by this measure. During the negotiations, the CFCs and derivatives were included into this list of risky gases because of their very high long-term stability, as were SF_6, CO_2, CF_4, N_2O, and perfluorocarbons (PFC). Each country sought to reduce GHG emissions in 2010, on the basis of 1992 emissions. However, each country was free to regulate this limitation, for example, reducing CO_2 emissions to moderately increase those in HFC.

It is clear that the objectives of these two programs are highly interconnected. In 2001, the United States had clearly announced their intention to ratify the Kyoto Protocol, arguing in particular that it did not limit the carbon emissions of countries like China and India. The European Union, which has adhered to the Kyoto directives, has embarked on a diplomatic path to the United States to invest in the process of limiting emissions. Russia and China have since ratified the Kyoto protocol. When all committed countries correspond to more than 55% of CO_2 emissions, the Kyoto Protocol will enter into implementation. Due to the European Union's "leading" position, actions taken now in the power generation, industrial processes, waste rationalization, and agriculture sectors are enabling many countries to take lead in developing cleaner technologies. In August 2003, the European Commission proposed to reduce the influence of GHGs mainly used in refrigeration and air-conditioning equipment, particularly automobiles. One of the key points of this proposal is the removal of one of these gases, HFC-134a, largely used for air conditioning in new vehicles. The Commission also planned to reinforce the restrictions for monitoring emissions of fluorinated gases, in order to improve their containment and to restrict their use if not fully justified. The objective was to reduce these gases in the EU by around 23 Mt CO_2-equivalent by 2010 and to achieve even greater reductions thereafter. According to the Commission's figures, emissions of such gases amounted to 65 Mt CO_2-equivalent in 1995 and, due to the lack of restrictions, they were expected to increase to around 98 Mt by 2010. On May 17, 2006, a new European Directive strengthened the control of fluorinated greenhouse gases and emissions of air-conditioning systems of motor vehicles. A visualization of the evolution of the ozone depletion is given in a video from NASA.[410]

The European Parliament has adopted two drastic directives to ban CFCs, HCFCs, and other HFCs starting from 2015: the MAC (for mobile air conditioning) directive, and the F-gas Regulation, which covers all other applications where fluorinated gases are involved. The current phase of the MAC Directive started on January 1, 2017, and requires the use of gases with GWP not exceeding 150 for automotive refrigeration in all cars and trucks produced from that date. Among the banned phases, let us mention HFC R-134a, of formula CH_2FCF_3, which has been used for several decades for air conditioning in automobiles. Today, it is hydrofluoroalkene or hydrofluoroolefins (HFO) which are the best candidates to replace this molecule,

[410]https://www.youtube.com/watch?v=lBu3vltczRw.

in particular 2,3,3,3-tetrafluoropropene, of formula $CH_2{=}CFCF_3$ with nomenclature R-1234yf or HFO-1234yf (GWP=4). Similarly, for firefighting, nonbrominated extinguishing products are proposed and fluorotelomer-based foams are also used, because fluorinated PFOS-based surfactants (perfluorooctane sulphonate) are banned. Various information documents on the use of halocarbons are reported in.[411]

The latest report from European Union on fluorinated greenhouse gases 2017 brings together all data reported by companies on the production, import, export, and destruction of fluorinated greenhouse gases in the European Union.[412] The EU policy is illustrated by the implementation of new European F-gas Regulation (No 517/2014), aiming to reduce F-gas emissions by two-thirds of the 2010 level by 2030. The progress and predictions of the EU in terms of HFC phase-down is given in Fig. 67. The 2016 EU total supplies of fluorinated gases by supply types, groups of fluorinated gases, and types of applications are compared in Fig. 68 in % CO_2 equivalents.

2.4.2 Control and limitation of fluorinated greenhouse gases (GHG)

In the troposphere, the impact of greenhouse gases (GHG) is to increase the percentage of heat returned to the Earth by absorption of the re-emitted IR radiation. According to the Kyoto Protocol, GHGs are grouped into five categories: CO_2, CH_4, N_2O, SF_6, and fluorinated gases (PFCs, CFCs, and derivatives), as shown in Figs 66 and 68 and Table 28. Among these species, we note several families of halogenated gases whose contribution is rather weak (3%) because of the lower volume involved; these are halocarbons such as chlorofluorocarbons (CFCs) and hydro (chloro)-fluorocarbons (HFCs and HCFCs) that were proposed as substitutes for CFCs, perfluorocarbons (PFCs), sulfur hexafluoride (SF_6), and derivatives. The CO_2 gas emissions are relative to energy sources of fossil fuels, as used in many countries (electricity, heating, transportation, industrial sector) and tropical deforestation; for CH_4 it is relative to the exploitation and distribution of gas and coal, agricultural activity, and the fermentation of organic waste. The N_2O emissions are relative to the storage and use of nitrogen fertilizers, and the production of nitric acid. With regard to fluorinated gases, it is refrigerants, electrical insulators, and aerosols that are mostly concerned. The direct contribution of a GHG to the balance of the planet is expressed in GWP (Global Warming Potential). For example, $GWP_{100} = 1000$ means that over a period of 100 years, the emission of 1 kg of the product will be identical to that of 1000 kg of CO_2. CFCs have a significant direct contribution to the heat returned to the Earth's surface. Table 26 groups the atmospheric lifetime and potential for ozone depletion of several fluorinated GHGs. Given the very low rate of these fluorinated species in the atmosphere, it is possible that their contribution will remain extremely limited, even by 2050.

[411] See for example documents proposed by Halocarbone Corp. (www.halocarbon.com/literature).

[412] https://www.eea.europa.eu/publications/fluorinated-greenhouse-gases-2017.

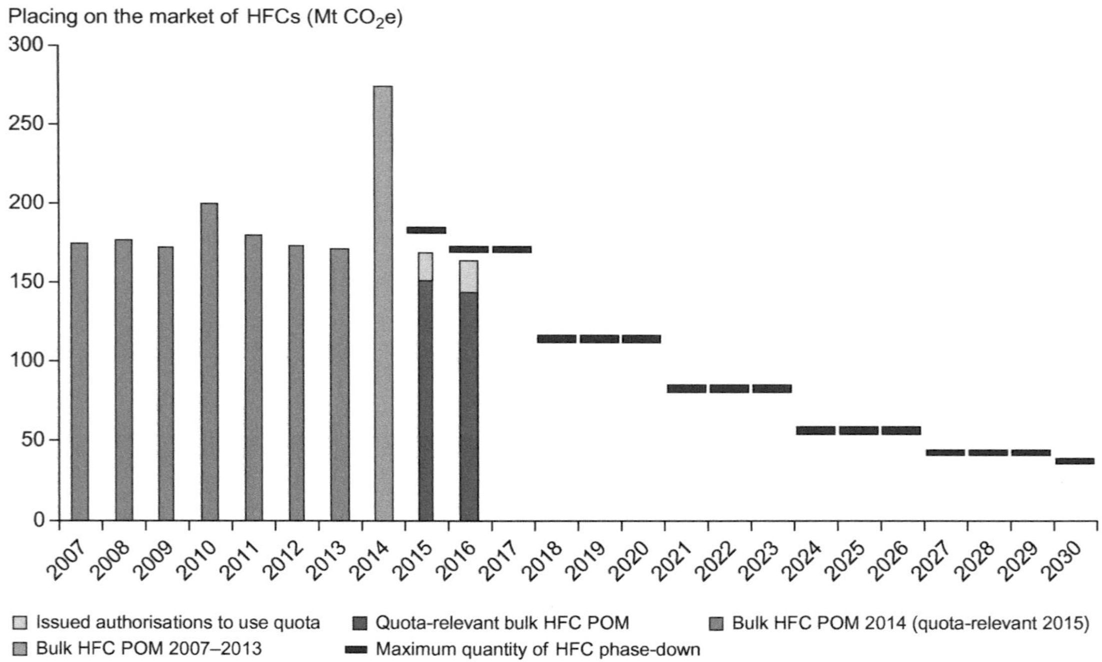

Notes: POM, placing on the market.

Values from 2007 to 2013 are based on the reporting obligations of the old F-Gas Regulation (EC) N0 842/2006 and are therefore not fully comparable to data from 2014 onwards (based on obligations of the new F-Gas Regulation No 517/2014). Similarly, the maximum quantities of the EU HFC phase-down will be recalculated for 2018 and are therefore for indicative purposes omly.

The quantity of bulk HFCs placed on the market in 2014 was used to determine 2015 quotas.

Sources: EC, 2011 and 2017a: EEA, 2016 and 2017b.

Fig. 67 Progress and predictions of the European Union for HFC phase-down (credit: https://www.eea.europa.eu/publications/fluorinated-greenhouse-gases-2017).

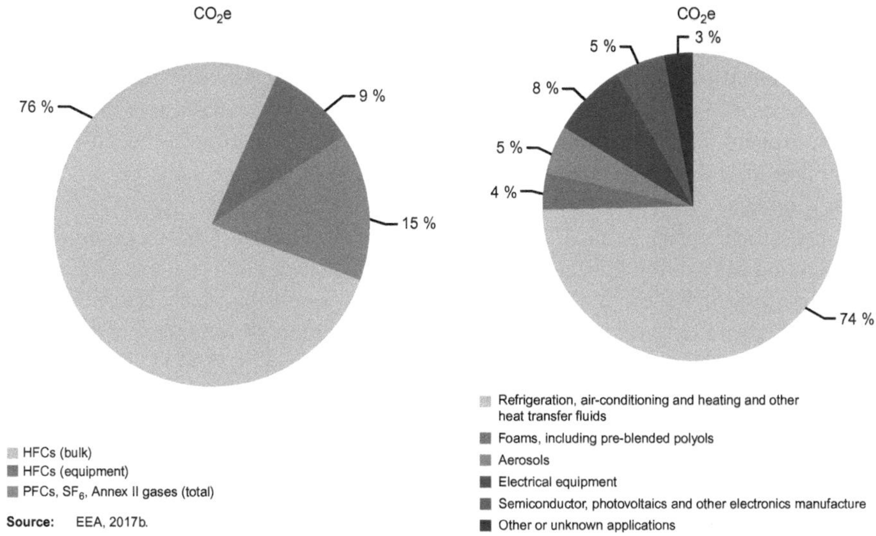

Fig. 68 The 2016 EU total supplies of fluorinated gases by supply types, groups of fluorinated gases and applications in % CO_2 equivalents (credit: https://www.eea.europa.eu/publications/fluorinated-greenhouse-gases-2017).

Table 28 Contribution of various GHGs, including fluorinated gases, and breakdown by sector (according to the Interministerial Mission for the Greenhouse Effect, France)

Type of gas	CO_2	CH_4	N_2O	CFC, HCFC, HFC	SF_6	PFC
(%)	69	13	16	1.7	0.4	0.3
Fields	Transportation	Industries	Agriculture	Building	Energy	Waste
(%)	26	21.3	19.35	18.6	12	2.5

2.4.3 *Contribution of fluorinated greenhouse gases to radiative forcing*

The concept of radiative forcing (RF) is fairly straightforward. Energy constantly flows into the atmosphere in the form of sunlight. About 30% of this radiation is reflected back to space, and the rest is absorbed by the planet, and some energy is always radiating back out into space as invisible infrared light. Radiative forcing or climate forcing is the difference between insolation by sunlight absorbed by the Earth and energy radiated back to space. Positive forcing (incoming energy exceeding outgoing energy) warms the system, while negative forcing (outgoing energy exceeding incoming energy) cools it. Causes of radiative forcing include changes in insolation and the concentrations of RF gases, such as GHG and aerosols. Radiative forcing is quantified at the tropopause, that is, at the border between troposphere and stratosphere. At this place, the temperature is stable, while a decrease is observed in the atmosphere from the ground and an increase in the stratosphere because of the absorption of UV rays by ozone. RF is measured in in units of watts per m^2 of the Earth's surface.

Among the anthropogenic GHGs, dichlorodifluoromethane (CFC-12) used for car refrigeration has was the third largest one for several decades before it was phased out. The evolution of RF from all halocarbons ($0.36\,W\,m^{-2}$) this last decade[413] has a reduced RF from chlorofluorocarbons (CFCs) but increases from many of their substitutes. Four of the halocarbons (trichlorofluoromethane (CFC-11), CFC-12, trichlorotrifluoroethane (CFC-113), and chlorodifluoromethane (HCFC-22)) account for around 85% of the total halocarbon RF. The first three have been declining over the last 5 years, but their combined decrease is compensated for by the increase from HCFC-22. The HFCs' RF has nearly doubled but still only amounts to $0.02\,W\,m^{-2}$, illustrating that their growth rate is smaller over the last decade than in the 1970s and 1980s.

Ozone-depleting substances (ODSs). Most of those compounds do not have natural emissions, and, because of the implementation of the Montreal Protocol and its amendments, total emissions of ozone-depleting substances have sharply decreased since the 1990s.[414] For CFCs, perfluorocarbons (PFCs), and SF_6, the main loss is through photolysis in the stratosphere. The CFC substitutes (hydrochlorofluorocarbons (HCFCs) and hydrofluorocarbons (HFCs)) are destroyed by OH oxidation in the troposphere. Their global concentration has steadily risen over the recent past.

The CFCs and HCFCs contribute approximately 11% of the global RF. Although emissions have been drastically reduced for CFCs, their long lifetimes mean that reductions take substantial time to affect their concentrations. The RF from CFCs has declined since 2005 (mainly due to a reduction in the concentrations of CFC-11 and CFC-12), whereas the RF from HCFCs is still rising (mainly due to HCFC-22). The RF of HFCs is $0.02\,W\,m^{-2}$ and has doubled compared to 2005 concentrations. HFC-134a is the dominant contributor, with an $RF = 0.01\,W\,m^{-2}$. Because the lifetimes of perfluorocarbons and SF_6 are thousands to tens of thousands of years long, their emissions essentially accumulate in the atmosphere. CF_4 has a natural source with a concentration of 35 ppt evaluated for the year 1750. These gases contribute $0.01\,W\,m^{-2}$ of the global RF values. NF_3 is used in the electronics industry

[413] Myhre, G., D. Shindell, F.-M. Bréon, W. Collins, J. Fuglestvedt, J. Huang, D. Koch, J.-F. Lamarque, D. Lee, B. Mendoza, T. Nakajima, A. Robock, G. Stephens, T. Takemura and H. Zhang, (2013), Anthropogenic and Natural Radiative Forcing. In: Climate Change 2013: The Physical Science Basis. Contribution of Working Group I to the Fifth Assessment Report of the Intergovernmental Panel on Climate Change (Stocker, T.F., D. Qin, G.-K. Plattner, M. Tignor, S.K. Allen, J. Boschung, A. Nauels, Y. Xia, V. Bex and P.M. Midgley (eds.)). Cambridge University Press: New York, NY, USA.

[414] Montzka, S. A., E. J. Dlugokencky, J. H. Butler, Non-CO_2 greenhouse gases and climate change, Nature, **476** (2011) 43–50.

and SOF_2 is used as a fumigant, but they only contribute around $0.0002\,W\,m^{-2}$ and $0.0003\,W\,m^{-2}$ to anthropogenic RF, respectively.[415]

2.4.4 What substitutes for CFCs, HCFCs, and HFCs?

For 15 years, many molecules have been tested as substitutes for CFCs. HFC arose as an option in the simple observation that the C–H bond in an organic molecule may be more easily attacked by $OH^{\bullet}$ hydroxyl radicals present in the troposphere than are C–Cl or C–F bonds, thus reducing the lifetime of these species, and thus, their GWP. The main properties that guide this choice are HFC's physicochemical properties which are consistent with a respect for the environment (in particular, a limited stability of long-term molecules), flammability, nontoxicity, and thermodynamic characteristics similar to those of CFCs, including the maximum operating pressures. Some examples will be given in the next section.

The impact of the CFC problem on science policies, that is, the global awareness of official health agencies and governments on the environmental persistence of potentially risky products, has been more recently illustrated in several other areas, in particular, by the phasing out of the production of perfluorooctanyl sulfonates, PFOS (see Chapter 8.3.4 in Section 1).

There is no panacea for finding new substitutes, and non-fluorinated substitutes for CFCs and derivatives are not without health and risk issues. The European Technical Committee on FluoroCarbones (EFCT)[416] has indeed underlined certain risks related to the replacement of HFCs by more "natural" fluids that are not compatible with the safety standards. For example, sulfuryl fluoride (SO_2F_2) was considered as a substitute for methyl bromide (CH_3Br) for post-harvest fumigation of crops; however, the National Research Council of Canada asked whether it was safe to maintain the use of this product, because metabolized residues could be found in nuts or grapes as toxic inorganic fluorides.

Now, HFCs are potent greenhouse gases with high global warming potential (GWP) and long lifetimes. So, they face international replacement by either hydrofluoroolefins (HFO) or "natural" refrigerants under the October 2016 Kigali Amendment to the Montreal Protocol.

[415] Weiss, R., J. Muhle, P. Salameh, and C. Harth, (2008). Nitrogen trifluoride in the global atmosphere. Geophys. Res. Lett., **35**, L20821; Muhle, J., et al., (2009). Sulfuryl fluoride in the global atmosphere. J. Geophys. Res. Atmos., **114**, D05306; Andersen, M., D. Blake, F. Rowland, M. Hurley, and T. Wallington, (2009). Atmospheric chemistry of sulfuryl fluoride: Reaction with OH radicals, CI atoms and O3, atmospheric lifetime, IR spectrum, and global warming potential. Environ. Sci. Technol., **43**, 1067–1070; Arnold, T., et al., (2013). Nitrogen trifluoride global emissions estimated from updated atmospheric measurements, Proc. Natl. Acad. Sci. U.S.A, **110**, 2029–2034.

[416] The European FluoroCarbon Technical Committee (http://www.fluorocarbons.org/) brings together the main European manufacturers of fluorinated gases. He is taking part in the climate change debate, notably following the changes proposed by the EU according to the Montreal Protocol and Kyoto, fighting illegal imports from countries that are still allowed to produce CFCs, providing governments with assistance in the risk assessment of products such as HCFC-22.

Many destruction programs for CFCs and HCFCs stocks are in progress, for example, the photo decomposition under UV.[417] However, it should be made clear that, given the very low rate of these fluorinated gases in the atmosphere, that is to say, $<1\%$ of all GHGs of anthropogenic origin, their direct contribution to global warming is extremely limited, and will remain so by 2050.[121] In conclusion, the "CFC effect," that is to say, the global awareness of official health agencies and governments regarding the environmental persistence of potentially risky products, has been more recently generalized in other areas, for instance, by phasing out the production of perfluorooctanyl sulfonates, PFOS and derivatives (see next section).

3 Fluorine and the lithosphere

3.1 Natural fluorinated compounds present in our environment and the various sources of fluorine

3.1.1 Discovery of fluorine gas (F_2) on earth

Because fluorine gas (F_2) is so incredibly reactive, it had been claimed it did not occur in nature and could be found only in combination with other elements, although it is relatively common in the Earth's. For a long time, among the numerous forms of fluorite (CaF_2) crystals, some dark varieties of a dark violet or even black color were known as antozonite, also called fetid fluorite or stinkspar, because they gave off a characteristic smell when crushed. There were suggestions that ozone or fluorine gas was responsible for the smell, but the exact cause of the odor remained a mystery. After isolating fluorine, Henri Moissan was very motivated to solve this problem, and he convinced his colleague Henri Becquerel to try characterizing the gas which was imbedded as bubbles inside the dark antozonite minerals from Quincié-en-Beaujolais, France. Although they could not find a definitive answer to the question, they came to the conclusion that elemental fluorine, and not ozone, was occluded in the mineral.[418] More recently, M. Reichling and coworkers observed the formation of sub-surface gas bubbles on electron irradiated CaF_2 crystals and assumed F_2 to be present.[419]

Wanting more proof, Florian Kraus, a chemist from Technical University of Munich, decided in 2011 to investigate with his team the story of these "fetid fluorites." He drove to a disused mine in Wölsendorf and picked up some of these fluorite rocks. He first tried to characterize the gas which came from the crushed crystals using a conventional mass spectrometer, but could not find F_2 and concluded that the gas must have reacted with parts of the mass spectrometer before the detection. Then they carried out ^{19}F solid-state NMR on pea-sized samples of antozonite

[417] Nishiumi H., K. Sato, R. Kato. Production of Second- or Third-Generation Fluorine-based Refrigerants from (Photo)-Dechlorination of Fluorocarbon Wastes, 133–166, Fluorine and the Environment, Vol. **1**, (ed.) A. Tressaud Elsevier (2006).

[418] H. Becquerel, H. Moissan, Etude de la fluorine de Quincié, Bull. Soc. Chim. Fr., **5** (1891), 154–157.

[419] R. Bennewitz, D. Smith, M. Reichling, Bulk and surface processes in low-energy-electron-induced decomposition of CaF2, Phys. Rev. **B59** (1999) 8237–8246.

crystals, and finally, they found a peak at 425 ppm that corresponds to the range expected for difluorine F_2. For the first time, free fluorine gas was discovered on Earth's surface.[420]

But how can F_2—the most reactive of the elements—be produced in these rocks and come to stay for millennia? The antozonite from Wölsendorf was known to contain uranium, mainly as UF_4. Tiny amounts of radioactive ^{238}U decay into β-emitting daughter nuclides. The β radiation strips fluoride anions of their extra electron. The resulting fluorine atom quickly forms a difluoride anion that is highly mobile. The anion loses an electron at a crystal defect and forms difluorine molecules, F_2, which are protected by CaF_2 layers. In some cases, bubbles up to 200 nm in size were seen with scanning electron microscopy. Clues from previous experiments confirmed how fluorine might be formed in the rocks. The experiments exposed artificial calcium fluoride to β- and γ-radiation, and high-energy electron beams. The ionizing radiation was splitting calcium fluoride apart to form clusters of calcium, which gave antozonite its dark-violet color. Subsequent tests showed that bubbles of fluorine gas were also forming in the lenses, and that high-quality synthetic fluorite used as optical components could be affected by radiation damage.

The same process could explain the stench of antozonite, says Kraus. The rocks have been lying around for 100 million years, which is enough time for the radioactive decay to produce the same effect as seen in the artificial fluorite experiments.

3.1.2 The fluorinated minerals

As was said in the introduction, it is essentially in the minerals of the earth's crust that the majority of natural fluoride is present in our environment. In the depths of the earth's crust, magmas contain dissolved gases. As they rise to the surface, these form a separate phase due to the pressure decrease. In the earth's crust, 1% fluorine concentrations are not unusual in alkaline-containing felsic rocks. Fluoride can, therefore, be considered one of the most important volatile components of magma and granitic rocks. The level of fluorine in the magma considerably modifies the properties of the latter, transport, viscosity, and eruptive properties, and so, influences the compositions of the resulting rocks: 360 ppm of fluorine in basalts, 245 ppm in andesites, and 740 ppm in rhyolites.[8] These data explain why fluoride is found in many silicates or aluminosilicates, especially in amphiboles and micas. In the Earth's crust, there are more than 300 minerals that contain up to 5% of fluorine from the griceite (LiF), which contains 57 wt% until fersmanite, a pyrosilicate (sorosilicate) of the family of cuspidine, with formula $((Ca,Na)_4(Ti,Nb)_2Si_2O_{11}(F,OH)_2)$. Fluorine is present in gems such as topaz, apatite, or fluorite. The presence of fluorine often gives these minerals quite unique properties. New fluorinated minerals are regularly discovered, for example, Rosenbergite, a hydrated aluminum trifluoride $AlF_3{\cdot}3H_2O$,[421] and

[420] Schmedt auf der Günne J., Mangstl M., Kraus F., Occurrence of difluorine F_2 in nature–in situ proof and quantification by NMR spectroscopy, Angew. Chem. Int. Ed. Engl., **51** (31) (2012) pp. 7847–7849.

[421] Rosenberg P. E., Aluminium fluoride hydrates, volcagenic salts from Mt. Erebus, Antarctica, Amer. Mineral. **73** (1988) 855–860.

simmonsite, derived from the cryolite Na_2LiAlF_6.[422] Although most fluoride-containing minerals are distributed throughout the globe, complex aluminum fluorides were concentrated in the Ivigtut mine in Greenland (see inset above), and also in St. Peter's Dome, El Paso, Colorado, in the United States. In addition, the majority of bastnaesite-type fluorocarbonates are extracted from the only Baotou mine in Inner Mongolia, China. Fluorinated phases that form around the craters and result from the condensation of species expelled during volcanic phenomena in contact with the atmosphere are of particular interest; they usually contain aluminum, silicon, and/ or boron (Table 29).

3.1.2.1 Main families of fluorinated minerals

The families of fluorinated minerals most used for by their industrial or lapidary uses include:

- Fluorite, fluorspar (CaF_2).
- Aluminum fluorides of the family of cryolite (Na_3AlF_6), in particular from the Ivigtut mine, Greenland.
- Mixed anions compounds with:
- Hydroxyfluoro silicates and aluminosilicates, such as fluoro-micas, muscovite $[KAl_2[Si_3AlO_{10}](F,OH)_2]$, or lepidolite $[K(Li,Al)_3\ [Si_3AlO_{10}](F,OH)_2]$. Among these silicates, there are a number of semi-precious stones such as tourmalines, $NaX_3Al_6\ (Si_6O_{18})(BO_3)_3(OH,F)_4$, topazes, $Al_2SiO_4(F,OH)_2$, or jadeites-nephrites, $Ca_2(Mg,Fe)_5Si_8O_{22}(F,OH)_2$, which have been mentioned previously. Fluoridation treatments can modify the morphology of these minerals, either by fixing additional fluorine atoms in the lattice, or by covering the particles of these minerals with protecting perfluorinated layers.[423]
- Fluoro-phosphates: apatites $Ca_{10}(PO_4)_6(OH,Cl,F)_2$ and carbonato-apatites, such as francolite, $Ca_{10}(PO_4)_{6-x}(CO_3)_x(F,OH)_2$. The hexagonal structure of fluorapatite $Ca_{10}(PO_4)_6F_2$, in which all the hydroxyls have been replaced by F^- ions, is shown in Fig. 69. This material is used in both the composition of biomaterials (see Chapter 6.3 of Section 2) and as a dispersant in organic chemistry for the oxidative dehydration of terpenes.[424] A special mention must be made for the bastnaesite-type minerals, $(Ce,La)(CO_3)F$, which are rare earth fluorocarbonates from which are extracted all the elements of the rare earth family. Most of these minerals are extracted from the Baotou mine, which supplies nearly 90% of the world's demand for rare earths. Each year, 15,500 t of rare earth oxides are exported to an estimated world production of 86,000 t.

Will we not further detail data on fluorinated minerals that can be found in many specialized works, particularly Materials Data Inc. software[425] and Dana's New Mineralogy book.[8]

[422] Foord E. E., O'Connor J. T., Hughes J. M.,Sutley S. J., Falster A. U., Soregaroli A. E., Lichte F. E., Kile D. E., Discovery of a new aluminofluoride mineral: simmonsite, Amer. Mineral. **84** (1999) 796–772; Ross K. C. Mitchell R. H., Chakhmouradian A. R., The crystal structure of synthetic simmonsite, J. Solid State Chem. **172** (2003) 95–101.

[423] L. P. Demyanova, A. Tressaud, Fluorination of alumino-silicate minerals: the example of lepidolite, J. Fluor. Chem. **130** (2009) 799–805.

[424] Ichihara J., Y. Sasaki, Catalysis Today **117** (2006) 120–125; Actes du Congrès ICFC, mai 2009, Kyoto, Japon.

[425] Nickel E.H., Nichols M.C., *Mineral Database* (2003), Materials Data Inc. 1224 Concannon Bvd, Livermore CA 94550, USA, mdi@materialsdata.com, Url: www.materialsdata.com.

Table 29 Mineral fluorides occurring around volcanic zones

Mineral	Formula (with Dana's notation)	Environment and place of discovery	Morphology and color	Chemistry: hardness (Mohs), D_{obs}, reactivity	Crystallography: Symmetry, Parameters $(\widetilde{\text{Å}}, \circ)$ D_{calc}, ICSD Sheets, JCPDS-ICDD-PDF
Rosenbergite	$AlF_3.3H_20$ AlF $[F_{0.5}(H_2O)_{0.5}]_4\cdot H_2O$	Erebus Mt., Antarctica, associated with monohydrate β-$AlF_3\cdot H_2O$ and ralstonite. Possible aussi during the eruption of Mt. Hekla, Iceland	Transparency, glassy luster	$D=2.10$	Tetragonal, *P4/n*: $a=7.715$, $c=3.648$; $Z=2$, $D=2.111$ PDF 35–0827; 46–1459
Carobbiite	KF (9.1.1.4)	Tiny crystals and stalactites in lava, Mt. Vesuvius, Italy	Semi-transparents	H = nd, D_{obs} = nd Soluble in water	Cubic, *Fm-3m*: $a=5.40$, $Z=4$ ICSD 44,280, 53,824, 64,686
Avogadrite	$(K,Cs)BF_4$ (11.2.2.1)	Sublimation product in fumaroles at Mt. Somma, Mt. Vesuvius, Italy; Kamchatka Peninsula, Russia	Tiny translucent to transparent translucent crystals, white to yellowish	$D_{obs}=2.6$. Soluble in water; easily fused	Orthorhombic, *Pnma*: $a=8.659$, $b=5.480$, $c=7.030$, (synthetic K); $Z=4$, $D=2.507–3.0$ ICSD 9875; 21,081; 22,260/PDF 16–0378
Ferruccite	$NaBF_4$ (11.2.3.1)	In the fumerolles of Mt. Somma and Mt. Vesuvius, Italy	Tiny translucent to transparent tabular crystals, forming aggregates	$H=3$, $D_{obs}=2.49$ Soluble in water; easily fused	Orthorhombic, *Cmcm*: $a=6.837$, $b=6.262$, $c=6.792$, $Z=4$, $D=2.507$. ICSD 30,349; 30,435; 36,067/PDF 11–671 (synth)

Continued

Table 29 **Continued**

Mineral	Formula (with Dana's notation)	Environment and place of discovery	Morphology and color	Chemistry: hardness (Mohs), D_{obs}, reactivity	Crystallography: Symmetry, Parameters $\left(\widetilde{\text{Å}}, \circ\right)$ D_{calc}, ICSD Sheets, JCPDS-ICDD-PDF
Barberiite	NH_4BF_4 (11.2.4.1)	In the fumaroles of the Fossa crater at Vulcano, Aeolian Islands, formed at $T \leq 600°$, associated with sassolite, malladrite, and various rare sulfides	Tiny translucent to transparent tabular crystals. Vitreous glaze	$H=1.5$, $D_{obs}=1.89$. Very unstable in air	Orthorhombic, *Pnma*: $a=9.062$, $b=5.673$, $c=7.267$ $Z=4$, $D=1.90$ ICSD 9400; 9918; 21,082; 24,017/PDF 15–745{syn); 47–1762
Hieratite	K_2SiF_6 (11.5.1.1)	In fumaroles of Vulcano volcano, Aeolian Islands	Transparent, glassy, white or gray Octahedra and stalagmitic concretions	$H=2.5$, $D_{obs}=2.668$ Soluble in water	Cubic, *Fm-3m*: $a=8.133$, $Z=4$, $D=2.719$ ICSD 29,407; 64,763; 73,722/PDF 7–217
Cryptohalite (dimorph with bararite)	$(NH_4)_2SiF_6$ (11.5.1.2)	Mt. Somma, Mt. Vesuvius, Italy	Transparent, glassy, white or gray. Massive gold mammillary arborescences	$H=2.5$, $D_{obs}=2.00$ Soluble in water	Cubic, *Fm-3m*: $a=8.395$, $Z=4$, $D=2.00$. ICSD 19,065; 38,545; 44,790; 18,033 PDF 7–0013

Bararite	$(NH_4)_2SiF_6$ (11.5.2.2)	Barari, Bengal, India, In fumaroles of Vesuvius, Italy	Transparent, glassy, white or gray. Tabular crystals and mammillary arborescences	$H = 2.5$, $D_{obs} = 2.15$ Soluble in water. Sublimates without residue	Hexagonal, *P*-3*m*1: $a = 5.77$, $c = 4.78$, $Z = 1$, $D = 2.144$. ICSD 18,027; 38,113 PDF 44–1424
Malladrite	Na_2SiF_6 (11.5.2.1)	In fumaroles of Vesuvius and Vulcano volcanos, Aeolian Islands	Prismatic vitreous crystals, pale pink to white	$H = 2.9$, $D_{obs} = 2.755$. Soluble in hot water	Hexagonal-rhombohedral, *P*321: $a = 8.866$, $c = 5.043$, $Z = 3$, $D = 2.729$. ICSD 16,598; 30,348; 40,917; 61,274 PDF 33–1280(synth)

Crystallographic data from R. V. Gaines *et al*., Dana's New Mineralogy, 8th Edition, John Wiley & Sons, New York (1997).

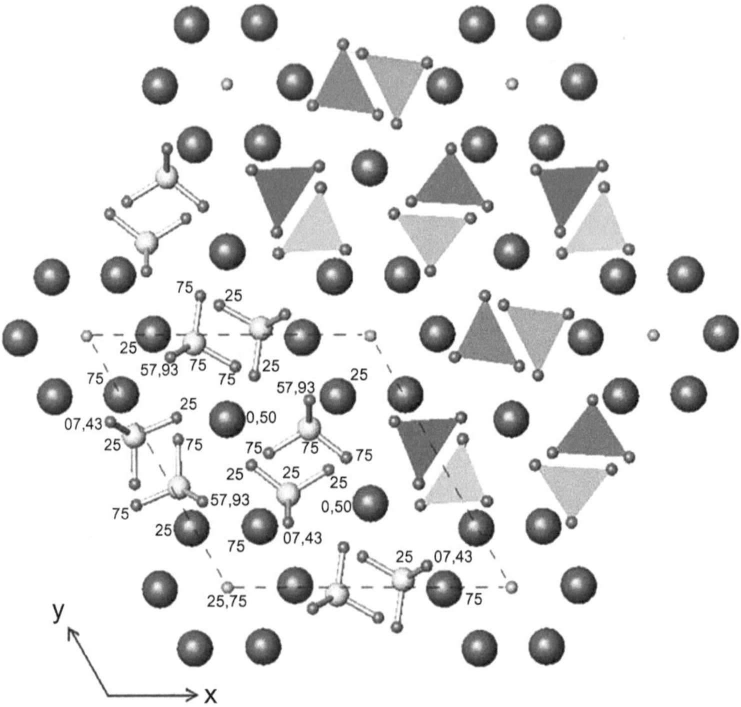

Fig. 69 The hexagonal structure of fluorapatite $Ca_{10}(PO_4)_6F_2$ in which all the hydroxyls have been replaced by F^- ions (at the vertices of the lozenge).

3.1.2.2 Recovery of byproducts of mineral fluoride industries

The application fields of fluorinated minerals and other inorganic fluoride derivatives are listed in Table 8. Among the material cited above, global CaF_2 resources (Table 9) are estimated at 500 Mt, with 80% of the reserves in China. At the rate of 6 Mt per year, we are heading toward their exhaustion within one century, and the recovery of some byproducts should be considered in order to allow sustainable development of fluoride resources:

- Sparfluor as a byproduct of limestone mines.
- Fluosilicic acid fertilizer plants using phosphates, because phosphate reserves are still very important and around 2×10^{10} t.
- HF and fluorides from aluminum blast furnaces.
- Fluorine of nuclear enrichment plants, etc.

3.1.3 The case of perfluorinated compounds of geological origin

Only a few years ago, it seemed evident that halomethanes and perfluorinated compounds emitted into the atmosphere were exclusively of anthropogenic origin. The main source of CF_4, for example, is due to the electrochemical production of aluminum, and SF_6 comes from high-voltage metallurgical and electrical installations. In 1966, Kranz provided some evidence that these last two compounds could also have a geological origin.[426] However, mass spectrometry data were not reliable enough for this finding to be universally recognized by the scientific community. Subsequently, high levels of CFC-11 ($CFCl_3$) were reported in crystals of halite (NaCl) or sylvite (KCl).[427] A controversy around the subject lasted for several decades, but today, the scientific evidence has confirmed that fluorocarbons and perfluorinated compounds of geological origin have been trapped in crystals (fluorite or granite) or directly emitted by volcanoes. The measurements are, nevertheless, particularly delicate given their very low rates. Assays of CF_4 in ice cores taken from Monte Rosa at the Italian-Swiss border and accurately dated with ^{210}Pb showed that this gas already existed in the atmosphere at the preindustrial time, long before the discovery of CFCs. For sections of ice corresponding to the period 1790–1860, the CF_4 base rate is 40 pptv.

The geological origin of these gases was recently confirmed by Harnisch and Eisenhauer, who showed that the emission rates from natural fluorite crystals were 40 ppt (mole) and 0.01 ppt (mole), respectively, for CF_4 and SF_6.[428] These authors evaluated the geological reservoir of CF_4 and SF_6 in fluorite minerals of the Earth's crust. Fluorite minerals, present as secondary minerals in granitic rocks constitute the main source, followed by the carbonato-fluorites present in the sediments, deposits, and ores of hydrothermal fluorites, can be neglected at first approximation. Using an average value of 10–100 wt ppt of CF_4 in the 2.4×10^{25} g of the Earth's crust, the estimated value for the geological CF_4 reservoir would be $0.5–5 \times 10^{14}$ g. This reservoir would, therefore, be 80–800 times greater than that of our atmosphere (which is about 6×10^{11} g, corresponding to 40 molar ppt). The factors causing the emission of these geological perfluoro-compounds to the atmosphere include erosion, metamorphism of some rocks, and diffusion losses.[429] The evaluation of this type of emissions at 0.1–1 t $year^{-1}$ for CF_4 and 0.006–0.06 t $year^{-1}$ for SF_6 corresponds well to the orders of magnitude that allow us to take into account the natural atmospheric levels of these gases, that is, 40 ppt (mole) and 0.01–0.05 ppt (mole).[430] It should be noted that the lifetime of CF_4 in the atmosphere is evaluated to be at least 170,000 years,[431] and that of SF_6 at 3200 years.

[426] Kranz R., Organische Fluorverbindungen in den Gaseinschlüssen der Wölsendorfer Flussspäte, Naturwiss. **53** (1966) 593–600.

[427] Isodorov V.A., Prilepsky E.B., Povarov V.G., Photochemically and optically active components of minerals and gas emissions of mining plants, J. Ecol. Chem. **1** (1993) 201–207.

[428] Harnisch J., Frische M., Borchers R., Eisenhauer A., Jordan A., Natural fluorinated organics in fluorite and rocks, Geophys. Res. Letters **27** (2000) 1883–1886.

[429] O'Nions K., The continents, in Understanding the Earth, (ed.) G.C. Brown, Cambridge Univ. Press, pp. 145–163 (1992).

[430] Maiss M., Steele L.P., Francey R.J., Fraser P.J., Langenfelds R.L., Trivett N.B.A., Levin I., Sulfur hexafluoride: a powerful new atmospheric tracer, Atm. Environ. **30** (1998) 1621–1629.

[431] Ravishankara A.R., Solomon S., Turnispeed A.A., Warren R.F., Atmospheric lifetimes of long-lived halogenated species, Science **259** (1993) 194–199.

Thus, although the volume of geological reservoirs is of the same order of magnitude as that of anthropogenic origin, the emission into the atmosphere of these perfluorinated geological compounds is by several orders inferior to those produced by man, which are, respectively, 15,000 and 5700 t $year^{-1}$ for CF_4 and SF_6, leading to a ratio of 104–105 between anthropogenic sources and natural sources.[432] Note that 700 ppt molar CF_4 levels were also found in natural gas of Russian origin. In addition, relatively high levels of CFC-11 ($CFCl_3$) were found in natural crystals of NaCl and KCl, as well as a variety of perfluorinated compounds: CF_4 (CFC-14), CF_3Cl (CFC-13), CF_2Cl_2 (CFC-12), $CFCl_3$ (CFC-11), CHF_3 (CFC-23), SF_6, and NF_3 in inclusions present in a large number of igneous or metamorphic rocks. Various halomethanes, CF_4 (CFC-14), CF_3Cl (CFC-13), CF_2Cl_2 (CFC-12), $CFCl_3$ (CFC-11), and CHF_3 (CFC-23), could be extracted from rocks of igneous or metamorphic origin. In addition, molecules of CH_3X (X = Cl, Br, I) were found by gas chromatography coupled to mass spectrometry in the fumaroles of Kilauea volcanoes and Mauna Loa of Hawaii. In one of the catches, the CH_3Cl level was even 100 times higher than the atmospheric average of 5.2×10^{-2} Tg $year^{-1}$. Other CFCs have been identified among these gases, such as (CFC-11) or CF2C12 (CFC-12), but at levels well below the atmospheric average.[409] It is clear, however, that due to the very low emission rate, all these gases have only a very small impact on the greenhouse effect or climate change.[433]

3.1.4 *Fluorinated organic molecules present in plants*

On Earth, fluorides are mostly of mineral nature, and there are very few natural organofluorine molecules in nature; however, fluoroacetate, fluorocitrate, fluoroacetone, and fluoro-oleic acid can be mentioned as shown below from left to right.[434]

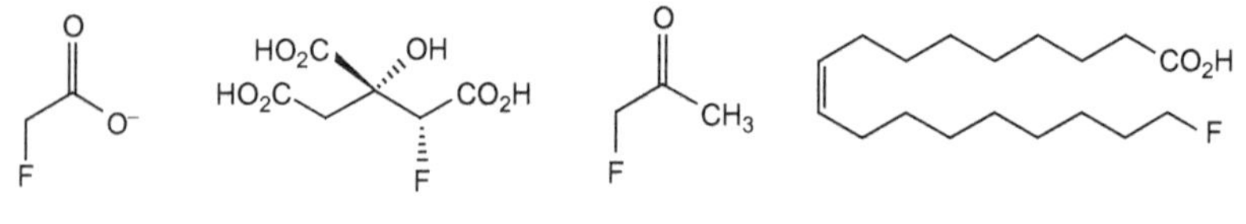

The best known molecule is monofluoroacetate. It occurs in a wide variety of plants in the southern hemisphere (Table 30), but it concentrates the most in acacia-type Dichapetalum or Acacia georginae trees (Fig. 70a) in South Africa and South Australia, where, because of its toxicity, it causes the death of many cattle. Very low levels are also present in guar gum (Fig. 70b), a vegetable grown in India and Pakistan and

[432] Harnisch J., Borchers R., Fabian P., Gäggeler H.W., Schotterer U., *Effect of natural tetrafluoromethane*, Nature **384** (1996) 32; Harnisch J., Borchers R., Fabian P.,Maiss M., Tropospheric trends for CF_4 and C_2F_6 since 1982 derived from SF_6 dated stratospheric air, Geophys. Res. Lett. **23** (1996) 1099–1102.

[433] Jordan J., Harnisch J., Borchers R., LeGuern F., Shinhara H., Volcagenic halocarbons, Environ. Sci. Technol. 34 (2000) 1122–1124.

[434] A.W. Davison, Weinstein, L.H. Some problems relating to fluorides in the environment: effects on plants and animals,—"Fluorine and the Environment: Atmospheric Chemistry, Emissions & Lithosphere ", Vol. **1**, (ed.) A. Tressaud, Elsevier (2006).

Table 30 Types of plants known to contain toxic organofluorinated molecules, particularly in the form of monofluoroacetate[388]

Type	Family	Country
Gastrolobium, Oxylobium	Fabaceae	Australia
Acacia georginae	Mimosaceae	Australia
Cyamopsis tetragonolobus (guar gum)	Fabaceae	India, Pakistan, grown in many countries
Dichapetalum	Dichapetalaceae	Africa, Borneo, China, Philippines, Brazil, Mexico
Palicourea marcgravii	Rubiaceae	Brazil
Arrabidea bilibiata	Bignoniaceae	Brazil
Spondianthus preussii	Anacardiaceae	Cameroon, Niger, Ivory Coast

Fig. 70 Fluorine concentrates (A) in acacia-type Dichapetalum or Acacia georginae trees; (B) in Guar beans.

used to maintain the texture of ice creams and as a stabilizer in some cheeses. It seems that the very low content of fluoroacetic acid, 0.07–1.42 $\mu g\, g^{-1}$,[435] has no impact on health. Note that in 2009, Rhodia Co. launched the marketing of a polymer, "Jaguar C500," based on guar gum, deemed to bring beauty and health to the hair.

We will see later that the major contaminations by fluoride (fluorosis) occur mainly via an accumulation of mineral fluorides in the body. In some plants such as camellias or tea trees (*Camellia sinensis*), a large accumulation of mineral fluorides occurs in the lower parts of the shrub, depending on the nature of the soil. Thus, the numerous cases of fluorosis observed in certain regions of China are due mainly to the massive use of tea bricks in culinary compositions (see below).

Recently, David O'Hagan discovered that bacteria, *Streptomyces cattleya*, was able to biosynthesize organofluorine molecules. The enzyme responsible for this

[435]Vartiainen, T. and J. Gynther, Fluoroacetic acid in guar gum, Food Chem. Toxicol., **22**, 1984, 307–308.

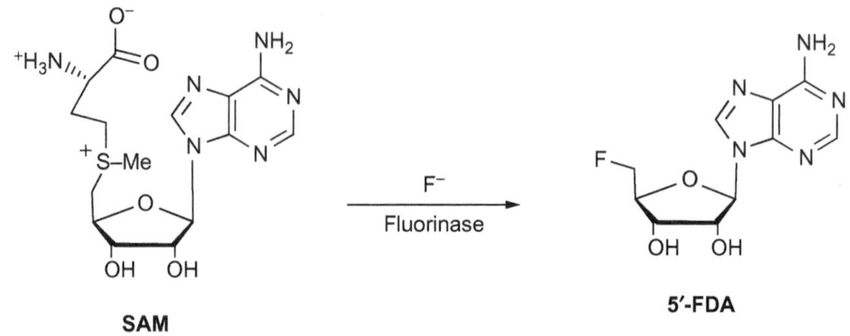

Fig. 71 First "natural" fluorination. The fluorinase catalyzes the reaction between *S*-adenosyl-L-methionine (SAM) and a fluoride ion to yield 5′-fluoro-5′-deoxyadenosine (5′-FDA) [from O'Hagan (436)].

natural fluorination has been identified from this microorganism. This fluorinase catalyzes the reaction between *S*-adenosyl-L-methionine (SAM) and a fluoride ion to yield 5′-fluoro-5′-deoxyadenosine (5′-FDA), as shown in Fig. 71.[436] It is sometimes very difficult to remove fluorinated substances from living organisms because of the very high stability of these molecules. For example, nature does not seem to offer many defluoridation reactions of monofluoroacetic acid, which is a violent poison. In addition, the aromatic fluorinated compounds can be biodegraded only under aerobic conditions, that is to say, in the presence of oxygen.[437] However recently, tests on a set of bacteria (*Actinobacteria*) have shown that ethyl difluoroacetate could be accumulated and converted into difluoroacetic acid by these bacteria, and the latter product can subsequently degrade into fluoride ion F^-, which is easily removable.[438]

3.2 Emissions of fluorinated gases of volcanic origin

Although the most common gases emitted by volcanoes are water vapor and CO_2, other species are also released into the atmosphere: F_2, CO, SO_2, H_2S, HCl, and Cl_2. It should be noted that volcanoes do not need to be erupted to emit gases, and that even in the waking state, they emit continuously. The quantities involved are estimated at >10 Mt of chlorine and several Mt of fluorine per year (between 5 and 9 Mt) coming mainly from alkaline volcanoes, the emitted fluorine being mostly

[436] O'Hagan, D., Harper D.B., J. Fluorine Chem., **100** (1999) p. 127; Dong C. J., Huang F. L., Deng H., Schaffrath C., Spencer J. B., O'Hagan D., Naismith J. H., Crystal structure and mechanism of a bacterial fluorinating enzyme, Nature, **427** (2004) 561.

[437] N. Iwai, Y. Watanabe, T.Kitazume, Retransformation of fluoride ion generated from the biodegradation of benzotrifluoride to calcium fluoride, J. Fluor. Chem. 135 (2012) 350–353.

[438] Morii Y., F. Hasumi, T. Kitazume, Accumulation of 2,2-difluoro-ethanol by bacteria. J. Fluor. Chem. **125** (2004) 731–733. Iwai N., R. Sakai, S. Tsuchida, M. Kitazume, T. Kitazume, Screening of fluorinated materials degrading microbes, J. Fluor. Chem. **130** (2009) 434–437.

transformed into hydrofluoric acid.[439,440] The annual rate of fluorine emitted between 1982 and 1984 during volcanic eruptions is of the same order of magnitude ($0.06–6 \times 10^{12}$ g) as that from industrial production of halocarbons and CFCs, which is around 0.273×10^{12} g.[441]

It has sometimes been suggested that fluorine and chlorine are an important source of stratospheric HF and HCl. Over the last 200 years, six volcanic explosions have had enough power to project gases into the stratosphere. The two most recent ones are El Chichon, Mexico, in 1982 and the terrible eruption of Mount Pinabuto in the Philippines in June 1991. During the eruption of the latter, during which tons of material were released into the atmosphere, measurements of stratospheric HF and HCl were made at the 1 m Fourier-Transform Macbath Telescope at Kitt Peak Observatory in Arizona while the eruption cloud was at its zenith. No substantial increase in the rate of these gases was observed, with a value as low as 6.3×10^{14} mol cm^{-2}, a value that can be compared with the minimum value measured during the same year: 6.0×10^{14} mol cm^{-2}.[442] Emissions of HF and HCl thus had no catalytic effect in the stratosphere, with 99% of the molecules being absorbed onto water droplets.[443] The contribution of even eruptions as spectacular as those of Mount Pinabuto or Mount Erebus in Antarctica in stratospheric fluorine or chlorine thus appears to be very limited.[444,445] It has been confirmed that fluorinated compounds do not interact with stratospheric ozone, with 99.99% being leached as they ascend into the troposphere.

3.2.1 Risks associated with volcanic eruptions with halide/fluoride emissions

What are the types of risks associated with volcanic eruptions? Chemicals from volcanoes travel either as aerosols (micro-droplets of fluid floating in the air), as solid products attached to the surface of particles or rocks ejected into the air, or as small particles, which may be crystallized or not. Fluorine-rich gases also deposit on the

[439] Halmer M.M., Schminke H.U., Graf H.F., The annual volcanic gas input in the atmosphere, J. Volcan. Geotherm. Res. **115** (2002) 511–528.

[440] Anazawa K., Fluorine and Coexisting Volatiles in the Geosphere, pp. 189–226, Fluorine and the Environment, Vol. **1**, (ed.) A. Tressaud, Elsevier (2006); Stormer, J.C., Jr., and Carmichael. I.S.E. The free energy of sodalite and the behavior of chloride, fluoride and sulfate in silicate magmas. Am. Mineral., **56** (1971) 292–306.

[441] Symonds, R. B., W. I. Rose and M. H. Reed, Contribution of Cl- and F-bearing gases to the atmosphere by volcanoes, Nature, **334**, (1988) 415–418.

[442] Wallace L., Livingston W., The effect of the Pinabuto cloud on hydrogen chloride and hydrogen fluoride, Geophys. Res. Letters **19** (1992) 1209.

[443] Tabazadeh A., Turco R., Stratospheric Chlorine Injection by Volcanic Eruptions: HCl scavenging and Implications for Ozone, Science, **260** (1993) 1082.

[444] Zreda-Gostynska G., Kyle P.R., Chlorine, fluorine, and sulfur emissions from Mount Erebus, Antarctica, Geophys. Res. Lett. **20** (1993) 1959–1962.

[445] Allen A.G., Baxter P.J., Ottley C.J., Gas and particles from Soufrière Hills volcano, Montserrat, West Indies: characterization and health hazard assessment, Bull. Volcanol. **62** (2000) 8–19.

surface of particles ejected as minerals, such as calcium fluorosilicates. Fluorine reacts rapidly with atmospheric water droplets to give hydrofluoric acid (HF), which can cause damage to the eyes, skin, and respiratory systems of living beings as well as to plants and our overall environment. In sites where a few centimeters of ash have been deposited, poisoning of the livestock can occur as soon as the fluorine content exceeds 250 ppm of dried grass. The particles of HF and HCl, however, propagate less in the atmosphere than those of SO_2 for example, because of their high reactivity which allows a rather fast recombination with the soil. Table 19 shows the evolution of emitted gases (SO_2, HCl, HF) as a function of the distance to the eruptive crater during the eruption of Soufriere, Montserrat, Antilles, in 1995. However, there is no clear data on the long-term effect of volcanic fluoride emissions.[446] During eruptions that do not involve excessive gaseous emissions, but rather lava flows, gases are emitted at the ground level, and their effects are simply local, given their very high reactivity. On the other hand, when these events result in the projection of hot gases and particles to an altitude of up to several tens of kilometers, gases, aerosols and microparticles can affect the stratospheric layers and take months or even years to fall down on the earth. Alkaline-type volcanism, which is generally associated with fluoride and chlorine emissions, has been widely present in the geological past, as evidenced by the preponderance of such volcanoes in East Africa and Antarctica, but is much less common today.

Data on HF and HCl emissions are provided in Table 31; they include various sources of halogens related to human activities, oceans and volcanism.[447,448] From these estimates, it can be noted that the main contribution of atmospheric fluoride appears to be of volcanic origin, although the results are tainted with great uncertainty. In the second part of the table are given the rates of HF and HCl that were measured during recent volcanic eruptions, such as those of Erebus, White Island, Masaya,[449] Poas, Etna, Stromboli, Mt. St. Helens,[450] and Merapi.[451] Some of these eruptions are illustrated in Fig. 72. Among these various types of eruptions, Etna, Erebus, and Masaya, for example, are basaltic volcanoes that emit gases continuously, while

[446] Edmonds M., Pyle D. M., Oppenheimer C. M. M., Young S. R., Pearce N. J., A Study of Degassing Using Remote Sensing and Geochemistry. Understanding Sulphur, Chlorine and Fluorine Emission From Active Volcanoes, Proceedings IAVCEI General Assembly, Bali, Indonesia, (2000).

[447] Francis P., Burton M. R., Oppenheimer C., Remote measurements of volcanic gases compositions by solar occultation spectroscopy, Nature **396** (1998) 567–569.

[448] Symonds, R. B., W. I. Rose and M. H. Reed, (1988), Contribution of Cl- and F-bearing gases to the atmosphere by volcanoes, Nature, 334, 415–418.

[449] Stoiber R.E., Williams S.N., Huebert B.J., Sulfur and halogen gases at Masaya caldera complex Nicaragua: Total flux and variations with time, J. Geophys. Res. **91** (1986) 12215–12,231.

[450] McGee K. A., The structure dynamics and chemical composition of noneruptive plumes from Mt. St Helen (1980–1988), J. Volcan. Geotherm. Res. **51** (1992) 269–282.

[451] Tazieff H., Monitoring and interpretation of activity on Mt. Merapi, Indonesia, 1977–1980: an example of practical, civil defense, volcanology, in: "Forecasting Volcanic Events" (H. Tazieff & J-C Sabroux, Eds.), Elsevier (1983) 485–492. Kadarsetia E., Hirabayashi J.I., Nakada N., The behavior of fluorine, chlorine and sulfur in the magma of Merapi volcano, Merapi Decade Volcano International Workshop II, UNESCO-VSI,Yogyakarta (1997).

Table 31 **Comparison of HF emission rates (in kg sec^{-1}) from various sources. Rate of HF and HCl emissions from recent volcanic eruptions. Halogen emissions are estimated from F/S and Cl ratios (Adapted from refs. 444–455)**

	Various halogen sources			Volcano emissions							
Types of emissions	**Global volcanism**	**Oceans**	**Anthropogenic**	**Mt. Erebus, Antarctica**[a]	**White Island, New Zealand (1983)**	**Masaya, Nicaragua (1998 and average 1979–85)**	**Poas, Costa Rica (1982)**	**Etna, Sicily, Italy (average 1975–97)**	**Stromboli, Italy (1997)**	**Mt. St. Helens, United States (1980)**	**Merapi, Indonesie (1984)**
HF emission rates (kg/s)	1.9–190	13–30	15.7	1.3–1.9	0.15	0.65/0.18	0.04	2.2	nd[b]	0.37	0.02
HCl emission rates (kg/s)	13–350	16×10^4 to 47×10^4	95	2.2–4.2	4.8	6.7/9.6	0.46	8.2	6.2	1.44	0.35

[a]Emissions measured on the 1983–91 period.
[b]nd: non-determined.

Fig. 72 Volcanic eruptions with notable emissions of halogens: (A) Poàs, Costa Rica (photo 2005 by Peter Andersen), (B) Lascar, Chile (Eruption May 2006, photo by Neil K), (C) White Island acid lake, New Zealand (photo by JShook), (D) Eruption of the Icelandic volcano Eyjafjöll (17 April 2010, photo by Årni Friöriksson).

White Island is an andesitic volcano with fumarole activity only, whereas Mt. St. Helens presents explosive episodes. Quantitative analyzes of the gases present in the plumes such as SO_2, HCl, HF, were measured not so long ago either by on-site sampling, followed by chemical analyzes in the laboratory, or by extremely expensive spectroscopic measurements taken by plane. Today, such data can be directly obtained on site using portable FTIR spectrometer (Fig. 73).[452] The Masaya volcano, Nicaragua, is a very good example, as it is one of the few exhibiting Plinian-type basaltic activity (eruptions: −20,000 and −6500 years), with permanent emissions of SO_2, HF, and HCl, as well as intermittent activity from the lava lake. Spectroscopic measurements using the solar spectrum taken at Masaya and Etna make it possible to compare HF and HCl emissions and showed that with respect to SO_2 emissions, Masaya is richer in halogens than Etna (Fig. 74). Because the Masaya eruptions occurred in the late Pleistocene, the fluoride rate emitted during the eruption could be approximated by the analysis of the vitreous minerals and inclusions found at the cone (Fontana cone) and resulting from the event. The same amounts of fluorine, of the order of

[452] Sawyer G, Oppeheimer C, Volcanic fluorine emissions: observations by Fourier transform infrared spectroscopy, pp. 167–187, Fluorine and the Environment, Vol. **1**, A. Tressaud Ed., Elsevier (2006).

Fig. 73 Characterization by FTIR spectrometry of volcanic gases emitted from the crater of Mount Erebus. Active lava serves as an IR source. The aerosol and gas plume is visible on the right of the photo.
Reproduced with permission of Clive Oppenheimer, Cambridge Volcano's Group.

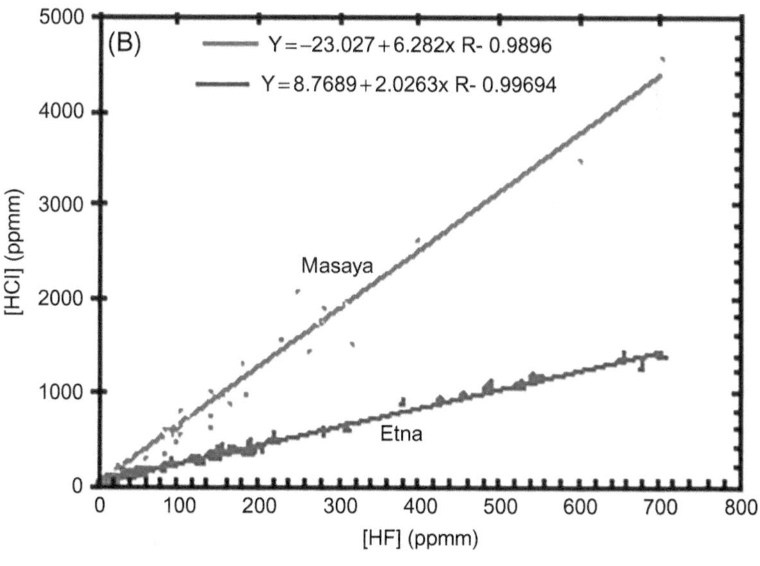

Fig. 74 Comparison of HCl and HF emissions at Masaya and Etna volcanoes using solar radiation absorption. Masaya is richer in halogens than Etna (values related to SO_2 emissions).

7000 ppm, which were measured on both types of samples (lava flow and vitreous inclusions), indicate that fluorine very little degassed during the eruption.[453]

Seminars are regularly organized by the Volcano Gas Chemistry Commission (CCVG) to monitor the knowledge and techniques in these areas. For example, the corrosion of materials such as steel, Cu, and Al, has been tested in a gaseous volcanic atmosphere containing SO_2, H_2S, HCl, and HF. The experiments took place near the orifices through which the gases escape in order to approach the corrosion mechanisms as well as possible. Protective layers were also tested at the crater edges of the Poas and Arenal volcanoes in Costa Rica, allowing researchers to obtain parameters representative of the corrosion such as wetting time, SO_2, Cl, and F deposition rates.[454] Thus, the natural sites of gas-emitting volcanoes could be easily transformed into experimental laboratories approaching the understanding of the acidification of soils exposed to substances emitted. The reaction between fluoride and Fe and Al hydroxides as well as with allophanes (phyllosilicates subgroup minerals) constitutes an important trapping mechanism for volcanic Masaya soils exposed to fluorinated particles transported by air. Although fluoride is not one of the most concentrated elements, it is the fluoride ions that dominate, regardless of the type of terrain. Prolonged exposure of soils to acid rain does not lower the pH much, but rather leaches most cations (Ca^{2+}, Mg^{2+}, K^+, and Na^+). During this leaching, fluorine accumulates as mineral fluoride, while chlorine is eliminated in the soil as soluble products. This process makes it possible to understand the preferential retention of fluorides with respect to chlorides and sulfates.[455]

Since the beginning of the 21st century, 151 volcanoes have erupted. The list can be found on the Internet (wikipedia.org). The types of eruptions and the risks associated with volcanic events in which fluorinated species have been clearly identified are summarized in Table 20. These data were either observed experimentally for the most recent eruptions, or deduced from analyzes of the minerals present on these sites, for older eruptions. The particular cases of eruptions of the Icelandic volcano Hekla and Erebus in Antarctica for which fluorine emissions have been followed in a very fine and regular manner are given below.

3.2.2 Two examples of volcanoes emitting gas and fluorinated solids

3.2.2.1 Recent eruptions of Hekla (Iceland)

During the 1947–48 eruption, which was the largest in Iceland during the 20th century, a Plinian-style column quickly reached a height of 30 km. Traveling at a speed of 50 km/h, the ash cloud reached Finland at 2800 km after 2 days. The Hekla magma

[453] Wehrmann H., Freundt A., Kutterolf S., Schmincke H., Strauch W., Volatile release and eruption dynamics of a basaltic plinian eruption from Masaya caldera, Nicaragua, Eos Trans. AGU, **84** (46), Fall Meet. Suppl., Abstract V42B-0358 (2003).

[454] Alvarez, J.F., Pridybailo, G., Calderón K., Duarte E., Malavassi E., Fernández E., Atmospheric corrosion at Poas volcano, 8th CCVG Workshop, Managua (2003); Alvarez J.F., Pridybailo G., Atmospheric corrosion in the vicinity of Arenal volcano, Abstracts, 8th CCVG Workshop, Managua (2003).

[455] Delfosse Th., Delmelle P., Strauch W., Malavassi E. Delvaux B., Impact of volcanic gases on volcanic ash soils at Masaya, Nicaragua, Abstracts, 8th CCVG Workshop, Managua (2003).

being particularly rich in fluorine, serious problems of cattle poisoning ensued. The last four eruptions of the Hekla took place in 1970, 1980, 1991, and 2000. In 1970, about 30 million m^3 of tephra (powder formed from fragments of rocks ejected during the explosion of magma) were produced in the first phase, which lasted 2 h. The maximum thickness in the vicinity of the volcano reached 18 cm. The fluorine was transported in the form of mineral fluoride adhering to the cloud particles that were formed. The high rate, about 2000 ppm, contaminated the pastures and the drinking water system in the northern part of the island.[456] In the Ytri-Ranga river, for example, the fluorine content was multiplied by 6 on the 5th day after the eruption, but reached its usual level by the 6th.[457] During the first days, up to 10 craters were active, but gradually the activity was located on a single outlet and after 2 months, the crater cone reached a height of 100 m slags. During the 1980–81 eruption, the first stage lasted 3 days and produced about 60 million m^3 of tephra and a lava river of 22.5 km of lava with 5 m thickness. The second stage lasted a week, and the lava produced from the only active crater covered an area of 6 km^2.

The 1991 eruption lasted 2 months, producing essentially andesitic lava. This lava covered 23 km^2 for a total volume of 0.15 km^3. On the other hand, the initial short-term phase of the Plinian type produced only 0.02 km^3. Its cloud reached 11.5 km high in just 10 min. At the last eruption, in 2000, a fissure 6–7 km long appeared, and a column of steam rose up to 15 km. Seven hours after the eruption, the layer of ash reached 4–5 cm thick at its maximum thickness, 21 km north of the volcano. Fortunately, most of the ashes felt into uninhabited areas. Nevertheless, again, some grazing animals were found dead, presumably because of the high rate of fluoride in the ashes.

The differences in composition found between the eruptive materials constituting the Hekla itself and the lavas resulting from the eruption can be explained in terms of a stratification of the magma reservoir by materials of different densities. The silica-rich magma (andesite) is located rather at the top of this reservoir, and the sillica-poor magma (basaltic) rather at the bottom. The fissures and chimneys located directly at the level of the volcano are thus fed with magmas rich in silica and associated with more free fluorinated species.

3.2.2.2 Emission of fluorine at Erebus volcano (Antarctica)

In Antarctica, Mount Erebus is a volcano that erupts once or twice a day. It is one of the three great volcanoes in the world to have a permanent lava lake. The alkaline magmas that are released contain high concentrations of halogens and various metals. It has been argued that due to high levels of chlorine, SO_2, and fluorine, this volcano contributes to the depletion of the ozone layer. The high rate of fluorine can be used for example as a tracer of its plume. Using the average F/S ratio of 0.69 ± 0.49, the fluorine production can be estimated at 0.24 kg s^{-1} which gives about 0.54 kg km^{-2} $year^{-1}$ of fluoride deposited on the Antarctic snow. Taking into account snow accumulation

[456] Oskarsson N., The interaction between volcanic gases and tephra: Fluorine adhering to tephra of the 1970 Hekla eruption, J. Volc. Geotherm. Res. **8** (1980) 251–266.

[457] Gudmundsson A., Oskarsson N., Gronvold K., Saemundsson K., Sigurdsson O., Stefansson R., Gislason S.R., Einarsson P., Brandsdottir B., Larsen G., Johannesson H., Thordarson T., The 1991 eruption of Hekla, Iceland, Bull. Volcanol., **54**, (1992) 238–246.

rates at 10 g cm^{-2} year^{-1} for Central Antarctica, the amount of fluoride in the snow was estimated by Zreda et al. at 0.3 μeq L^{-1}. The marine contribution of fluorine is much lower, given the low rate present in seawater (see Table 31). From an environmental viewpoint, if the plume is moving toward central Antarctica during the austral summer, its high halogen content must significantly affect the chemistry of the snow deposited in this region of the globe. Due to the remoteness of the Antarctic continent from any other source of fluoride (natural or anthropogenic) this feature has been used to date Antarctic meteorites (see Chapter 1.5., this Section). However, we will see that such dating must be used with great care because of the possible temporal fluctuations of fluorine emissions. CO_2 emissions associated with volcano eruptions can be estimated at around 130–230 million tons per year. This is much less than the 27 billion tons of CO_2 emitted each year by human activities. Volcanic CO_2 emissions are, therefore, on average, 130 times lower than emissions from anthropogenic sources.

3.3 Emissions of anthropogenic origin

Fluorine is present in many rocks and minerals, the mineralurgy treatment is likely to lead to emissions of fluorinated products. The processing of many common products such as iron, steel, aluminum, glass, and ceramics are affected by this problem. The question can be illustrated by three examples: the ceramic industry, the case of coal-fired power plants, and the aluminum industry. Concerning the mining sites, where the ores are not heated, fluoride emissions are generally limited to the immediate vicinity of these sites and instead affect neighboring lands and streams.

Moreover, many industries use gases or fluorinated products in their processes, such as the conventional chemical industry (heavy chemistry, fluoropolymers, perfluorocarbons, pharmaceutical, and agro-food industries, etc.) and the nuclear industry, in which fluorine is used for completing uranium isotopic enrichment. Surface treatment units are also concerned: stripping of silicon wafers from electronic components, passivation of fuel tanks in the automotive industry, textile protection, etc. It should be noted that information on these emission rates and the methods used to reduce them are sometimes extremely difficult to obtain.

3.3.1 Example of the ceramic industries

3.3.1.1 Manufacturing processes

The raw materials of the ceramic industries (bricks, tiles, pottery) consist of natural materials and ores: clay, sand, feldspar, carbonate, etc. The three main pollutants of these industries are lead (which is one of the main constituents of varnishes), suspended particles, and fluoride. Hydrated clays and micas (muscovite, illite[458])

[458] Muscovite, (Dana's group 71.2.2a.1): mica de formulation $KAl_2[AlSi_3O_{10}](OH,F)_2$; Illite (Dana's group 71.2.2d.2): mica with formulation $K(H_3O^+)[AlSi_3O_{10}](OH)_2$. Micas such as muscovite, can contain high levels of fluorine [see Ottolini L., Camara F., Hawthorne F.C., Stirling J., SIMS matrix effects in the analysis of light elements in silicate minerals, Am. Miner. **87** (2002) 1477–1485; Ottolini L., Camara F., Bigi S., An investigation of the matrix effects in the analysis of fluorine in humite-group minerals by EPMA, SIMS, SREF, Am. Miner. **85** (2000) 89–102.

which constitute the raw material contain between 0.02% and 0.3% of fluorine, which is substituted for the hydroxyl groups in these structures. On a starting material containing, before heating, a fluoride level of about 200 mg kg^{-1}, 53 mg of gaseous fluorides are emitted per kg of heated product.[459] It should be noted, however, that no significant relationship could be found between the nature of the starting material and the amount of fluoride emitted. During the heating cycles at 500–600°C, fluorine present in these minerals begins to volatilize in the form of HF or SiF_4 or to a lesser extent of inorganic fluoride particles. The materials are finally heated to high temperatures around 1200°C. During the ceramics manufacturing process, 30%–95% of the fluorine will be released into the atmosphere. The two main parameters that will determine the rate of fluoride emissions to the atmosphere are the temperature and the duration of the heating cycle.

In Italy, which is the world's leading producer of ceramics and tiles, two cycles are used in glazed tile or roof tile manufacturing processes (Fig. 75). A number of

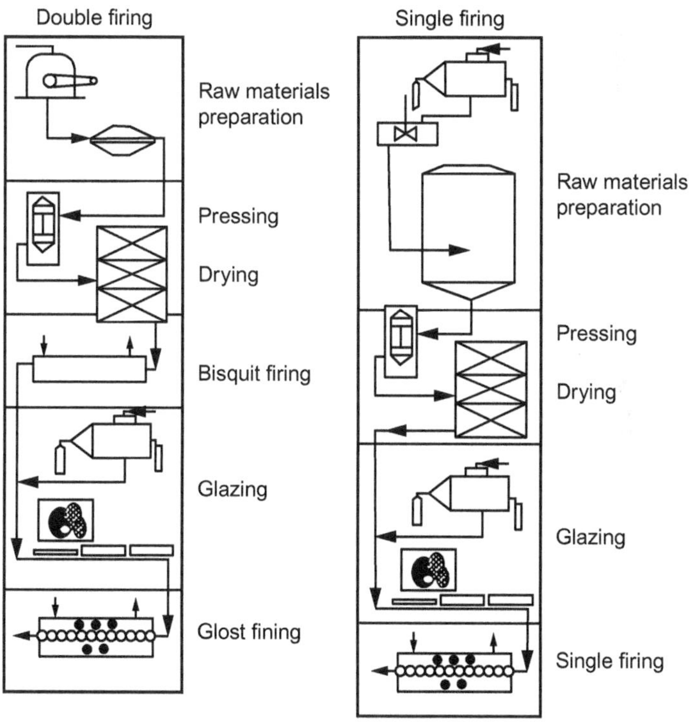

Fig. 75 Schematization of the two ceramic tile production routes at Italian Ceramics Center, Bologna, Italy [from Bonvicini et al. (460)].

[459] Dogeroglu T., Kara S., Fluoride emission factors in ceramic floor-tile kiln, Amer. Ceram. Soc. Bull. **81** (2002) 50–53.

operations first make it possible to reduce the particle size by grinding and to regulate the water content according to the desired shape and size:

- A double heating cycle can be performed before and after the application of varnish and decorations. In this case, dry grinding with successive water addition makes it possible to obtain the same humidity conditions (between 5% and 7%).
- A single heating cycle after application of the varnish and decorations on the dried model is used in 60% of Italian production. In this case, the starting material is subjected to wet grinding and successive spay drying of the engobe, allowing manufacturers to obtain a similar water content.[460] The three main types of emissions at each stage of the process are given in Table 32. An emission control during the various cycles showed that 30% was in particulate form and ~70% in gaseous form, predominantly HF.[461] The emission of fluorinated compounds (HF, H_2SiF_6, SiF_4, and, to a lesser extent, alkaline fluorides) occurs during the heating and sintering of the clay (see Table 32). The influence of the two parameters on the rate of fluorine emissions (temperature and duration of successive heatings) has been clearly demonstrated for three types of process used in manufacturing red ceramics. Indeed, reductions in the emission rate of about 12% are obtained when a single slow cycle is substituted to a double heating cycle (fluoride emission: $4\,g\,m^{-2}$) (fluoride emission: $3.5\,g\,m^{-2}$), and finally, up to more than 75% when this single cycle is carried out rapidly (fluoride emission: $1\,g\,m^{-2}$). The data on fluorine emissions measured at the outlet of factory chimneys are quite

Table 32 Three main types of emissions (fluorine, solid particles, lead) produced at each stage of glazed tile manufacture in Italy[461]

	Emission rate in the atmosphere (g of pollutant per m^2 of tile produced)					
	Emissions without purification			Emissions with purification process		
Production steps	**F**	**Particles**	**Pb**	**F**	**Particles**	**Pb**
Preparation of starting material	–	300–850	–		1.6–1.8	–
Pressing	–	340	–		2	–
Varnishing	–	20	4		1.4	0.5
Heating step	0.6–0.9	1–2	0.1–0.3	0.1–0.7	0.5–0.7	0.02–0.05
Sintering	0.8	2	0.8	0.03	0.2	0.03

[460] G. Bonvicini, A. Fregni, C. Palmonari, Fluorine compounds in gaseous emissions from industrial sources: the case of ceramic industries, pp. 227–252, Fluorine and the Environment, Vol. **1**, A. Tressaud Ed., Elsevier (2006).

[461] Palmonari C., Timellini G., Pollutant emission factors for the ceramic floor and wall tile industry, J. Air Pollut. Control Assoc. **32** (1982) 1095–1100; Palmonari C., Timellini G., Air pollution from the ceramic industry: control experiences in the Italian Ceramic-tile industry, Amer. Ceram. Soc. Bull. **68** (1989) 464–1469.

disparate and vary from source to 5–50 mg m^{-3}.[462] When single and fast heating cycles are implemented, for example in Turkey, these gaseous emissions are around 30 mg m^{-3}, and in the region of production, the average daily rate of gaseous fluorides is 20 μg m^{-3}.

3.3.1.2 Monitoring and purification system

Two purification systems are used, wet or dry. For the limitation of fluoride emissions, this is the first route that is most effective. It is water that is the most used liquid due to the solubility of most fluorinated salts. The gases, which come out of the furnaces at more than 150°C, are cooled in order to facilitate their absorption in the water. HF-charged gases, however, must not be cooled below the dew point to avoid condensation of hydrofluoric acid, which is corrosive for materials of the purification systems.[463] The efficiency of the purification systems is attested to in Table 32. Between 1975 and 1990, the concentration of particles decreased from 130 to 30 μg m^{-3} in the Sassuolo area, while at the same time the annual production of tiles increased from 150 to 330 million m^3. In addition, a close mesh of pollution control systems, as experienced for example in the Sassuolo area, makes it possible to quickly locate the origin of any excessive pollution. In this region of Emilia-Romagna, the main center of this industry, the daily average of the rate of fluorine present in the atmosphere in the 1980s was on the order of 10 μg m^{-3}.

Now, the results of regular pollution control close to the industries likely to emit fluoride are very often made available to local residents. For example, in Western Australia, near the Swan Valley Bricks Plant, the Department of Environmental Protection controls, in partnership with local residents, the respect of fluoride levels in the atmosphere. Acceptable daily values should be between 0.8 and 1.5 μg m^{-3} (average over 7 days), depending on the distance to the brick factory.

Thus, in the most industrialized countries, pollution related to the ceramic industries has been considerably reduced in 20 years thanks to extremely rigorous control and constant improvement of gas and effluent purification techniques.

3.3.2 Processes involving coal

3.3.2.1 Domestic coal heating

In most coal ores, fluorine is present as various mineral inclusions; the main one is fluorapatite, but in many other minerals such as kaolinite, amphibole, mica, and fluorite, the element fluorine is also present by substitution with hydroxyl groups.[464] The rate of fluorine varies in considerable proportions depending on the ore and the location. In most producing countries, Russia, the United States, Australia, the European community, India, and China, it is included between traces (<20 mg kg^{-1}) and a few

[462] Brosnan D.A., HF emissions in manufacturing traditional ceramics, Amer. Ceram. Soc. Bull. **76** (1997) 59–61; Timellini G., Cremonini F., Palmonari C., Air pollution from ceramic tile processes, Ceram. Eng. Sci. Proc. **14** (1993) 445–456.

[463] Palmonari C., Timellini G., Cremonini F., Maroncelli M., The control of gaseous emissions, pp. 83–109, in "Ceramic Tiles," 6th SIMCER, C. Palmonari, Timellini Eds., Centro Ceramico, Bologna, Italy (1984).

[464] Swaine D.J., Trace elements in coal, Butterworths, London (1990).

hundred mg kg^{-1}. In China, however, it can reach up to 2350 mg kg^{-1} for the most polluted samples, with an average value of 268 mg kg^{-1}.[465] A significantly lower value (82 mg kg^{-1}) was found more recently by Luo et al.[466] These authors evaluated the various factors conditioning the amount of fluorine in the coal, that they announced to be between 50^{-1} and 400 mg kg^{-1}, and correlated the rate in fluorine with the origin of the ore. The presence of strong volcanic activity associated with rising magma considerably increases this rate. In the southeastern Qinling Mountains (Daba area), as well as in southwestern China (Yunnan and Guizhou), this rate can reach 3 g kg^{-1}. The major problem is that these ores are made of anthracite, and are used as such by local people for cooking and heating.

3.3.2.2 Coal-based thermal power plants

Fluorine, mainly in the form of HF, is the third mineral substance most commonly emitted by thermal power plants after chlorine, as HCl, and sulfur, as H_2SO_4. In China, coal used for thermal power plants consists essentially of non-cokable coal, such as bituminous coal of permo-carboniferous origin. Coal, with 800 Mt produced annually, is the major source of energy (84%) in this country. Of this total, 60%, or 500 Mt, are burned in thermal power plants that are of two types: the high-temperature plants (1200–1500°C) in which the emission of fluoride into the atmosphere is, on average 90 g t^{-1}, corresponding to a 96% yield, while in those for which the boilers heart is at 800–1200°C, the emission is only g t^{-1}. By taking an average rate of 100 mg kg^{-1}, Luo et al. thus evaluated the fluorine emission in China to be 66×10^3 t $year^{-1}$ [467] (Fig. 76). For the United States, the Environment Protection Agency (EPA website: http://www.epa.gov/) has assessed such an emission at around 32×10^3 t $year^{-1}$.

3.3.2.3 Technologies allowing a reduction of fluorine emissions in coal-fired power plants

The first solution to significantly reduce the rate of emitted fluoride was the development of pulverized coal plants with limestone or lime injection to fix sulfur, which is unfavorable in terms of CO_2 emissions. Their yield, which peaked at 39%, should reach up to 47% in "super critical" plants under development. The purification columns mainly used to remove SO_2 have proved very effective in retaining fluorinated residues. The generation of fluidized bed boilers brings more safety to emissions. In the fluidized bed, the fuel and the desulfurization additive (limestone) injected into the furnace have a particle size corresponding to a coarsely ground coal. The fuel is kept in suspension in the furnace by a current of primary air injected under a distribution grid; the particles burn in suspension, and the dust (totally or incompletely burned) leave the hearth by its upper part to be recovered in a cyclone. At its base, the solids maintained

[465] Zheng B., Ding Z., Huang R., Zhu J., Yu X., Wang A., Zhou D., Mao D., Su H., Issues on health and diseases relating to coal use in southwestern China, Intern. J. Coal Geol. **40** (1999) 119–132.

[466] Luo K., Ren D., Xu L., Dai S., Cao D., Feng F., Tan J., Fluorine content and distribution patterns in Chinese coals, J. Coal Geol. **57** (2004) 143–149.

[467] Luo K., Xu L., Li R., Xiang L., Fluorine emission from combustion of steam coal of North China plate and Northwest China, Chinese Sci. Bull. **47** (2002) 1346–1350.

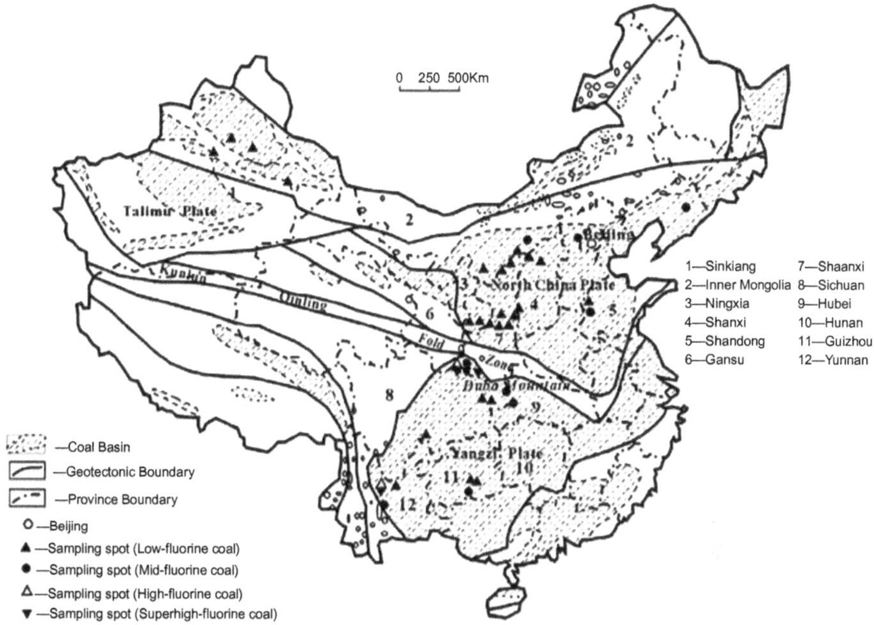

Fig. 76 Distribution of fluorine content in Chinese coal ores, according to K. Luo et al.[467]

in fluidization are re-injected into the hearth. They are, therefore, recycled continuously (from 20 to 35 passages in the fireplace), thus, allowing a perfect combustion at a lower temperature: <900°C, instead of 1300°C in a conventional boiler.

In addition to the advantages of flexibility and efficiency, the combustion technique generates almost absolute cleanliness; it leaves very few unburnt residues, allows a very effective desulphurization of the fumes, produces little NO_x, emits only small quantities of heavy metals and traps in the ashes almost all chlorine and fluorine. A combustion at 850°C instead of 1300°C—one of the major features of the process—first guarantees a very low content of NO_x as well as heavy metals in the fume, because the used temperature is lower than the one of volatilization of these metals. Particle agitation and stirring, repeated solids reinjection and continuous limestone injection into the hearth allow an excellent combination of more than 90% sulfur dioxide, chlorine, and fluorine with calcium. The adoption of a bag filter finally ensures the dust removal of fumes with a yield of 99.9%. In addition, new types of cooling chimneys have allowed dilution of fluorides in such a way that analyses at ground level are generally satisfactory. In the Netherlands, particles of 1000 MW coal-fired power plants are converted into wet fallout by leaching the plume; the hydrofluoric acid thus formed is more easily removed.[468]

[468] Ten Brink H.M., et al., Plume wash-out near a coal-fired power plant: measurements and model calculations, Atmosph. Environm. **22** (1988) 177–188.

3.3.3 *Aluminum industry and overall assessment of the greenhouse effect*

3.3.3.1 Life cycle assessment of aluminum

Due to the very significant mineral resources associated with renewable hydroelectric power and considerable efforts in the recycling of finished products, aluminum can be considered as one of the key materials for sustainable development (see IAI "Sustainability Report").[469] Although a controversy has emerged because of fears that an increase in aluminum production in the future would be linked to the use of polluting energy such as coal-fired power plants, several ongoing projects continue to favor hydroelectric power; this is the case, for example, of the extension of Alcan smelters in Alma, Quebec. Over the next 5–10 years, this last source of energy will represent 55%, while the sources based on coal or gas will be of 30% and 15%, respectively.

Since 1997, IAI has initiated a discussion on the analysis and evaluation of the aluminum life cycle, which takes into account the advantages, but also the disadvantages and risks of the aluminum industry, in particular due to high energy consumption and greenhouse gas (GHG) emissions. This report concerns current aluminum production and its involvement in the automotive industry, a major contributor to CO_2 emissions. These prospects naturally take into account the consequences linked to an increase in production, as far as one knows that the production of 1 t of aluminum is associated with 0.30 kg of CF_4 emissions and 0.03 kg of C_2F_6, which correspond to 2.2 t CO_2 equivalents when applying the instructions of the Intergovernmental Committee on Climate Change (IPCC). It should be noted that the use of a light metal allows fuel savings of 5%–10%.

GHG emissions take place during the use of the vehicle, in relation to the emissions involved in the preparation and assembly of the various components. A vehicle that has traveled 200,000 km should save 6–13 L of fuel for each kilogram of aluminum that has replaced two kilograms of heavier metal, and in other words, should save 20 t CO_2 equivalent for each extra ton of aluminum incorporated in the vehicles. The life cycle assessment takes into account the entire chain from ore extraction to manufacturing, transportation, and use, but also recycling and energy saving. Such a complete consideration is increasingly being taken into account in political decisions important for our planet; however, it must be handled with great care, for how can we compare a decrease in the greenhouse effect and a decrease in the resources of raw materials?

3.3.3.2 Impact of the aluminum industry on PFC emissions

The impact of the aluminum industry on PFC emissions was estimated in 1997 at 110 Mt CO_2 equivalent, 45% of which comes from CF_4 and C_2F_6 emissions. Remember that these gases do not affect the ozone layer at all. On average, smelting operations produce per aluminum ton 1.7 t CO_2 per ton of carbon anode consumption and 2 tons of equivalent CO_2 per PFC emission.

[469] Reports International Aluminium Institute (IAI), "Aluminium for Future Generations Sustainability/ 2007 Update" (Site Internet: http://www.world-aluminium.org/Sustainability). C. Gentaz, « Aluminium, développement durable et énergie », SMIA -03, Congrès International "Sustainable management in action", Genève, sept. 2003.

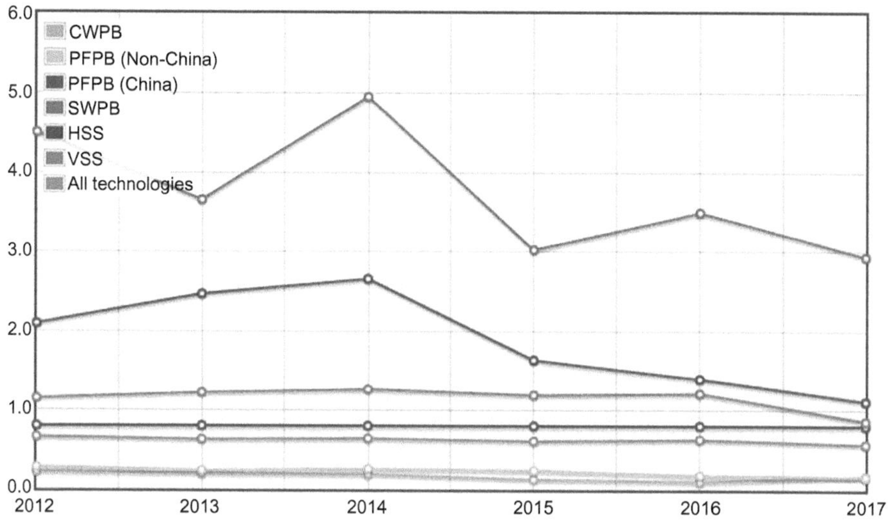

Fig. 77 Global Mean PFC Emissions Intensity for 2012 to 2017 in t CO_2e/t Al . PFC Emissions are aggregated by different technologies as follows: CWPB: Centre Worked Prebake - Bar broken centre feed / PFPB: Centre Worked Prebake - Point centre feed / SWPB: Side Worked Prebake / HSS: Vertical Stud Soderberg / VSS: Horizontal Stud Soderberg.
From IAI's annual Anode Effect Survey Report, based on reported data - http://www.world-aluminium.org/publications/tagged/PFC/

Over the past 15 years, technological improvements at all process levels, including the anode effect, have resulted in a substantial reduction (60%) in CF_4 emissions per ton of produced aluminum, and this for 63% of the world's aluminum production. A 46% reduction in PFC emissions was achieved in 1998 by US industry, or 2.2 Mt CO_2 equivalents (Fig. 77).

This environmental approach is accompanied by the rehabilitation of bauxite mining sites, a better use of residues, especially red mud, and the improvement of dry pickling, which avoids 96%–99% of fluoride emissions in the atmosphere. Thus, 1.1 kg of fluoride are emitted per ton of aluminum today on average, instead of 4 kg as in 1974. These emissions are either particulate (NaF, AlF_3, Na_3AlF_6) or gaseous (HF, PFC). The most significant efforts now must come from Eastern Europe and China. A policy of metal recycling and replacement of old types of blast furnaces is in place. It must not be forgotten, however, that aluminum production gives rise to significant pollution. Each ton of aluminum results in the release of 4 tons of harmful red mud with emissions of fluoride in the air and water. Highly oxidizing fluorine vapors attack vegetation; the waters are loaded with fluoride and toxic aluminum salts. This negative ecological balance should lead to reserving aluminum for permanent uses (aircraft parts, vehicles, etc.) and to avoid ephemeral uses. In these areas too, it should be noted that a protocol to harmonize the main international GHG standards, the Greenhouse Gas Protocol Initiative, has been established by the profession and allows for a significant reduction beyond 2010 (website: http://www.ghgprotocol.org/).

3.4 Fluorine in the water cycle

3.4.1 Hydrochemistry of fluorine in groundwater

In water, the fluorine concentration depends essentially on the state of saturation with CaF_2:

$$CaF_2(s) \leftrightarrow Ca^{2+}(aq) + 2F^-(aq); \qquad Ks = 10^{-10.57} \text{ at } 25^{\circ}C.$$ [470]

If the water is under-saturated, the fluorine content will reflect the amount present in the leached rock. On the other hand, if the water is saturated with CaF_2, the fluorine concentration will be limited by the solubility constant Ks of the fluorine dissolution reaction,[471] as shown by Travi for the Senegalese and Tunisian aquifers. Indeed, although these two reserves are located in phosphated sedimentary formations, and therefore, saturated with fluorine, only Senegal's aquifers have high levels of fluorine ($13\,mg\,L^{-1}$). In the case of Tunisia, the presence of gypsum ($CaSO_4$) provides a significant Ca concentration and therefore limits fluorine levels below $2\,mg\,L^{-1}$.[472] Groundwater with high fluorine content is often of the sodium-bicarbonate type, with low levels of magnesium and calcium and generally with pH values >7, which favors the substitution between hydroxyl and fluoride ions. The hydrodynamics of groundwater also plays an important role, because a weak current will allow a longer contact between the water and the rock, increasing the quantities of dissolved substances.

Example of Lake Titicaca—The hydroclimatic and hydrogeological study of the Lake Titicaca environment is of great importance because of the serious problems related to the high salinity of the groundwater resources of this region of the Andean Altiplano.[473] This lake with a volume of $8400\,km^3$ has one of the highest rates of fluoride, with an average of $0.475\,mg\,L^{-1}$. This rate is related to the nature of the soils close to the lake and rocks that are crossed by the groundwater. The phenomenon of evaporation which concentrates the ions present in the water is also drastic. A study of the chemical speciation of fluorine in the groundwater of the Altiplano (i.e., the nature of the chemical species present) shows that the F^- ions predominate (>95%), however, with 5% of MgF^+ and CaF^+ species. Because these waters are not saturated with fluorine, the fluorine content is not limited.

[470] Smyshlyaev S.I., Edeleva N.P., Determination of the solubility of minerals. Solubility product of the fluorite, Izv. Vysshikh Uchebn. Zavedenii Khim. Teknol., Vol 5 (1992).

[471] D. K. Nordstrom, E. A. Jenne, Fluorite solubility equilibria in selected geothermal waters, Geochimica et Cosmochimca Acta. (1977). Vol. **41**. pp. 175–188.

[472] Travi Y., Hydrology and hydrochemistry of Senegal's aquifers; Hydrogeochemistry of fluorine in groundwater/Hydrologie et hydrochimie des aquifères du Sénégal; Hydrogéochimie du fluor dans les eaux souterraines, (N° 95) Institut de Géologie, Université Louis Pasteur, Stasbourg (1993).

[473] Coudrain A., Loubet M., Condom T., Talbi A., Pouyaud B., Quintanilla J., Dieulin C., Dupre B., Isotopic data ($^{87}Sr/^{86}Sr$) and hydrological changes since 15,000 years on the Andean Altiplano/ Données isotopiques ($^{87}Sr/^{86}Sr$) et changements hydrologiques depuis 15,000 ans sur l'Altiplano andin, Hydrol. Sci. J. **47** (2002) 293–306.

3.4.2 Fluorine content of groundwater

The fluorine content of all groundwater in the world is generally between 10 and 4 $mg L^{-1}$, but can sometimes exceed 10 $mg L^{-1}$. It should be noted that these high rates of fluorine are due to soils rich in this element that are in contact with groundwater.[474,475] Many solutions are proposed to reduce these rates.

In many countries, fluorine-containing groundwater is often in excess of the WHO's proposed maximum rate of 1.5 $mg L^{-1}$, either statistically distributed in countries such as the United States, (or more locally, in France, for example). Some carbonated sparkling waters (Vichy, Célestins, Saint Amand, Badoit) contain a slightly larger amount of fluorine.

Despite these high rates, few or no cases of fluorosis can be reported in technologically advanced countries where adapted treatments, or even alternatives, are offered to the population. On average, 40% of drinking water is slightly fluoridated with $1 \pm 0.15\, \mu g L^{-1}$.[476] In the United States, a search for trace elements in water of various origins has shown that it was of two types: either relatively fluorinated with $\sim 1\, mg L^{-1}$ or having a practically undetectable fluorine level ($<0.5\, mg L^{-1}$).

3.4.3 Influence of hydro-climatic conditions in countries with high risk of fluorosis

It must be made clear that while the application of such standards is feasible without too much difficulty in industrialized countries (those least affected by the problems of fluorosis), it is, on the other hand, very delicate to implement in developing countries, which are those most affected by this disease, for multiple reasons (cost of implementation, secular traditions, lack of training of the population, etc.). By comparing various hydroclimatic data, it is possible to deduce that fluorosis primarily affects arid or semi-arid zones for two main reasons.[477] Evaporation concentrates residual water, and the very rare precipitation reduces the flow of water, thereby increasing the water's contact time with the rocks. Three types of land give rise to waters rich in fluorine:

- Sedimentary rocks in which fluorine is the most common halogen, for example in fluorspar, fluoapatite, or mica deposits (Turkey, Maghreb, and Sahel).[478]

[474] Näslund J., I. Snell, Fluoride contaminated Groundwater in Nakuru & Baringo districts, Kenya, Master's Thesis, Luleå Univ. Technol., Division Sanitary Engin., Suède (2005).

[475] I.E. Nezli, S. Achour, M. Djidel, S. Attalah, Presence and Origin of Fluoride in the Complex Terminal Water of Ouargla Basin (Northern Sahara of Algeria), Amer. J. Appl. Sci. **6**, 876–881, 2009.

[476] Miller-Ihli N.J., Pehrsson P.R., Cutrifelli R.L., Holden J.M., Fluoride content of municipal water in the United States: what percentage is fluoridated? J. Food Comp. Anal., **16** (2003) 621–628.

[477] Mazet P., Les eaux souterraines riches en fluor dans le monde, USTL, Montpellier (2002). Site Internet: http://horizon.documentation.ird.fr/exl-doc/pleins_textes/divers4/010031017.pdf.

[478] Koritnig S., Zur geochemie des fluor den sedimenten, Fortsch. Geol. Rheinld Westf., **10** (1963) 231.

- Volcanic zones (East-Africa Rift, Mexico), where the fluoride comes either from the leaching of alkaline magmatic rocks containing fluorinated minerals, or volcanic emissions of fluorides.[479,480]
- Granite reservoirs (India, China, South Africa, and Australia).

4 Fluorine, friend or foe for human health?

Is fluoride an essential element for human health according to the criteria defined by the World Health Organization?[481] The answer is obviously yes if we consider the benefit it can bring for the improvement and strengthening of dental hygiene (see previous chapter). On the other hand, it can be rather negative if one takes into account the risks of fluorosis linked to a consumption too rich in fluorine.

The knowledge of trace elements in our bodies and their effects on our health is obviously related to the advancement of characterization techniques, and the physiological consequences are often to be taken with great caution. Fluoride is a very good example of the duality of these trace elements on health issues: a low intake (<2.5 mg/day) has a significant positive prophylactic impact, for example, by reducing tooth decay. However, taken at higher doses (>8 mg/day), it can induce pathologies of dental or bone fluorosis, and for doses approaching 1 mg/kg, signs of acute fluoride intoxication may appear.[482] This element perfectly illustrates the famous phrase of Paracelsus, "All is poison, nothing is poison, what makes the poison is the dose."[483]

What are the various origins of our fluoride exposure and the consequences? Except in exceptional cases, fluorine is ingested as F^- fluoride ion. The recommended rate is 0.05 mg day^{-1} kg^{-1}. Drinking water, beverages, and toothpastes are the main source of fluoride in the body, except for particular dietary habits or regular contact with this element.[484] We have previously listed the various sources of fluoride in the atmosphere, both natural and anthropogenic, and transported to humans via dust and aerosols. The levels of fluoride in the air are generally very low: more than 98% of the analyses carried out between 1966 and 1968 on 9170 samples taken in urban environment and 2160 outside the cities showed rates lower than 0.05 μg/m^3, with maxima of

[479] Gaciri S.J., Davies T.C., The occurrence and geochemistry of fluoride in some natural waters of Kenya, Hydrolog. Sci. J., **143** (1993) 395–412.

[480] Chernet T., Travi Y., Valles V., Mechanism of degradation of the quality of natural waters in the Lakes region of the Ethiopian Rift valley, Water Res. **35** (2001) 2819–2832.

[481] Expert Consultation of World Health Organization/Food and Agriculture Organization of the United Nations/International Atomic Energy Agency (WHO/FOA/IAEA), Trace Elements in Human Nutrition and Health, OMS, Genève, (1996).

[482] Janssen P.J.C.M., Janus J.A., Knaap A.G.A.C. Integrated criteria document fluorides—effects. Appendix. Bilthoven, The Netherlands, National Institute of Public Health and Environmental Protection, 1988 (Appendix to Report no. 75847005).

[483] Paracelsus, Swiss alchemist (1493–1541) *"Alle Ding sind Gift, und nichts ohn Gift; allein die Dosis macht, daß ein Ding kein Gift ist".*

[484] R.J. Thompson, T.B. McMullen, and G.B. Morgan, Fluoride concentrations in the ambient air, J. Air Pollut. Control Assoc., (1971), **21**, 484–487.

1.9 and 0.16 μg/m^3, depending on whether the test was made in urban areas or not. Concentrations in rainwater are negligible. In freshwater (rivers and lakes), the rates depend on the presence of fluorinated sediments in soils and, of course, human sources of pollution. On average, the rates vary from 0.01–0.3 mg L^{-1} and can reach 1 mg L^{-1} for geological zones rich in mineral fluorides. Groundwater contains a little more fluoride, especially at the foot of high mountain ranges and in areas where the soil is composed of marine sediment.

4.1 Levels of fluorine in food

Fluoride is present in the vast majority of foods, obviously, as traces (see M. Ponikvar (485)). In fruit juices, soft drinks or sodas (Coca-Cola, Sprite, etc.) the rate is usually <1 mg L^{-1}. In vegetables, rates vary greatly depending on the nature of the land; the highest are found in spinach (up to 70 mg), cabbage (up to 40 mg), and beets (up to 20 mg) per kg of fresh product. The average levels of fluoride in certain foods and beverages as well as in various parts of the human body are given in Table 33. Special mention should be made of fish (0.1–30 mg kg^{-1}, with maxima observed of 60 mg kg^{-1})[486] and especially tea: 3–300 mg kg^{-1} depending on the source and type of tea, with an average of 100 mg kg^{-1} of dried leaves, corresponding to 1–15 mg L^{-1}, that is, about 0.4–0.8 mg fluorine per two to three cups.[487,488,489]

The acceptable maximum levels of ingested fluoride are regularly regulated and modified, if necessary, by decision-making bodies of the various governments. Some examples are provided below. Remember that the maximum level in water for human consumption is fixed at 1.5 mg L^{-1}, to limit the daily intake dose to 0.05 mg kg^{-1} (Directive 98/83/European Community). The Cosmetic Directive prohibits the use of hydrofluoric acid, its hydrofluorinated salts and complex compounds as an ingredient in cosmetics with the exception of a number of exempted substances, for example, sodium, potassium, magnesium, and ammonium fluosilicates, limited to oral hygiene products for a maximum rate of 0.15% in the finished product. In the United States, NaF, sodium fluophosphate, and stannous fluoride (SnF_2) are limited to 1.15 g kg^{-1} of toothpaste. Even maximum levels for animal food have been defined at the European level (Table 34).

[485] Ponikvar M., Exposure of Humans to Fluorine and Its Assessment pp. 487–552, Fluorine and Health, A. Tressaud, G. Haufe Eds., Elsevier (2008).

[486] Slooff W et al., (eds.). Basis document fluoriden. Bilthoven, Netherlands, National Institute of Public Health and Environmental Protection, (1988) (Report no. 758474005).

[487] Asanami S., Tanabe Y., Koga H., Takaesu Y., Fluoride contents in tea and sakura shrimp in relation with other inorganic constituents, Shikwa Gakuho, **89** (1989) 1407–1412.

[488] OMS Fluorine and fluorides. Geneva: World Health Organization, (1984). (Environmental Health Criteria, No. 36).

[489] G.V. Iyengar, L. Tandon, Minor and trace elements in human bones and teeth. International Atomic Energy Agency, NAHRES-39 report, Vienna, 1999.

Table 33 **Mean fluorine content in foods and beverages and in various parts of the human body**

Food and drink	French mineral waters (mg/L)	Vichy Saint Yorre (mg/L)	Drinking water (EU) mg/L	Sodas (mg/L)	Tea (mg/kg)	Fish (mg/kg)	Spinach (mg/kg)	Kale (mg/kg)	Beet (mg/kg)	Table salt (mg/kg)	Tea bricks, Tibet, China (mg/kg)	Toothpaste (EU) (mg/kg)
Taux	0.1–8	8	1.5	<1	3–300	0.1–30	70 max.	40 max.	20 max.	250	400	1500

Localization of fluorine in the body	Bones[489]	Dental enamel[489]	Soft tissues	Blood
Average amounts of fluorine	600 à 6000 (mg/kg)	300–2600 (mg/kg)	<<1 (mg/kg)	0.028 (mg/L)

Table 34 Maximum levels for animal food with a moisture content of 12% (Directive 202/32/EC of 07–05-2002 (European Parliament))

	Raw materials			Complete foods			
Type of food	**General case**	**Food of animal origin**	**Phosphates**	**Cattle, sheep, goats**	**Pigs**	**Poultry**	**General case**
Amount of F (mg/kg)	150	500	2000	30–50	100	250–350	150

4.2 International standards for drinking water and water fluoridation

Historically, as early as 1958 and 1963, WHO has set an average tolerated fluoride value in drinking water of 0.8 and 1.5 $mg L^{-1}$, respectively for warm and temperate countries in International Standards for Drinking Water. These differences took into account the fact that, because in hot countries, people eliminate more by sweating, they are supposed to consume more water. Because these rates have been revised to 1 and 1.2 $mg L^{-1}$, the upper limit has been set at 1.5 $mg L^{-1}$ (a value beyond which there is competition between cariostatic benefit and the risk of dental fluorosis). Between 2002 and 2010, the "Guideline for Drinking Water Quality, USA" and "Health Canada" confirmed the upper limit of 1.5 $mg L^{-1}$ [490] (IPCS Directive),[491] (WHO Directive); www.hc-sc.gc.ca.

However, not all various government agencies completely agree on optimal levels of fluoride in water. Some standards for drinking water use are given in Table 35.[492] In 1986, the US Environmental Protection Agency (EPA) set the upper limit of fluoride in drinking water at 4 $mg L^{-1}$. With fluoride found acceptable in a new report from the administration in 1993, the EPA asked the National Research Council to possibly propose new standards. However, it should be noted that 0.2% of the US population consumes water at more than 2.0 $mg L^{-1}$ fluorine.[493] Such recommendations do not take into account multiple external factors, for example:

[490] Fluorine and fluorides, Environmental Health Criteria 227, IPCS International Programme on Chemical Safety (IPCS), WHO, (2002). Site Internet: http://www.inchem.org/documents/ehc/ehc/ehc227.htm.

[491] OMS: Guidelines for drinking-water quality, 3rd ed., (2006), Geneva: World Health Organization (WHO).
Site Internet: http://www.who.int/water_sanitation_health/dwq/gdwq3rev/en/.

[492] Jackson P.J., Harvey P.W., Young W.F., Chemistry and bioavailability aspects of fluoride in drinking water, Report N° 5037, WRC-NSF Ltd., (2002) Marlow, UK.

[493] Lantz O., Jouvin M.H., de Vernejoul M.C., Druet P., Fluoride-induced chronic renal failure, Am. J. Kidney Dis. **10** (1987) 136–139.

Table 35 Influence of pH on the nature of the main fluorinated species in water containing 1 mg of fluoride per liter in the presence of aluminum[492]

pH	F^-	AlF_3	AlF_2^+	AlF^{2+}
6	21.35	12.82	60.4	5.36
7	97.46	1.22	1.26	0.02
8	100	0	0	0
9	100	0	0	0

- Importance of diets: low calcium diet that promotes the retention of fluoride in the body, high consumption of tea that is harmful;
- Industrial or simply agricultural environment (polluting industries, use of coal for heating, phosphate fertilizers);
- Poor knowledge of possible changes in fluorine levels in groundwater and river water.

In France, fluoride levels in tap water remain low: 83% of tap water has a rate of $<0.3\,mg\,L^{-1}$. For example, in the water distributed in 2017 in the city of Pessac, Gironde, the rate was mostly $<0.3\,mg\,L^{-1}$, and additional intake is recommended for the prevention of cavities.[494] Areas that exceed this value are located in the East and North, and in the Pyrenees. The rate in mineral waters varies between 0.1 and $8\,mg\,L^{-1}$, but is particularly important for bicarbonated waters such as Vichy-Célestins and Saint-Yorre which have 5.5 and $8\,mg\,L^{-1}$, respectively, or calcium sulfates such as Wattwiller water ($2\,mg\,L^{-1}$). Such amounts may be the cause of some kidney disorders. Nevertheless, in most cases, fluoridation of drinking water supplies remains the most effective public health decision for preventing and treating dental caries. The French Agency for the Safety of Health Products (AFSSAPS) recommends a daily intake of 0.05 mg per kilogram of weight in children from 0 to 12 years of age.[495]

The controversy between supporters of fluoridation of drinking water and its detractors is still heated, especially in the United States. This can easily be seen by visiting dozens of "anti-fluoride/fluoridation" sites on the Internet, most of which often use fallacious arguments and documents taken out of context. On these sites, the situation is dramatized, even demonized, attributing to fluorine all the evils of the Earth, with even the evocation of conspiracies against the population. Although the problem is indeed of paramount importance, it should not be forgotten that in most industrialized countries, fluorosis is not a major risk, whereas dental caries are really a major health problem. Only one case of severe fluorosis was reported in a survey of

[494] Document DDASS, Departmental Directorate of Sanitary and Social Affairs of the Gironde, Ministry of Health, "Water Quality of the Public Network" / Direction Départementale des Affaires Sanitaires et Sociales de la Gironde, Ministère de la Santé, "Qualité des eaux du réseau public" (2018).

[495] In France, information can be found at the French Agency for Sanitary Safety of Health Products (AFSSAPS) and the National Agency for Accreditation and Evaluation in Health (ANAES).

6000 12-year-old adolescents conducted between 1998 and 1999 by the French Union for Oral Health. Control of the quantities of fluoride ingested, however, remains the rule to avoid any risk of overdose.

4.3 Importance of fluoride in oral health

The beneficial effect of fluoride on oral health is now widely recognized in both children and adults.[496,497] Its use is generally via weakly fluoridated drinking water and toothpastes; in addition, varnishes, mouthwashes, gels, tablets, chewing gums, or fillers may contain fluorine. These fluorinated supplements generally correspond to 0.25–1 mg of fluoride per intake in the form of NaF. The US Centers for Disease Control and Prevention (CDCPs) provide recommendations for preventing and controlling cavities.[498] Its cariostatic effect has been demonstrated before the appearance of teeth in young children,[499] as well as on caries once dentition is formed. A dose of 0.05 mg per day is recommended for infants. On the other hand, waters too rich in fluorine such as Vichy Celestin or Saint-Yorre must be banned until the age of 12 years. It should be noted that a survey conducted in the United States between 1988 and 1994 showed that 45% of children 5–17 years old had caries in their final teeth,[500] and that 25% of adults had carious tooth roots. In 10 years, however, the number of damaged teeth has decreased by 50% among French children aged 6–12 years. The reduction of the risk of dental caries by fluoridation results essentially from the substitution of F^- ions for OH^- hydroxyl ions in apatite crystals, thus forming a fluorohydroxyapatite, more resistant to solubilization in acidic medium (see previous chapter on bioceramics).[501] In addition, these treatments strengthen the enamel during the development of the tooth. The impact of fluoride occurs at three main levels:

- Inhibition of bacterial metabolism.
- Inhibition of demineralization when fluoride is present on the enamel surface in an acid medium.
- Strengthening of the mineralization by F/OH substitution in the apatite network.

For these last two points, the fluorine must be constantly in contact with the teeth, by topical application or by any other means able to maintain an adequate concentration in the mouth.[502]

[496] Burt B.A., Eklund S.A., Fluoridation of drinking waters, in "Dentistry, dental practice and the community", Burt & Eklund Eds., W.S. Saunders Co.: Philadelphia, (1999) pp. 297–314.

[497] Beltran-Aguilar E.D., Beltran-Neira R.J., Oral diseases and conditions throughout the lifespan, Gen. Dentistry, **52** (2004) 21–26.

[498] CDCP: *Recommendations for using fluoride to prevent and control dental caries in the United States*, MMWR Recomm. Rep. **50** (RR.14) (2001), 1–42.

[499] Chow LC. Tooth-bound fluoride and dental caries. J Dent Res., **69** (1990) 59–600.

[500] Kaste I.M., Drury T.F., Horowitz A.M., Beltran E., An evaluation of NHANES III estimates (Health & Nutrition Examination Survey) of early childhood caries, J. Public Health Dent., **59** (1999) 198–200.

[501] Hamilton IR. Biochemical effects of fluoride on oral bacteria. J Dent Res; 69, 660–667, (1990); Marquis RE., Antimicrobial actions of fluoride for oral bacteria. Can. J. Microbiol., Ottawa, **41**, (1995) p.955–964.

[502] Levi S.M., Zarei M.Z., *Evaluation of fluoride exposures in children*, J. Dentistry for children **58** (1991) 467–472.

Note that fluoride-based treatments can influence the mechanical properties of bones. An experiment conducted on rats having ingested doses of L-glutamine monofluorophosphate and calcium showed a significant increase in femoral head strength and compressive strength, with howevera decrease in flexural strength.[503]

4.4 Possible use of fluorinated salts in the treatment of osteoporosis

For the prevention of osteoporosis, hormone replacement therapy must be implemented with calcium and vitamin D supplementation. Additional fluoride-based treatments have also been proposed over the last 30 years, as they stimulate the formation of osteoblasts and therefore bone reinforcement.[504] Despite these seemingly attractive properties, contradictory results have been reported concerning the effect of such treatments on the (non) decrease in the number of fractures (vertebral and non-vertebral). These discrepancies are likely due to differences in dosage, formulation, calcium supplements, diet types, etc. and could even lead to an increase in non-vertebral fractures or undesirable gastrointestinal effects. As fluorinated salts promote trabecular bone formation, their use has been recommended for patients with moderate osteoporosis in which the trabecular meshwork was not impaired too much. An estrogen-based hormonal treatment that decreases the resorption of bone has been combined with an addition of fluorides that increase bone mass and trabecular thickness (+ carbonate supplement or calcium citrate). Fluorine was introduced as sodium monophosphate. A 3%–5% increase in bone mineral density was observed in the femoral neck or lumbar vertebrae when hormone replacement therapy (estrogens + progestins) and adequate fluoride treatment were added.[505] A comparable result, with 5% improvement in bone mineral density, was obtained in 110 patients with osteoporosis with vertebral fracture. In this case, fluorine was ingested as NaF.[506] It should be noted, however, that estrogen/progestin hormone replacement therapy is currently being seriously questioned as a result of the increased risk of developing breast cancer.

4.5 Bioassimilation of fluoride in humans and animals

The average fluoride intake for an adult is 1.1 mg per day for areas with non-fluorinated drinking water and can reach 3.8 mg for areas with fluoridated water, which corresponds to an average amount of 2.6 g of fluoride in the body. We can see in the inset below a particular case related to the consumption of tea bricks. The bioavailability of fluoride in the body is 100%, and there is no retention. All

[503] Giavaresi G., Fini M., Gnudi S., Mongiorgi R., Ripamonti C., Zati A., Giardino R., The mechanical properties of fluorine-treated bone in the ovariectomized rat, Cacif. Tissue Int. **65** (1999) 237–241.

[504] Eastell R., Treatment of postmenopausal osteoporosis, N. Engl. J. Med. **338** (1998) 736–746; Murray T.M., Ste-Marie L-G., Fluoride therapy for osteoporosis, Can. Med. Assoc. **155** (1996) 949–954.

[505] Negri A.L., Sarli M., Insua A., Zanchetta J.R., Hormone replacement therapy with and without fluoride for postmenopausal osteoporosis, Osteologia-Osteology **2** (1999) 58–63.

[506] Pak C.Y., Sakhaee K., Adams-Huet B., et al., Treatment of postmenopausal osteoporosis with slow-release sodium fluoride: final report of a randomised controlled trial, Ann. Int. Med. **123** (1995) 401–408.

studies have shown that the fluoride ion is very easily absorbed in the gastrointestinal transit, and only very high levels of calcium can lead to the formation of insoluble species. The bio-assimilation of fluorides in the body also depends on the other components of the diet, especially the presence of cations likely to form complex fluoride ions. It is reduced to 60%–80% if there is simultaneous consumption of foods containing calcium, such as milk. A physical–chemical study of speciation in freshwater has shown that this trend also depends on the pH; in acidic media, fluorinated aluminum complexes will predominate, whereas in basic pH, it will be the F^- ions (Table 30). Once in the blood plasma, fluorine is rapidly distributed to all body parts, but is fixed only in the calcified tissues at a proportion of 50%, the rest being eliminated in the urine. Absorption via inhalation and the epidermis is extremely limited and only concerns direct contact with fluorinated gases or products such as hydrofluoric acid.

The same factors as those determined for humans remain valid for all animals (herbivores, omnivores, carnivores). For example, a study of rats fed for 1 week with diets rich in aluminum shows that 99% of the fluorine is absorbed when it is in the form of NaF, while 86% or 16% are absorbed, respectively, when it is under the form of Na_3AlF_6 or $AlF_3{\cdot}H_2O$.[507] Similarly, when calcium is added to the livestock diet, the fluorine absorption rate is reduced to 40%.[508] It has also been shown that there is no real evidence of fluoride accumulation in the animal food chain (mammals, invertebrates, etc.), but that the two main factors favoring uptake and accumulation of fluoride in the body are the acidity of the digestive system's pH of the and the presence of calcified tissues.

5 Contamination and physiological effects of fluoride on living organisms

The risk of contamination depends on the degree of daily exposure to fluoride and fluorinated products. Drinking water content is of primary concern, but diet is also—especially if it contains tea or fish in high doses—as is the immediate environment: heating and cooking habits, presence of close industrial environment. Finally, excessive use of fluoride toothpastes can also have a negative influence. Bioaccumulation of substances such as PFOA and PFOS in organisms has also been reported.

5.1 *Water contamination and fluorosis*

The UNICEF reports count more than 20 countries affected by fluorosis.[509] Countries with endemic fluorosis are shown in Fig. 78. According to WHO and UNICEF data, this disease is due to an excessive intake of fluorine over a period of at least 5 years.

[507] Wright D.A., Thompson A., Retention of fluoride from diets containing materials produced during aluminium smelting, Brit. J. Nutrit. **40** (1978) 139–147.

[508] Shupe J.L., Miner M.L., Harris L.E., Greenmood D.A., Relative effects of feeding hay atmospherically contaminated by fluorine residue, Amer. J. Veterin. Res. **23** (1962) 77–787.

[509] UNICEF, "Fluoride in water: an overview" (1998). UNICEF, WES Section (Water Environment & Sanitation) Programme division, UNICEF 3, U.N. Plaza, NYC, 10017, USA.

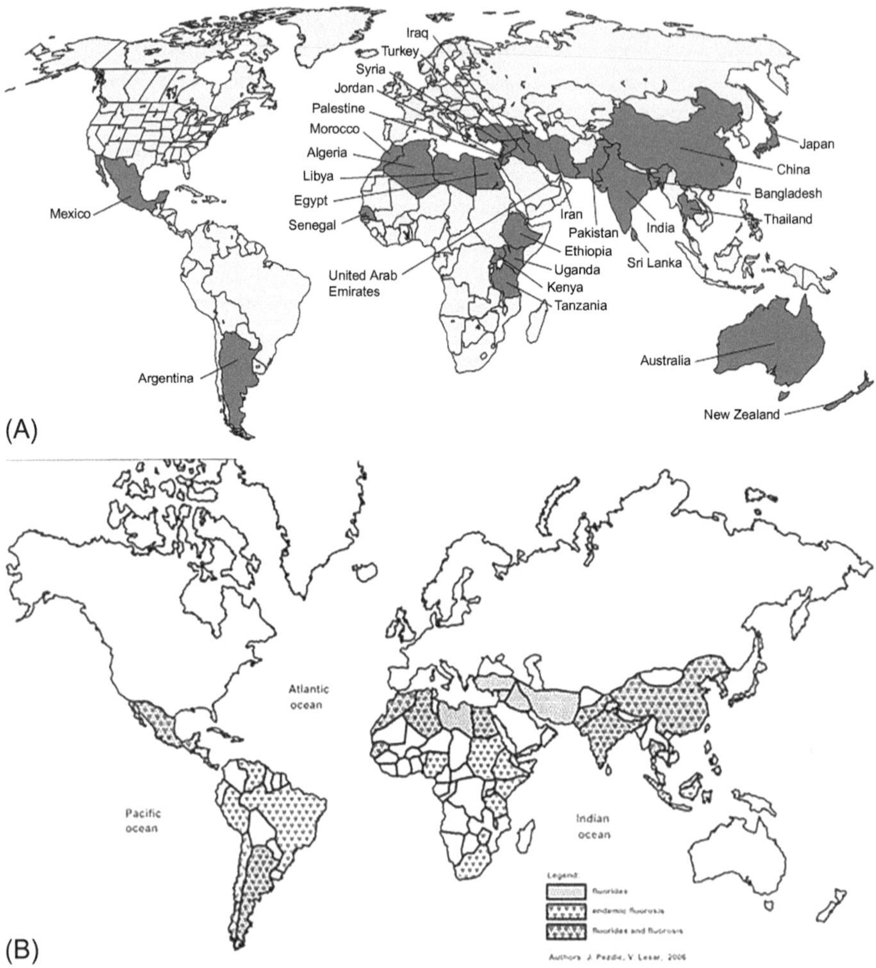

Fig. 78 (A) Countries in which fluorosis is endemic in relation to high levels of fluoride in drinking water, from UNICEF, "Fluoride in water: an overview" (1998). UNICEF, WES Section (Water Environment & Sanitation) Programme division, UNICEF 3, U.N. Plaza, NYC, 10017, USA, (B) Level of fluorosis, according to P. Mazet.[477]

The WHO has defined this rate as being of the order of 8 mg per day.[510] More recently, a highly documented bibliographic report conducted by P. Mazet at "Maison des Sciences de l'Eau" of Montpellier showed that the previous estimate must be revised upward in 30 countries concerned with the problems of fluorosis in the 1960s, for which the average rate of fluorine in water is $>1.5\,\mathrm{mg\,L^{-1}}$.[477] This work has attempted to identify the chemical and hydrological mechanisms leading to high levels of fluoride in freshwater, in relation to geological, climatological, industrial, and agricultural activities. These factors are sources of fluoride pollution, particularly in the Andean

[510] Office of Drinking Water. *Drinking water criteria document on fluoride*. Washington, DC, US Environmental Protection Agency, (2010).

Altiplano. The various water treatments used are also available with the corresponding bibliography.[477] Several websites provide information on issues related to water scarcity and purification, particularly in South Africa and India, where the majority of states face cases of endemic fluorosis. In India and China, the number of people suffering from dental fluorosis is about 25–38 million, with 1–2 million exhibiting more serious bone fluorosis. In India, where fluorine levels in water can reach $18\,mg\,L^{-1}$, many cases of fluorosis are reported in the Agra region. For example, the site http://www.twadboard.gov.in/twad/index.html is dedicated to these issues affecting the State of Tamil Nadu.[511] The main source of pollution is water from wells drilled in rocks with a high content of fluorinated minerals.[512] The relationship between the number of cases of fluorosis observed in Indian Rajasthan villages and the rate of fluoride in the water used has been clearly established.[513] In all countries, it can be estimated that 70 million people are affected to some degree by these diseases.

The first level of fluorosis, dental fluorosis, is aesthetic with characteristic spots on the teeth. Bone fluorosis, which is much more dangerous than dental fluorosis, first results in an increase in bone mass with fluoride concentrations that can reach two to five times normal, that is, $3–5\,g\,day^{-1}\,kg^{-1}$. The symptoms that follow are pain and hardening of the joints, followed by osteosclerosis that may ultimately lead to skeletal deformity by calcification of ligaments, osteoporosis of long bones, and neurological problems due to hypercalcification. In this final stage, the fluorine content in the bones can reach or even exceed $10\,g\,kg^{-1}$.[514] The US Food & Nutrition Board study has shown that such cases are extremely rare in the United States, even when water contains up to $20\,mg\,L^{-1}$ of fluoride.[515] There are also populations in northern Tanzania that have been using water containing $10–14\,mg\,L^{-1}$ of fluoride for generations without any of the symptoms of fluorosis appearing. Conversely, in the western part of Guizhou Province in China, signs of bone fluorosis have been clearly demonstrated although the level of fluoride in water is low and ranges from 0 to $0.18\,mg\,L^{-1}$. On the other hand, in some islands of the South Atlantic, the dental fluorosis often observed seems to be due, not to water ($0.2\,mg\,L^{-1}$), but to fish consumed (average $7\,mg\,kg^{-1}$).

5.2 Food contamination by tea bricks

The risk of fluorosis does not depend solely on the level of fluoride present in drinking water, but also on deficiencies and eating habits such as the use of tea bricks in Tibet[516] (see "Consumption of tea bricks and fluorosis in Tibet"). From all the raw materials that go into the composition of foods or beverages, tea (*Camellia sinensis*

[511] Mariappan P., Vasudevan T., Tamilnadu Water Supply and Drainages Board (2004), http://twadboard.gov.in/home.asp.

[512] Khandare A.J., G.S. Rao, J. Hum. Ecol. **19**, 283–288 (2006).

[513] S.L. Choubisa, Fluorosis in some tribal villages of Udaipur district (Rajastan), J. Environm. Biology, **19**, 341–352 (1998).

[514] Eble DM, et al., *Fluoride concentrations in human and rat bone,* J. Public Health Dent. **52** (1992) 288–291.

[515] Food and Nutrition Board, Institute of Medecine, USA; site: http://www.iom.edu/CMS/3788.aspx.

[516] M.P Whyte, W. G Totty, V. T Lim, G. M Whitford, Skeletal Fluorosis From Instant Tea, J. Bone Mineral Res. **23**, 759–769 (2008).

L.) is the plant that accumulates the most fluorine in its leaves and branches. As this shrub is evergreen, it is the older leaves (closer to ground) that contain the most fluoride. A study of commonly used teas shows that the rate of fluoride in the leaves varies from about $50\,\mathrm{mg\,kg^{-1}}$ to several $100\,\mathrm{mg\,kg^{-1}}$ of dry leaves depending on provenance, with the highest levels being measured on common mixtures (which contain most old leaves). These values correspond to $1–3.5\,\mathrm{mg\,L^{-1}}$ once the tea is infused in distilled water.[517] The amount of fluoride absorbed daily by tea drunk by the population of Tibet varies between 7 and 8 mg according to the regions of the country and corresponds to rates of fluorosis of 53%–83%.[518] An additional toxicological problem is due to the presence of aluminum, which has sometimes been associated with Alzheimer's disease.[519] Levels between 300 and $2700\,\mathrm{mg\,kg^{-1}}$ were analyzed in Chinese teas, corresponding to $1–6\,\mathrm{mg\,L^{-1}}$ infusion, well above the WHO limit of $0.2\,\mathrm{mg\,L^{-1}}$.[520]

Consumption of tea bricks and fluorosis in Tibet

For nearly 1000 years, many people in China have been conditioning tea in the form of 1.5 kg bricks. The leaves—usually the oldest leaves closer to the ground but also branches and fruits and sometimes the roots are compressed in the form of bricks so as to preserve the scent of the tea and facilitate its transport and preservation. These practices occur almost everywhere in China, but are more particularly developed by the Mongolian and Tibetan populations. Although tea bricks are less commonly produced in modern times, many post-fermented teas, such as *pu-erh*, are still commonly found in bricks, discs, and other pressed forms. Tea bricks can be made into beverages like tea or eaten as food, and were also used in the past as a form of currency. Does not one of their proverbs say "we can do without cereals for three days, but tea only for one day." The problem is that one of the special characteristics of the tea plant is to selectively absorb fluorine from the soil and store it in its lower leaves. What's more, in the old leaves that make up the bricks, the fluorine content is 10–100 times higher than for young leaves that are used for green or black tea and can reach $440\,\mathrm{mg\,kg^{-1}}$.[521]. Similarly, the Al content is two to eight times higher than teas prepared from young leaves. Drinking 0.3 L of infusion is equivalent to taking more than 4 mg. An epidemiological survey was conducted in Sichuan and Tibet to determine the causes of the high rate of fluorosis in the population.[522]. This has shown that the origin of fluoride came from the only use of tea bricks in many and frequent culinary compositions. In addition, in Tibet, fluorosis is aggravated by poor general nutrition and especially a lack of oxygen related to the very high altitude.

[517] Dabeka W.D., McKenzie, A.D., Survey of lead, cadmium, fluoride, nickel, and cobalt in food composites and estimation of dietary intakes of these elements by Canadians in 1986–1988, J. AOAC Int. **78** (1995) 897–909.

[518] Cao J., Zhao Y., Liu J., Fluoride in the environment and brick-tea fluorosis in Tibet, J. Fluor. Chem. **106** (2000) 93–97.

[519] McLachlan D.R.C., Aluminium and the risk for Alzheimer disease, Environmentrics **6** (1995) 233–275.

[520] Chen Z.M., The environmental pollutants in tea products, Chin. J. External Agric. **3** (1984) 1–10; Wong M.H., Zhang Z.Q., Wong J.W.C., Lan C.Y., Trace metal contents (Al, Cu, Zn) of tea, Environ. Geochem. Health **20** (1998) 87–94; Wong M.H., Fung K.F., Carr H.P., Aluminium and fluoride contents of tea with emphasis to brick tea and their health implications, Toxic. Lett. **137** (2003) 111–120.

[521] FSA Expert Group in Vitamins and Minerals, Food Standards Agency, UK (2001). https://cot.food.gov.uk/cotstatements/cotstatementsyrs/cotstatements2003/fluorine

[522] Cao, J.; Zhao, Y.; Liu, J.; Xirao, R.; Danzeng, S.; Daji, D.; Yan, Y. (2003). "Brick tea fluoride as a main source of adult fluorosis". Food and Chemical Toxicology. **41** (4): 535–42.

A change in the dietary habits of the populations concerned, with a preference for young leaves and a better selection of the types of tea (e.g., a variety of Assam plants which concentrate less F and Al[523]) considerably limits both F and Al levels in the tea bricks. Soil treatment with Ca salts, such as low-dose carbonate or chloride, increases the pH with a simultaneous decrease in F and Al, while maintaining the plant in good health.

5.3 Environmental contamination

In regions with high volcanic activity, fluoride intake for an adult can reach 30 mg per day, mainly by drinking water pollution. Risks incurred during volcanic eruptions with high fluoride emissions have been presented previously. This is also the case for cattle feeding in fields near areas with volcanic or industrial activities are emitting fluoride vapors. Exposure of plants to fluoride particles leads to a deposition layer on their outer surface with fluoride uptake. Beyond an intake of $0.2\,\mu g\,m^{-3}$, irremediable deterioration of the plant is observed. It should be noted that in the European Union, the permissible rates of certain fluorinated gases are set in the working places at $2.0\,mg\,m^{-1}$ over 8 h for F_2 and $2.5\,mg\,m^{-3}$ over 8 h for HF.

5.4 Industrial and domestic contamination: Example of the use of coal

In some parts of China and India, coal heating is also a source of fluoride pollution. While in developed countries, the use of coal occupies a decreasingly important place for domestic heating, in some parts of China and India, it is still the primary fuel used, and very serious problems may be caused in the populations by both by ingestion and food pollution.

In China, in particular, the combustion of domestic coal emits gaseous or aerosol pollutants that are either directly ingested by the people or are finally found in their food. Four-hundred million people depend on this heating medium, and many areas are affected by endemic fluorosis (Fig. 7).[524] The areas concerned can be connected to those corresponding to most fluorinated carbon ores (see Fig. 79). In Beijing, the average aerosol concentration is $0.61\,\mu g\,m^{-3}$, with differences between 0.08 and $1.61\,\mu g\,m^{-3}$. The highest concentrations ($1.61\,\mu g\,m^{-3}$) occur in winter and are 20 times greater than in summer. This difference is due to the use of coal for heating as well as to the increase of dry deposition in winter dry periods.[525] Other harmful emissions include heavy metals and metalloids such as As and Se, SO_2, aromatic

[523] Fung K.F., Zhang Z.Q., Wong J.W.C., Wong M.H., Fluoride contents in tea and their infusion, Environ. Pollut. **104** (1999) 197–205; Fung K.F., Zhang Z.Q., Wong J.W.C., Wong M.H., Al and F concentrations of three tea varieties growing at Lantau Island, Hong Kong, Environ. Geochem. Health **25** (2003) 219–232.

[524] Zheng B., Ding Z., Huang R., Zhu J., Yu X., Wang A., Zhou D., Mao D., Su H., Issues on health and diseases relating to coal use in southwestern China, Intern. J. Coal Geol. **40** (1999) 119–132.

[525] Feng Y. W., Ogura N., Feng Z. W., Zhang F. Z., Shimizu H., Concentrations and Sources of Fluoride in Atmospheric Depositions in Beijing, China Water, Air, and Soil Pollution, **145** (2003) 95–107.

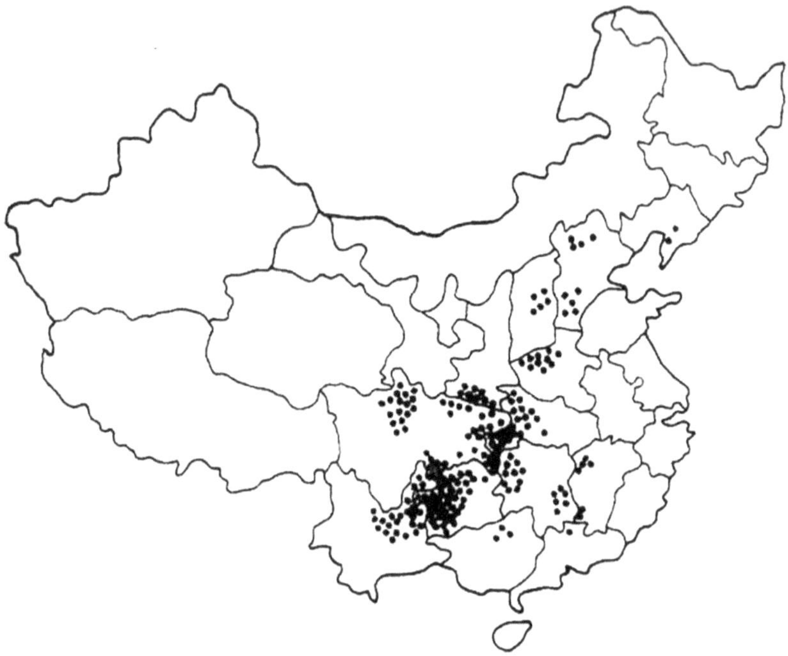

• Area of the endemic fluorosis related to indoor combustion of coal

Fig. 79 Regions of China in which endemic fluorosis is linked to domestic heating by coal, according to B. Zheng.[524]

hydrocarbons.[526] HF and SiF_4 constitute the two major species of fluorinated emissions, representing between 40% and 80% of the emitted inorganic fluorides and leading to rates of 10–150 $\mu g\,m^{-3}$ inside houses.[527] A study of the various sources of fluoridated pollution in areas with various degrees of pollution in Jiangxi Province, China, showed (Table 36) that the food issues were by far the most important, coming both from the very high levels of fluoride in the soil and pollution by gases emitted by heating or when cooking food.[528] In the most polluted areas, the most common cases of dental fluorosis were observed in children rather than adults: 70% of the population suffered from bone fluorosis. Up to 100% of children who could have dental fluorosis, with the aggravating fact that in the presence of arsenic, the rate of fluorine necessary to induce bone fluorosis is lowered.

[526] An D., He Y., Xu Q., Poisoning by coal smoke containing arsenic and fluoride, Fluoride, **30** (1997) 29–32.

[527] Zhang Y., Cao S.R., Coal burning induced endemic fluorosis in China, Fluoride **29** (1996) 207–211.

[528] Ando M., Tadano M., Yamamoto S., Tamura K., Asanuma S., Watanabe T., Kondo T., Sakurai S., Ji R., Liang C., Chen X., Hong Z., Cao S., Health effects of fluoride pollution caused by coal burning, Sci. Total Environ. **271** (2001) 107–116.

Table 36 Fluorine concentrations of various origins and daily intake (mean values) in areas with variable fluoridated pollution rates in Jiangxi Province, China[528]

Origin	Coal (mg/kg)	Soils (mg/kg)	Air (mg/m^3)	Water (mg/L)	Harvests (mg/kg)	Vegetables, including peppers (mg/kg)	Daily intake of fluoride (mg/person)
Heavily polluted areas	656	620	0.070	0.19	70	195	43.2
Medium polluted areas	212	4807	0.076	0.10	3.6	129	14.4
Non-polluted areas	152	285	0.005	0.57	1.3	1.7	3

5.5 The ban of persistent organic pollutants (POP): The case of PFOS and PFOA

It was noted that humans and large proportion of wildlife could be contaminated with environmentally persistent long-chain perfluoroalkyl compounds. In December 2002, following a large-scale pollution, 3M Co. stopped making PFOA perfluorooctanoic acid ($C_8HF_{15}O_2$) and perfluorooctane sulfonyl PFOS ($C_8HF_{17}O_3S$) that they produced at the plant in Cottage Grove, Minnesota. These products were sold to DuPont for use as emulsifiers in the synthesis of fluoropolymers. They allowed production of a large range of products, including nonstick cookware, grease-proof food wrappers, stain-resistant carpets, water repellent clothing, etc. In particular, they were the raw materials for manufacturing Scothchgard products. All carboxylates and sulfonates with long-chains C6–C12 were implicated. Because of the stability of these molecule in the atmosphere and their bioaccumulation in the human body (liver and blood) and in certain species such as bees, the main producing companies agreed to stop production on the basis of agreements proposed by the Environmental Protection Agency (EPA-USA). In December of 2006, the production of these molecules was prohibited by the European Parliament. This bioaccumulation, whose half-life is 100 days for rats and from 4 to 5 years for men, has been related to the length of the alkane chain. Long chains >4 carbons are now prohibited. A first detailed account for the global production history, types of emissions and fate of PFOAs have shown that between 1950 and 2004, 6900 t of PFOAs were directly emitted from chemical plants.[529]

The PFOA and PFOS molecules have been found in surface water and tap water in both developed and developing countries around the world including in North America, Europe, and Asia. In most countries, rivers are the source of tap water, which is one of the important pathways in which PFCs reach humans. Following the Kunacheva study, the evaluations were conducted during 2004–2010 in 15 countries, and 539 samples were collected from the rivers in 41 cities. A solid phase extraction coupled with HPLC-ESI-MS/MS were carried out in industrialized and non-industrialized areas demonstrating that these compounds undergo long-range transportation in the environment. The average concentration of PFOS ranged from not detected to 70.1 ng/L and of PFOA ranged from 0.2 to 1630.2 ng/L.[530]

New alternative shorter chain molecules such as perfluorobutane sulfonyl fluoride ($C_4F_9SO_2F$) without risk to health, have been proposed as substitutes. These new short-chain molecules, although they are still persistent, are not bioaccumulative and have a better toxicity profile. The impact of phase-out of long-chained PFOSs program is already evident, because its level in US adult blood donors dropped by 60% and PFOAs by 25% from 2000 to 2006.[531] The EPA has established health

[529] K. Prevedouros, I. T. Cousins, R. C. Buck, S. H. Korzeniowski Sources, Fate and Transport of Perfluorocarboxylates, Environment Sci. Technol. **40** (2006) 32.

[530] Kunacheva C, Fujii S, Tanaka S, Seneviratne ST, Lien NP, Nozoe M, Kimura K, Shivakoti BR, Harada H., Worldwide surveys of perfluorooctane sulfonate (PFOS) and perfluorooctanoic acid (PFOA) in water environment in recent years. Water Sci Technol., **66**(12) (2012) 2764–71.

[531] G. W. Olsen et al., Decline in Perfluorooctanesulfonate and Other Polyfluoroalkyl Chemicals in American Red Cross Adult Blood Donors, 2000–2006, Environment Sci. Technol. **42** (2008) 4989.

advisories for PFOA and PFOS based on the agency's assessment of the latest peer-reviewed science to provide drinking water system operators. In addition, the agency is committed to supporting states and public water systems as they determine the appropriate steps to reduce exposure to PFOA and PFOS in drinking water.[532] GenX is a trade name for chemicals that were proposed around 2010 as an alternative to persistant perfluorooctanoic acids (PFOA or C8) in the synthesis of PTFE. GenX (trade names as FRD-903, FRD-902, E1) have fewer carbon atoms in the chain, but have many similar physical and chemical properties as their predecessors, as they both repel oil and water. These materials are therefore essential for the production of common household products such as non-stick pans, firefighting foam, and common outdoor fabrics such as. Gore-Tex. See article "*Confronting Fluoroethers*" that appeared in C&E News in the Feb. 12, 2018 issue (C&EN 96, 8, 8–8).

6 Improving the quality of drinking water by defluoridation

For industrialized countries, controlling the fluoride content in drinking water must correspond primarily to a better health quality. But for many countries in which fluorosis is endemic, the cost and difficulty of implementing the operation remain limiting factors. Other factors to take into consideration are the efficiency that should lead to fluoride levels of 1–2 $mg L^{-1}$, the quality, and adapted pH of the treated water, not to mention the religious taboo issues concerning, for example, the origin of the calcined bones used to obtain activated charcoal. It should be noted that, given the importance of the issue, meetings devoted to this crucial problem are held regularly throughout the world.[533]

In Africa, cases of fluorosis are found in countries rich in phosphate deposits (Tunisia, Morocco, Algeria, and Senegal). Studies show that from 2 $mg L^{-1}$, all children are affected and that 60% of them have severe fluorosis for levels close to 4 $mg L^{-1}$. For larger exposures or longer duration, for example, when the percentage in the water is >7 $mg L^{-1}$, the damage affects the bones and joints (see previous chapters).

A large number of physicochemical or physical-type techniques have been proposed.[534] Four major methods of defluoridation are used:

- Co-precipitation, sometimes called the Nalgonda technique, with aluminum salts (sulfates or chlorides) mixed with water that give rise to insoluble fluorine compounds. These

[532] 2016 - Supporting Documents for Drinking Water Health Advisories for PFOA and PFOS, EPA website: https://www.epa.gov/ground-water-and-drinking-water/supporting-documents-drinking-water-health-advisories-pfoa-and-pfos.

[533] Water Engineering Development Center Conferences/International Workshops on Fluorosis Prevention and Defluoridation of Water/Water & Wastewater Asia Conferences/Afriwater/Asia Pacific Training Workshop on Ecohydrology/Water Framework Directive (WFD): Ecological Implications/Environment 2003: Innovation in Environmental Practice/Water Resource and Water Quality Management for Sustainable Drinking Water Supply, India.

[534] Pontié M., Rumeau M., Ndiaye M., Diop C.M., Sur le problème de la fluorose au Sénégal: bilan des connaissances et présentation d'une nouvelle méthode de défluoration des eaux de boisson, Cahiers Santé **6** (1996) 27–36.

methods may also be based on calcium salts; in this case, the F^- ions react with lime or calcium chloride to form highly insoluble CaF_2.
- Adsorption on all kinds of substrates: clay, activated alumina, calcined bone;
- Ion-exchange resins based on sulfonated compounds derived from coconut; it should be noted that the purchase of these columns is too expensive for most poor countries where fluorosis is prevalent;
- Membrane processes such as reverse osmosis have proven to be the most effective, even if their cost is rather high.

6.1 Precipitation processes

Given the low insolubility of fluorides, it is difficult to carry out precipitation treatment of insoluble fluorides for the treatment of drinking water. In practice, few compounds can diminish the fluoride concentration to truly acceptable levels. Complexation is necessary; once complexed with the fluoride ions, aluminum and calcium salts precipitate because of their low solubility, so, a decanting operation is also necessary. However, a lime treatment [$Ca(OH)_2$] can be used to treat industrial effluents containing significant levels of hydrofluoric acid (HF).

Other methods may be reported, such as electrocoagulation using aluminum bipolar electrodes,[535] distillation of water in contact with brushite and calcite,[536] or "Krass" technology with very low residual aluminum levels.[537]

In areas with a high risk of endemic fluorosis, a variety of low-cost methods have been implemented to bring the fluoride content in drinking water to acceptable levels, often through initiatives of humanitarian organizations under WHO or UNICEF patronage.[538] Prior to any technical intervention, detailed information should be provided to modify the population's habits, as in the three-phase program in Nagaon District, Assam, India, with the assistance of the Fluorosis Research and Rural Foundation of India (http://aphe.gov.in/). The so-called Nalgonda precipitation technique[539] proposed by a Danish group and developed in the Nagpur region of India in the 1970s has been transposed to a pilot scale in Kenya, Senegal, and Tanzania.[540] It consists of adding alum (aluminum sulfate or aluminum and potassium sulfate) and lime to the polluted water. The proportions of these two additives are determined so that the final pH is between 6.5 and 7. After stirring, flocculation occurs because of the formation of

[535] Mameri N., Yeddou A.R., Lounici H., Belhocine D., Grib H., Bariou B., Defluoridation of septentrional Sahara water of North Africa by electrocoagulation process using bipolar aluminium electrodes, Water Res. **32** (1998) 1604–1612.

[536] Larsen M.J., Pearce E.I.F., Defluoridation of drinking water by boiling with brushite and calcite, Caries Res. **36** (2002) 341–346.

[537] Argawal K.C., Gupta S.K., Gupta A.B., Development of new low cost defluoridation technology (krass), Water Sci. & Technol. **40** (1999) 167–173.

[538] Näslund J., I. Snell, Fluoride contaminated Groundwater in Nakuru & Baringo districts, Kenya, Master's Thesis, Luleå Univ. Technol., Division Sanitary Engin., Sweeden (2005).

[539] Nawlakhe W.G., Kulkami D.N., Pathak B.N., Bulusu K.R., Defluoridation of water by Nalgonda technique, Ind. J. Environ. Health **17** (1975) 26.

[540] Dahi E., Mtalo F., Njau B., Bregnhj H., *Defluoridation using the Nalgonda Technique in Tanzania*, 22nd Water Engineering Development Center Conference, New Delhi, India (1996).

aluminum hydroxides. During the process, an exchange takes place between the hydroxyl and fluoride ions, and the latter are trapped in the flocculated solid, where they can be removed by simple sedimentation. The experiment carried out in the village of Ngurdoto in Tanzania on a water containing between 8.8 and 12.5 m mg L^{-1} of fluoride allowed a post-treatment reduction of this rate to 2.1 ± 0.7 mg L^{-1}. Although this rate is slightly higher than current standards, the great advantage of the method is that it can be implemented and maintained at lower cost by the villagers.

6.2 Adsorption and ion exchange processes

The use of extremely porous and highly adsorbent solids such as activated alumina, activated carbon, or tricalcium phosphate makes it possible to fix the F^- ions. The lack of selectivity is often a handicap of the method. The adsorption of fluorine on products derived from hydrotalcite $[Mg_6Al_2CO_3 (OH)_{16}]\cdot 4H_2O$ was, for example, studied with the aid of adsorption isotherms. The influence of pH, material crystallinity and the presence of exchangeable anions has been optimized for application in rural areas of northern China. The use of locally available materials is particularly important for obvious economic reasons. These include activated carbon derived from wheat residues,[541] clays (aluminates and silicates),[542] soils derived from volcanic ash (soils of the Ando type) allowing effective defluoridation in rural Kenya and the East African Rift.[543] In southern India, iron-rich clays (hematites) can reduce the fluoride level from 4 to 1 mg L^{-1}.[544] The adsorption of fluoride ions onto new types of materials is

Table 37 Defluoridation capacities of different chemical absorbents

Material	Aluminium hydroxyde	Charcoal	Apatite	Bone charcoal	Activated alumina
Useful capacity (g L^{-1} of F^- per g of material)	0.05	0.2	0.3	1	3.3

[541] Singh R.B., Singh Y., Swaroop D., *Defluoridation of groundwater in Agra city using low-cost adsorbents*, Bull. Environ. Contamin. Toxicol. **65** (2000) 120–125.

[542] Wang H., Ji J., Chen J., Defluoridation of drinking water with hydrotalcite-like compounds, Paper 83–18, Geol. Soc. Amer. Denver Annual Meeting, (Oct. 2002); Defluoridation of drinking water by natural and synthetic clays, Goldschmigt Conference (2003).

[543] Zevenbergen C., van Reeuwijk L.P., Frapporti G., Louws R.J., Schuiling R.D., A simple method for defluoridation of drinking water at village level by adsorption on Ando soil in Kenya, Sci. Total Environ. **188** (1996) 225–232.

[544] Padmasiri J.P., Fondeta W.S.C.A., Liyanapatabendi T., Low cost fluoride removal by an upward flow household filter, The IWSA International Specialized Conf. on Advanced Treatment and Integrated Water System Management into the 21st Century, Osaka, Japon, Blackwell Scientific Publ. Oxford, UK (1995).

tested continuously,[545] for example, on carbon nanotube bundles, which at pH = 7 adsorb 4.5 $mg\,g^{-1}$ of fluoride from solutions concentrated at 15 $mg\,L^{-1}$.[546]

The treatment by precipitation with aluminum, adsorption on activated alumina, or on bone charcoal (containing phosphates) are part of these techniques. As can be seen in Table 37, the best adsorbent is activated alumina. Zirconia should give even higher results because of the stability of the formed complexes. It would remain, then, to study the cost of the operation and the possibilities of regeneration of the adsorption columns. We can also consider the use of adsorbents available on site: mineral adsorbents such as carbonized coconut, attapulgite (a kind of amphibole clay) or even fly ashes from thermal power plants. Although the fixation power for the fluoride ion is low, the abundance of these materials reduces the cost of the operation. The use of local adsorbents makes it possible to produce defluoridation columns which are very simple in both their design and their use and to carry out small-scale treatments at very low cost.

6.3 Nanofiltration (NF)

Nanofiltration is a separation technique using a membrane that allows fractioning of solutes whose molar mass is between 100 and 1000 $g\,mol^{-1}$, while reverse osmosis concerns solutes with molar masses $<100\,g\,mol^{-1}$. Both processes work by permeation under pressure. In NF, it is possible to carry out a partial desalination of a brackish water avoiding a remineralization after treatment and to determine the level of reduction the salt by modifying the operating parameters: transmembrane pressure, conversion rate, and ionic strength. Moreover, it is interesting to note that, due to a greater opening of the pores in NF, the hydraulic permeability is about five times higher than that obtained in reverse osmosis.

For polluted waters containing both As and F, the use of highly selective nanoadsorbents or nanofunctionalized membranes are able to remove both pollutants.[547]

6.4 Reverse osmosis

This technique uses a semi-permeable membrane that separates a concentrated solution from another, whether pure or diluted. In general, the osmotic pressure of water allows it to cross the membrane to dilute the concentrated solution. Conversely, in the case of reverse osmosis, a hydrostatic pressure (>30 bar) greater than the osmotic pressure of the water is applied to the part containing the concentrated solution, which allows the pure water to pass through the membrane, whereas suspended or dissolved solids cannot. Membranes commonly used are based on cellulose acetate or polyamide composite film. New nanoscale membranes allow the use of lower pressures

[545] Onyango M. S., H. Matsuda Fluoride Removal from Water Using Adsorption Technique, pp. 1–48, Fluorine and the Environment, Vol. **2**, A. Tressaud Ed., Elsevier (2006).

[546] Li Y.H., Wang S., Zhang X., Wei J., Xu C., Luan Z., Wu D., Adsorption of fluoride from water by aligned carbon nanotubes, Mat. Res. Bull. **38** (2003) 469–476.

[547] S. V. Jadhav, E. Bringas, G. D. Yadav, V. K. Rathod, I. Ortiz, K. V. Marathe, Arsenic and fluoride contaminated groundwaters: A review of current technologies for contaminants removal, J. Environ. Manag. **162** (2015), pp. 306–325.

Fig. 80 Mobile reverse osmosis unit established in Senegal in 1994 in the Mbour region, according to M. Pontié.
From Pontié M., C. Diawara, A. Lhassani, J.C. Schrotter, Treatment of water intended for human consumption, L'actualité chimique—N° **301–302** pp. 106–112 (2006).

(<20 bar) and can thus reduce the cost of the method.[548] An installation of this type is shown in Fig. 80. Thus, reverse osmosis and the use of adsorbents (such as activated alumina) seem to be the most effective defluoridation methods and therefore, have the most promising development.[549]

[548] Pontié M., C. Diawara, A. Lhassani, J.C. Schrotter, Treatment of water intended for human consumption, L'actualité chimique—N° **301–302** pp. 106–112, (2006); Pontié M., C. Diawara, A. Lhassani, H. Dach, M. Rumeau, H. Buisson, J.C. Schrotter, Water Defluoridation Processes: Nanofiltration for Future Large-Scale Pilot Plants, pp. 49–80, Fluorine and the Environment, Vol. **2**, A. Tressaud Ed., Elsevier (2006).

[549] U.S. Department of the Interior, "Fluoride fact sheet" (2001); U.S. Department of the Interior, Water Treatment Engineering & Research Group, Denver CO 80225, USA.

Conclusions: What future for fluorine and fluorine products?

We have seen that fluorine, with its long history, is present today in various application fields: in chemistry, of course, but also in biology, medicine, agrochemistry, new technologies, energy transformation, etc. Thanks to its specific properties (very high reactivity and electronegativity, high oxidizing power), fluorine has given rise to a remarkable chemistry and applications. Today, there are more than 600,000 compounds containing at least one fluorine atom, and the chemistry of fluorine and fluorinated products has led to advances in a wide variety of applications.

- *In chemistry and applied sciences*
  - New molecules with highly selective properties (sulfur-based fluoride compounds used as steroids, triflic acid derivatives, sugars, vitamins).
  - Fluoropolymers derived from Teflon, whose remarkable resistance to corrosion has revolutionized many domains: the packaging of highly reactive products, nonstick culinary containers, cardiovascular surgery materials, ion exchange membranes, tissues which "breathe" after treatment with Gore-Tex.
  - Superacid media used for the production of high-octane fuels.
  - Surfactants used for the protection of surfaces (fabrics, carpets, leather) and for firefighting.
  - Fluorinated materials with specific properties (catalysts, colored pigments, chemical sensors, biomaterials).
  - Nanocomposites based on fluorinated silanes and silica, metal and fluorinated polymers with ultrahydrophobic properties, selective membranes for gas filtration.
  - Surface treatments for the protection of our cultural heritage, antigraffiti and antireflective layers, UV absorbers, car protection.
- *In the storage and conversion of energy*, fluorine can be considered an irreplaceable element of the nuclear cycle. Fluorinated products are used as components of batteries and solid electrolytes based on ionic conductors, as well as perfluorinated membranes of the nafion type at the heart of PEM fuel cells.
  - *In photonics and micro-electronics*, fluorine and fluorinated gases occupy a strategic place in the production chain of silicon components because they allow the elimination of any trace of impurities on the surface of the semiconductor that would be unacceptable for the functioning of our computers. Ceramics, glasses, and fluorinated glass ceramics are used in the composition of laser fibers, optical amplifiers, and waveguides for telecommunications microlasers. We can add color pigments, UV absorbers for protection of wood furniture, phosphors for television, heavy metal bases for fiber

lasers, transparent conducting films (TCF), dye-sensitized-perovskite-quantum-dot-sensitized solar cells, LEDs, OLEDs, etc.

- *In agrochemistry*, fluorine is present in half of the molecules with herbicidal, fungicidal, or insecticidal properties.

– *In environmental protection*—After the discovery of the involvement of CFCs, HCFCs, and other PFCs in the depletion of the ozone layer (followed by their final ban) spectacular results have been achieved within fewer than 30 years. The result of research by industrial and academic groups has made it possible to propose new formulations and substitutes which have thus greatly slowed down this type of atmospheric degradation.

- *In biology, medicine, biotechnologies, phytosanitation*—The potential of fluorinated compounds is multifaceted, with fluorinated molecules exhibiting anticancer, antiinflammatory, antibiotic, neuroleptic, antidepressant, anesthetic, or antihypertensive properties. Biomimetic materials use the properties of fluoroapatite, one of the main constituents of our skeleton. Fluorinated blood substitutes could be used for emergency transfusions. Perfluorocarbons are used in vitreoretinal surgery during retinal detachment operations. In the early detection of cancers, positron emission tomography (PET) using the ^{18}F fluorine radioisotope, allows the production of high-quality scans of our organs, tissues, or cells. Over the last 20 years, new tools have become available to fight against certain diseases, thanks to the rapid progress in the organic and bio-organic chemistry of fluorine. Increasingly powerful synthesis methodologies, associated with original and varied fluorinated structural units, make it possible to access the design of new drugs or phytosanitary products with increasingly sophisticated structures. Moreover, the understanding of the multiple effects produced by fluorine atoms on the biological behavior of a molecule is in full development. Thanks to the rapid progress in fluorine chemistry over the last 20 years, powerful new synthesis methodologies and original and varied fluorinated structural motifs are available, and as a result, the design of new drugs with increasingly sophisticated structures is facilitated.

It can be argued that the impact of fluorine and fluorinated products on our daily lives is part of the general context of confrontation between scientific advances and societal benefits, that is, between benefits and prejudices for humans and their environment. The presentation of some major discoveries and technological improvements in fluorine and fluoride products that have punctuated the 20th and early 21st centuries can illustrate this point. Among these advances, some had to be called into question because of serious disturbances caused to humans or the environment. For each of these areas, we have endeavored to show that, as soon as it turned out that these products were likely to create major problems, scientists put everything into play to propose the most appropriate solutions. At the beginning of the century, the benefits of fluorinated compounds are evident in many areas, and the future of this element is very encouraging because of the promising contributions of fluorinated molecules and materials in many fields of science and technology. Nevertheless, drawbacks to our environment caused by some of these products should not be underestimated. Perfluorinated compounds such as PFOA which bioaccumulate in the body are classified as endocrine disruptors, as are other species such as persistent organic pollutants (POPs), pesticides (DDT), polychlorinated biphenyls, polychloroterphenyls, bisphenol A, or phthalates. They may be harming the health of wildlife; the potential health effects of PFOA and other PFCs to humans are under investigation in the United States by the C8 Science Panel [C8 for perfluoro*octa*noic acid] (http://www.c8sciencepanel.org/).

In addition, solutions must be proposed to further limit the emission of gases or fluorinated pollutants into our atmosphere and environment.

However, certain aspects of the history of fluorinated products evoked in this book encourage us to remain optimistic. It can be considered that if there had not been a rapid worldwide awareness following the scientific discovery of the ozone layer depletion, the problem could have become rapidly insoluble. This has led researchers and companies to surpass themselves in the search for technological breakthroughs to find new molecules with higher safety, less harmfulness, and more efficiency standards. After the global decisions concerning the stratospheric ozone depletion, it can be emphasized that the strong increase in the stratospheric chlorine values observed in the 20th century is anticipated to be reversed in the next decade and possibly return to pre-1980 values around 2050 in the mid-latitudes, and even possibly return to pre-1980 values in polar regions around 2065.

Other optimistic examples quoted in this book should be pointed out, such as the discovery of new types of molecules with low GWP, such *as* hydrofluorophenols (e.g., HFO-1234yf) for automotive refrigeration, new alternative shorter chain molecules such as perfluorobutane sulfonyl fluoride for replacing phased-out PFOA, the intense efforts for reducing nocive fluoride emissions in aluminum, ceramics, or other polluting industries, and also for limiting the level of fluorine in drinking water in areas where fluorosis occurs (mostly due to geological reasons). This last point is strongly helped by the awareness of the political decision makers and populations who face these risks because of particular geological conditions and bad consumption habits. In all these cases, scientists must be highly encouraged in their quest for finding the best solutions to overcome all these problems and proposing the most adapted solutions.

Books on fluorine chemistry and fluoride products (2000–2018)

1. Generalities on fluorine chemistry and inorganic fluorinated materials
   - Progress in Fluorine Science: a series, A. Tressaud Ed. Elsevier: Vol. 1 - Photonic and Electronic Properties of Fluoride Materials, A. Tressaud, K. Poeppelmeire, Eds. (2016). Vol. 2 - New Forms of Fluorinated Carbons, O. Boltalina & T. Nakajima, Eds. (2016).
   - New Materials for Catalytic Applications, V. I. Parvulescu, E. Kemnitz, Elsevier (2016).
   - Surface Modified Carbons as Scavengers for Fluorides from Water, V. Sivasankar, Ed., Springer (2016).
   - Special Issue "Fluorine," Dalton Transactions. J. A. Love Ed., ISSN 1477-9226, RSC (2015).
   - Fluorinated materials for energy conversion, T. Nakajima, H. Groult Eds, Second edition, Elsevier (2015).
   - Fluorine Chemistry, a thematic issue, V. Gouverneur, K. Seppelt, Eds. Chemical Reviews, 115 (2015).
   - Efficient preparations of fluorine compounds, H.W. Roesky, Wiley (2012).
   - Le Fluor: Histoire, Applications & Paradoxes, A. Tressaud, CNRS-Editions (2011).
   - Functionalized Inorganic Fluorides: Synthesis, Characterizations & Properties of Nanostructured Solids, A. Tressaud, (Ed.), Wiley-Blackwell, Oxford (2010).
   - Special issue « Fluorine and fluorinated products at dawn of the XXIe century, A. Tressaud, Coordinator, *"Actualité Chimique", French Chemical Society*, Paris, N° 301-302, (2006).
   - Fluorine and the Environment: Vol.1-Atmospheric Chemistry, Emissions & Lithosphere, Vol.2-Agrochemicals, Archaeology, Green Chemistry & Water, A. Tressaud, (Ed.), Elsevier B.V., Amsterdam (2006).
   - Fluorinated materials for energy storage, T. Nakajima, H. Groult, (Eds.), Elsevier B.V. Amsterdam (2005).
   - Fluorides in the Environment: effect on plants and animals, L. H. Weinstein, A. Davison, CABI Publishing, Cambridge, Mass. (2004).
   - Chemistry of Tantalum and Niobium Fluoride Compounds, A. Agulyansky, Elsevier, Amsterdam (2004).
   - Advanced Inorganic Fluorides: Synthesis, Characterization and Applications, T. Nakajima, B. Zemva, A. Tressaud (Eds), Elsevier, Amsterdam (2000).
   - Fluorine Chemistry at the Millennium (partly in 100th Volume, J. Fluorine Chem., 1999), R. E. Banks, (Ed.), Elsevier Science & Technology, Amsterdam (2000).
2. Chemistry of fluoro-organics, fluoropolymers, phytosanitary products, medicine, biomedical technologies
   - Fluorinated Polymers: Vol. 1, Vol. 2, B. Ameduri, H. Sawada, RSC Publishing (2016, 2017).

- Progress in Fluorine Science: a series, A. Tressaud Ed. Elsevier: Vol. 3 Processes and Reactivity of Fluorinated Compounds, H. Groult, F. Leroux & A. Tressaud, Eds. (2016); Vol. 4 Fluorine & Life Sciences, G. Haufe, F. Leroux. Eds (to appear 2018).
- Organometallic Fluorine Chemistry, T. Braun Ed., Springer (2015).
- Fluorine in Heterocyclic chemistry V. G. Nenajdenko Ed., Vol. 1 and 2, Springer (2014).
- Fluorine in Pharmaceutical and Medicinal Chemistry, V. Gouverneur, K. Muller, Eds. World Scientific Publ. Co. (2012).
- Fluorine in Medicinal Chemistry and Chemical Biology, Iwao Ojima, (Ed.), Wiley-Blackwell, Oxford (2009).
- Bioorganic and Medicinal Chemistry of Fluorine, J.-P. Bégué, D. Bonnet-Delpon, John Wiley & Sons, USA (2008).
- Fluorine and Health: Molecular Imaging, Biomedical Materials and Pharmaceuticals, A. Tressaud, G. Haufe, (Eds.), Elsevier, Amsterdam (2008).
- Organofluorine chemistry, K. Uneyama, Blackwell Publ., Oxford (2006).
- Chimie bioorganique et médicinale du fluor, J-P. Bégué, D. Bonnet-Delpon, EDP-Sciences CNRS, Paris (2005).
- Fluorine-containing syntheses, V. A. Soloshonok, ACS Symposium, Washington DC (2005).
- Fluorine in Organic Chemistry, R. Chambers, Blackwell Publ., Oxford (2004).
- Well-Architectured Fluoropolymers, B. Ameduri, B. Boutevin, Elsevier, Amsterdam (2004).
- Handbook of Fluorous Chemistry, J. A. Gladysz, D. P. Curran, I. T. Horvath Eds., Wiley New York (2004).
- Modern Organofluorine Chemistry - Synthesis & Applications, P. Kirsch, Wiley-VCH New York (2004).
- Fluorinated Surfactants and Repellents, E. Kissa, Marcel Dekker Inc. New York (2001).
- Fluorinated Surface, Coatings and Films, D.G. Castner, D.G. Grainger (Eds.) ACS Symposium Series 787, Washington-DC, (2001).
- Organo-fluorine Compounds, 4th Ed., Houben-Weyl, Collectif, Vol. E10 Thieme, Stuttgart (2000).
- Fluorine Chemistry for Organic Chemists: Problems and Solutions, M. Hudlicky, Oxford Univ Press, Oxford (2000).
- Organofluorine Compounds: Chemistry & Applications (2000), T. Hiyama, H. Yamamoto, Springer-Verlag Berlin (2000).

Index

Note: Page numbers followed by *f* indicate figures, *t* indicate tables, and *b* indicate boxes.

D

E

Acronyms and definitions quoted in the text

Acronyms

AGAGE	advanced global atmospheric gases experiment (sponsored by NASA's atmospheric composition focus area in earth science)
AD	Anno Domini
AES	auger electron spectroscopy
AMS	accelerator mass spectrometry, radiocarbon dating
AFEAS	alternative fluorocarbons environmental acceptability study
BP	Before Present
BC	Before Christ
BCE	before the common (or current) era
CDCP	US Centers for Disease Control and Prevention
CFC	chlorofluorocarbons
COP	conference of the parties
DSSC	dye-sensitized solar cell
DABCO	1,4-diazabicyclo[2.2.2]octane
EPMA	electron probe microanalysis
EPA	US Environmental Protection Agency
ESR	electron-spin resonance
EFCT	European Technical Committee on FluoroCarbons
FAP	fluoroapatite
FTIR	Fourier-transform infrared
GHG	greenhouse gases
GWP_{100}	global warming potential (calculated over 100 years)
FTO	fluorine-doped tin oxide
F-TCO	fluorine-doped transparent conducting oxides
HAP	hydroxyapatite
HCFC	hydrochlorofluorocarbons
HFC	hydrofluorocarbons
HPLC/ESI-MS/MS	high-performance liquid chromatography/electrospray ionization tandem mass spectrometry
ISE	ion selective electrode
IBA	ion beam analysis
ICOM	International Council of Museums
IAI	International Aluminum Institute
IPCC	Intergovernmental Panel on Climate Change
IISD	International Institute for Sustainable Development
LIBS	laser-induced breakdown spectroscopy

LED	light-emitting diode
MCMB	mesocarbon microbeads
MNHN	National Museum of Natural History, Paris
mtpa	million tons per annum
MRI	magnetic resonance imaging
NRA	nuclear reaction analysis
NMR	nuclear magnetic resonance
NLO	nonlinear optics
ODP	ozone depletion potential
OLED	organic light-emitting diodes
ODS	ozone-depleting substances
PIXE	particle-induced X-ray emission
PIGE PIGME	proton-induced gamma-ray emission
PFC	perfluorocarbons
PTFE	polytetrafluoroethylene
PFOA	perfluorooctanoic acid
PFOS	perfluorooctanesulfonic acid
PFAS	*per-* and polyfluoroalkyl substances
PEMFC	proton exchange membrane fuel cell
PEFC	polymer electrolyte fuel cell
PET	positron emission tomography
PDT	photodynamic therapy
PEF	plasma-enhanced fluorination
POP	persistent organic pollutants
PAH	polycyclic aromatic hydrocarbons
QDSSC	quantum dot-sensitized solar cell
RF	radiative forcing
RBS	Rutherford backscattering
RE, REE	rare earth, rare earth elements
SPECT	single photon-emission computed tomography
SEB	scanning electron microscopy
SEM-EDX	energy-dispersive X-ray spectroscopy analysis via scanning electron microscopy
SCQD	semiconductor quantum dots
SPS	spark plasma sintering
TSCA	Toxic Substances Control Act
TFE	tetrafluoroethylene
TEM	transmission electron microscopy
TSL	thermostimulated luminescence
TCO	transparent inorganic conducting oxides
TTB	tetragonal tungsten bronze
USGS	US Geological Survey
UNFCCC	United Nations Framework Convention on Climate Change
UCNP	upconversion nanoparticles
VUVS	vacuum ultraviolet spectroscopy
VSEPR	valence shell electron pair repulsion
WMO	World Meteorological Organization

WHO	World Health Organization
XPS	X-ray photoelectron spectroscopy
XANES	X-ray absorption near-edge structure

Definitions

Allophane	mineral composed of hydrated aluminum silicate, poorly crystallized
Amphiboles	silicate types including nephrite and tremolite
Andesites, rhyolites	magmatic and volcanic rocks
Basalt	volcanic rock from a magma cooled quickly in contact with water or air. It is the main constituent of the upper layer of the oceanic crust.
Cariogenic treatment	treatment limiting caries
Chondrites	meteorites characterized by smooth swellings (chondrules), resembling a nodule. The chondrule melts while the rest of the meteorite remains more or less solid, and then solidifies
Compomer	glass ionomer modified with resin
Corticosteroids	natural steroid hormones secreted by the adrenocortical gland
Cytotoxicity	the property of a chemical or biological agent to alter or even destroy cells
Enzyme inhibitors	play an important role in the control of biological mechanisms, including the regulation of metabolic pathways
Felsiques	silicate magmatic rocks enriched in elements such as oxygen, aluminum, sodium, or potassium; for example quartz, muscovite, feldspar orthoclase. The name comes from the combination of "feldspar" and "silica"
GABA	Γ-aminobutyric acid, the main inhibitory neurotransmitter of the central nervous system
Half-life	corresponds to the moment when half of the radioactive nuclei of a source have disintegrated
Holocene	geological period covering the last 10,000 years
Ionomer glass cement	used in dentistry as a sealing cement especially for all-ceramic crowns
Ionomer glass	material used in odontology (to seal dental cavities). Prepared using a powder composed of glass particles and quartz or silicates mixed with an acidic liquid. Ionomer glass adheres very strongly to enamel and dentin. It releases fluoride ions, increasing the anticaries power
Lipophilic	affinity for fat
Lithosphere	the most superficial terrestrial rigid envelope (stone sphere)
L-DOPA	for 3,4-dihydroxyphenylalanine (or levodopa), able to cross the blood-brain barrier. Substance used in the treatment of Parkinson's disease by increasing dopamine levels
Osteoblasts	cells responsible for the synthesis of bone tissue. They secrete collagen and alkaline phosphatase which releases inorganic phosphate which, by binding to calcium, gives calcium phosphate

Osteoporosis	frequent disease for women after menopause, characterized by excessive fragility of the skeleton due to decreased bone mass and deterioration of the bone microarchitecture
Plagioclases	types of alkaline feldspars, containing calcium and no potassium
Pleistocene	geological period starting 1,8M yr BP, ending 11,400 yr BP, and covering most of the last ice ages
Plinian	type of eruption in which the lava is extremely pasty because very rich in silica. The gases cannot be released and, as the pressure increases in the magma chamber, explosions spray the lava by throwing ashes tens of kilometers high, thus reaching the stratosphere (named after Pliny the Elder, who died during the eruption of Vesuvius in 79)
Radioelements	atoms whose nucleus is unstable and therefore radioactive. They exist naturally but can also be artificially produced by a nuclear reaction (as ^{18}F)
Sorosilicates	minerals of the family of silicates whose $[SiO_4]$ tetrahedra are grouped two by one of their vertices to form $[Si_2O_7]^{6-}$
Tephra	all materials blown into the air or water during explosive volcanic eruptions
Trabecula	a fine extension that separates from a wall of an anatomical tissue and protrudes into a cavity
Vitreo-retinopathies	affection affecting vitreous humor and retina
Wafer	designates a fairly thin disc of semiconductor material such as silicon
$[^{18}F]$ FDG	2-$[^{18}F]$fluoro-2-deoxy-D-glucose
$[^{18}F]$ DCFBC	N-[N-[(*S*)-1,3-dicarboxypropyl]carbamoyl]-4-$[^{18}F]$fluorobenzyl-L-cysteine

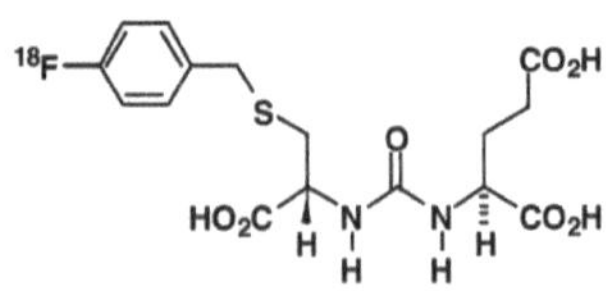

$[^{18}F]$DCFBC

Acknowledgments

These years spent tracking down new fluorinated materials could not have been possible without two great scientists, Paul Hagenmuller (1921–2017), one of the initiator of solid-state chemistry in France and worldwide, who welcomed me in his group in Bordeaux, one day of September 1964, and Neil Bartlett (1932–2008), discoverer of noble gas compounds, in whose laboratory I spent my postdoctoral time at U. C. Berkeley and who remained my mentor until his early disappearance.

I am indebted to CNRS for allowing me to realize my passion for research since 1966, and more recently to the National Program: "*French Fluorine Network*", to have made me aware of new scientific horizons.

Many thanks to all my colleagues of ICMCB-CNRS, from the oldest times: R. de Pape, J. Portier, J. Grannec, L. Lozano, L. Fournès, J. Darriet, J.P. Chaminade, J. Claverie, J. Etourneau, J.M. Dance (†), to more recent ones: A. Demourgues, M. Gaudon, B. Chevalier, E. Durand, who shared my joys of discovery, to all my post-docs and students who have helped to keep intact my energy and my scientific curiosity.

Also to all my French and foreign colleagues who, through constant and fruitful exchanges, opened me to new perspectives by abolishing borders, in particular H. Groult (Paris), but also D. Babel (Marburg), K. Christe (Los Angeles), L.P. Demyanova (St Petersburg), J. B. Goodenough (Austin), M. Feist (Berlin), I.N. Flerov (Krasnoyarsk), R. Hagiwara (Kyoto), E. Kemnitz (Berlin), A. Kharitonov (†) (Moscow), M. Leblanc (Le Mans), D. Lutar (†) (Ljubljana), D. Meshri (Catoosa), V.N. Mitkin (Novosibirsk), T. Nakajima (Toyota), P. Nuñez (La Laguna), H. Roesky (Göttingen), G. Schrobilgen (Hamilton), H. Touhara (Ueda), N. Watanabe (†) (Kyoto), J. Winfield (Glasgow), B. Žemva (Ljubljana).

So many thanks to my wife Viviane, to Laurent and Mathieu, and all my family who have often had to suffer from my passion for research. And finally to everyone who has made my life more enjoyable.

Kolibri 360 GmbH
Ferdinand-Jühlke-Straße 7
99095 Erfurt, Deutschland
produktsicherheit@kolibri360.de